Based on the 2009 Canadian Electrical Code®

ELECTRICAL WIRING
RESIDENTIAL

Fifth Canadian Edition

Ray C. Mullin

Gary Miller

Paul Stephenson

Tony Branch

Sandy F. Gerolimon

Bill Todd

Craig Trineer

NELSON / EDUCATION

NELSON / EDUCATION

Electrical Wiring: Residential, Fifth Canadian Edition

by Ray C. Mullin, Gary Miller, Paul Stephenson, Tony Branch, Sandy F. Gerolimon, Bill Todd, and Craig Trineer

Vice-President and Editorial Director:
Evelyn Veitch

Editor-in-Chief:
Anne Williams

Executive Marketing Manager:
Dave Ward

Developmental Editor:
MyEditor Inc.

Permissions Coordinator:
Nicola Winstanley

Senior Content Production Manager:
Natalia Denesiuk Harris

Proofreader:
June Trusty

Production Coordinator:
Ferial Suleman

Creative Director:
Angela Cluer

Cover Design:
Katherine Strain

Cover Image:
© Charles O'Rear/Corbis

Compositor:
Pamela Clayton

Printer:
Edwards Brothers

Library and Archives Canada Cataloguing in Publication

Electrical wiring : residential / Ray C. Mullin ... [et al.]. — 5th Canadian ed.

"Based on the 2009 *Canadian Electrical Code*".
Includes index.
ISBN 978-0-17-650215-7

1. Electric wiring, Interior.
2. Dwellings—Electric equipment.
I. Mullin, Ray C.

TK3285.M84 2009
621.319'24   C2009-903679-7

ISBN-13: 978-0-17-650215-7
ISBN-10: 0-17-650215-7

CONTENTS

Introduction xi

Preface xii

Acknowledgments xiii

Unit 1   GENERAL INFORMATION FOR ELECTRICAL INSTALLATIONS **1**
    Safety   1
    Codes and Standards   1
    Personnel Safety   3
    Testing and Accreditation   3
    Units of Measure   5

Unit 2   DRAWINGS AND SPECIFICATIONS **9**
    Technical Drawings   9
    Visualizing a Building   10
    Building Views   11
    Symbols and Notations   12
    Dimensions   12
    Scale   18
    The Working Drawings   18
    Specifications   20

Unit 3   SERVICE ENTRANCE CALCULATIONS **23**
    Size of Service Entrance Conductors and Service Disconnecting Means   23
    Service Calculations for a Single-Family Dwelling,
        Rule 8–200: Calculating Floor Area (Ground Floor)   24
    Service Calculations for Apartments   29

Unit 4   SERVICE ENTRANCE EQUIPMENT **33**
    Overhead Service   34
    Mast-Type Service Entrance   35
    Underground Service   37
    Main Service Disconnect Location   37

Service Entrance Conduit Sizing 44
Meter 45
Grounding—Why Ground? 47
Grounding Electrode Systems 47
Grounding the Service When Non-Metallic Water Pipe Is Used 49
Summary—Service Entrance Equipment Grounding 49
Bonding 51
Multiple Meter Installations 53
Branch Circuit Overcurrent Protection 55
Interrupting Ratings for Fuses and Circuit Breakers 58
Panels and Load Centres 60
Reading the Meter 63
Caution: Series-Rated System... 63
Cost of Using Electrical Energy 63

Unit 5   ELECTRICAL OUTLETS 71
Electrical Outlets 71
Lighting Fixtures and Outlets 72
Switches 75
Junction Boxes and Switch (Device Boxes) (Rules 12–3000 to 12–3034) 75
Non-Metallic Outlet and Device Boxes 78
Ganged Switch (Device) Boxes 82
Boxes for Conduit Wiring 83
Special-Purpose Outlets 83
Number Of Conductors in a Box 86
Selecting a Box When All Conductors Are the Same Size 87
Selecting a Box When Conductors Are Different Sizes (Rule 12-3034(4)) 89
Box Fill 90
Locating Outlets 91
Positioning Receptacles 91
Bonding 94

UNIT 6   DETERMINING THE NUMBER AND LOCATION OF LIGHTING
AND RECEPTACLE BRANCH CIRCUITS 99
Lighting 99
Receptacles 99
Basics of Wire Sizing and Loading 100
Voltage 103
Lighting Branch Circuits 104
Receptacle Branch Circuits 106
Preparing a Lighting and Receptacle Layout 107

Unit 7   CONDUCTOR SIZES AND TYPES, WIRING METHODS,
WIRE CONNECTIONS, VOLTAGE DROP, NEUTRAL SIZING
FOR SERVICES 112
Conductors 112
Voltage Drop 118

Approximate Conductor Size Relationship 121

Calculation of Maximum Length of Cable Run Using *Table D3* 121

Non-Metallic-Sheathed Cable (*Rules 12–500* to *12–526*) 123

Armoured Cable (*Rules 12–600* to *12–618*) 126

Installing Cables through Wood and Metal Framing Members (*Rule 12–516*) 129

Installation of Cable in Attics 131

Installation of Cables through Ducts 132

Connectors for Installing Non-Metallic-Sheathed and Armoured Cable 133

Electrical Metallic Tubing (*Rule 12–1400*), Rigid Metal Conduit

   (*Rule 12–1000*), and Rigid PVC Conduit (*Rules 12–1100* to *12–1122*) 134

Rigid PVC Conduit (*Rules 12–1100* to *12–1122*) 134

Flexible Connections (*Rules 12–1000* and *12–1300*) 138

Flexible Metal Conduit (*Rules 12–1002* to *12–1014*) 138

Liquidtight Flexible Metal Conduit (*Rules 12–1300* to *12–1306*) 141

Liquidtight Flexible Non-Metallic Conduit (*Rules 12–1300* to *12–1306*) 142

Unit 8 SWITCH CONTROL OF LIGHTING CIRCUITS, RECEPTACLE BONDING, AND INDUCTION HEATING RESULTING FROM UNUSUAL SWITCH CONNECTIONS **146**

Conductor Identification (*Rule 4–036*) 146

Toggle Switches (*Rules 14–500* to *14–514*) 149

Bonding at Receptacles 155

Nonbonding-Type Receptacles 158

Induction Heating 159

Unit 9 GROUND-FAULT CIRCUIT INTERRUPTERS, ARC FAULT CIRCUIT INTERRUPTERS, TRANSIENT VOLTAGE SURGE SUPPRESSORS, AND ISOLATED GROUND RECEPTACLES **165**

Electrical Hazards 165

Code Requirements for Ground-Fault Circuit Interrupters (GFCIS) 166

Precautions for Ground-Fault Circuit Interrupters 168

Ground-Fault Circuit Interrupter in Residence Circuits 168

Feedthrough Ground-Fault Circuit Interrupter 171

Identification, Testing, and Recording of GFCI Receptacles 172

Replacing Existing Receptacles 173

Ground-Fault Protection For Construction Sites 176

Arc-Fault Circuit Interrupters (AFCIs) 176

Transient Voltage Surge Suppression (TVSS) 178

Isolated Ground Receptacle 180

Unit 10 LIGHTING FIXTURES AND BALLASTS **185**

Types of Lighting Fixtures 185

CEC Requirements for Installing Recessed Fixtures 187

Ballast Protection 191

Lighting Fixture Voltage Limitations 191

Unit 11   BRANCH CIRCUITS FOR THE BEDROOMS, STUDY, AND HALL **195**

Grouping Outlets 195

Cable Runs 196

Estimating Loads for Outlets 197

Symbols 199

Drawing the Wiring Diagram of a Lighting Circuit 199

Lighting Branch Circuit A16 for Front Bedroom and Study 201

Branch Circuit A24 Outside Receptacle 202

Determining the Wall Box Size 203

Bonding of Wall Boxes 203

Positioning of Split-Circuit Receptacles 204

Positioning of Receptacles Near Electric Baseboard Heating 204

Fixtures in Clothes Closets 204

Lighting Branch Circuit A14 for Master Bedroom 204

Sliding Glass Doors 210

Selection of Boxes 210

Lighting Circuit for Study and Main Bathroom 211

Valance Lighting 211

Paddle Fans 211

Unit 12   BRANCH CIRCUITS FOR THE LIVING ROOM AND FRONT ENTRY **220**

Lighting Circuit Overview 220

Track Lighting 222

Dimmer Controls 223

Incandescent Lamp Load Inrush Currents 226

Dimming Fluorescent Lamps 228

Branch Circuit for Front Entry, Porch 229

Unit 13   BRANCH CIRCUITS FOR BATHROOMS **237**

General Comments on Lamps and Colour 240

Hanging Fixtures in Bathrooms 241

Hallway Lighting 241

Receptacle Outlets in Hallways 241

Bonding Requirements for a Bathroom Circuit 241

Bathroom Ceiling Heater Circuits (▲)$_K$ (▲)$_J$ 243

Hydromassage Tub Circuit (▲)$_A$ 245

Unit 14   LIGHTING BRANCH CIRCUIT AND SMALL APPLIANCE CIRCUITS
FOR THE KITCHEN **249**

Lighting Circuit B7 249

Kitchen Lighting 249

Small Appliance and Branch Circuits for Convenience Receptacles in Kitchen 253

Split-Circuit Receptacles and Multiwire Circuits 256

General Bonding Considerations 257

Unit 15 SPECIAL-PURPOSE OUTLETS FOR RANGES, COUNTER-MOUNTED COOKING UNITS ▲G, WALL-MOUNTED OVENS ▲F, FOOD WASTE DISPOSALS ▲H, AND DISHWASHERS ▲I **26**

Counter-Mounted Cooking Unit Circuit ▲G 263
Temperature Control 264
Wall-Mounted Oven Circuit ▲F 265
Branch-Circuit Requirements for Ranges, Ovens, and Countertop Cooking Units 266
Freestanding Range 266
Branch Circuits Supplying Separate Built-in Cooking Units 267
Food Waste Disposal ▲H 267
Bonding Requirements 270
Dishwasher ▲I 270
Portable Dishwashers 271
Cord Connection of Fixed Appliances 271

Unit 16 BRANCH CIRCUITS FOR THE LAUNDRY, WASHROOM, AND ATTIC **276**
Dryer Circuit ▲D 276
Receptacle Outlets—Laundry 279
Lighting Circuit 279
Humidity Control 280
Attic Lighting and Pilot Light Switches 281
Attic Exhaust Fan Circuit ▲L 283

Unit 17 ELECTRIC HEATING AND AIR CONDITIONING **291**
General Discussion 291
Types of Electric Heating Systems 292
Separate Circuit Required 292
Control of Electric Heating Systems 292
Circuit Requirements for Baseboard Units 294
Location of Electric Baseboard Heaters In Relation to Receptacle Outlets 295
Circuit Requirements for Electric Furnaces 296
Heat Pumps 298
Bonding 298
Marking the Conductors of Cables 298
Room Air Conditioners 298
Receptacles for Air Conditioners 299
Central Heating and Air Conditioning 299
Special Terminology 299
Loads Not Used Simultaneously 301

Unit 18 OIL AND GAS HEATING SYSTEMS **306**
Principles of Operation 306
Major Components 306
Self-Generating (Millivolt) System 312
Supply Circuit Wiring 313
Control Circuit Wiring 313

...N ROOM **320**

...Room Lighting 320

...les and Wet Bar 322

...RANCH CIRCUITS FOR WORKSHOP AND UTILITY AREA **331**

Workshop 331

Workbench Lighting 331

Receptacle Outlets 331

Cable Installation in Basements 333

Multi-Outlet Assembly 333

Empty Conduits 334

Utility Room 335

Water Pump Circuit (▲)$_B$ 335

Jet Pump Operation 335

Submersible Pump 337

Water Heater Circuit (▲)$_C$ 338

Heating Elements 340

Speed of Recovery 341

Scalding From Hot Water 341

What about Washing Dishes? 342

Sequence of Operation 342

Metering and Sequence of Operation 344

Water Heater Load Demand 345

Effect of Voltage Variation 346

Unit 21   HEAT AND SMOKE DETECTORS, CARBON MONOXIDE DETECTORS,

AND SECURITY SYSTEMS **353**

The Importance of Heat and Smoke Detectors 353

Types of Smoke Detectors 354

Installation Requirements 355

Manufacturers' Requirements 356

Combination Direct/Battery/Feed-Through Detectors ... 358

Carbon Monoxide Detectors 358

Security Systems 358

Unit 22   EXTRA-LOW-VOLTAGE, REMOTE-CONTROL

AND SIGNAL SYSTEMS **363**

Remote-Control Switches 364

Extra-Low-Voltage Relay 365

Transformers 366

Conductors 367

Installation Procedure 367
Typical Wiring Installation 368
CEC Rules 368
Signal System (Chimes) 370

Unit 23   SWIMMING POOLS, SPAS, HOT TUBS **379**
Pool Wiring (*Section 68*) 379
Electrical Hazards 379
Code-Defined Pools 381
Grounding and Bonding of Swimming Pools 381
Lighting Fixtures Under Water 384
Electric Heating of Swimming Pool Decks 384
Spas and Hot Tubs (*Rules 68–400* to *68–408*) 385
Hydromassage Bathtubs (*Subsection 68–300*) 386
Fountains 387
Summary 387

**Unit 24   TELEVISION, TELEPHONE, DATA, AND HOME
AUTOMATION SYSTEMS** **391**
Television (Symbol TV) 391
Satellite Antennas 393
CEC Rules for Installing Antennas and Lead-in Wires (*Section 54*) 395
Telephone Wiring (Symbol ▶|) (*Section 60*) 397
Data Systems 401
Home Automation 405

**Unit 25   LIGHTING BRANCH CIRCUIT FOR THE GARAGE
AND OUTDOOR LIGHTING** **411**
Lighting Branch Circuit 411
Lighting a Residential Garage 411
Receptacle Outlets 413
Outdoor Wiring 413
Underground Wiring 414
Overhead Garage Door Opener ▲E 419
Central Vacuum System ▲Q 420

APPENDIX A: Specifications for Electrical Work—Single-Family Dwelling **424**
APPENDIX B: Jack Pin Designations and Colour Codes **430**
APPENDIX C: Formulas and Equations **431**

INDEX **432**

RESIDENTIAL PLANS (INSERTED AT BACK OF BOOK)

Sheet 1 of 10   Site Plan: South, North, East, West Elevations
Sheet 2 of 10   First Floor Electrical Layout Drawing
Sheet 3 of 10   Basement Floor Plan
Sheet 4 of 10   Basement Electrical Layout Drawing
Sheet 5 of 10   First Floor Plan
Sheet 6 of 10   Code Requirements for Swimming Pool
Sheet 7 of 10   Section A-A
Sheet 8 of 10   First Floor Electrical Plan
Sheet 9 of 10   Interior Elevations and Schedules
Sheet 10 of 10   Basement Electrical Plan

INTRODUCTION

This fifth Canadian edition of *Electrical Wiring: Residential* is based on the 2009 edition of the *Canadian Electrical Code, Part I: The Safety Standard for Electrical Installations (CEC)*. The authors welcome readers' suggestions for improvements that may be included in subsequent editions of the text.

Electrical Wiring: Residential provides an entry-level text that is both comprehensive and readable. It is suitable for colleges, technical institutes, vocational/technical schools, and electrical programs in high schools, and for readers who wish to study the subject on their own.

The wiring in the home illustrated in *Electrical Wiring: Residential* incorporates more features than are absolutely necessary to present as many *CEC* rules as possible to give students the information they need to complete a safe installation. The text focuses on the technical skills required to perform electrical installations. Topics include calculating conductor sizes, calculating voltage drop, sizing services, connecting electrical appliances, grounding and bonding equipment, and installing recessed fixtures. These are critical skills that can make the difference between an installation that "meets code" and one that is exceptional. The electrician must understand that the reason for following *CEC* regulations is to achieve an installation that is essentially free from hazard to life and property.

The *CEC* is the basic standard for the layout and construction of all electrical systems in Canada. However, some provincial and local codes may contain specific amendments that must be adhered to in all electrical wiring installations in those jurisdictions.

The authors encourage the reader to develop a detailed knowledge of the layout and content of the *CEC,* which must be used in conjunction with a comprehensive study of *Electrical Wiring: Residential* to derive the greatest benefit from the text.

The *CEC* is divided into numbered sections, each covering a main division. These sections are further divided into numbered rules, subrules, paragraphs, and subparagraphs. All references in the text are to the *Section* or to the *Section rule number.* For example, *Rule 8–200(1) (a) (i)* refers to *Section 8, rule 8–200, subrule 1, paragraph a, subparagraph i.* This explanation should assist the student in locating *CEC* references in the text.

Thorough explanations are provided throughout, as the text guides the student through the steps necessary to become proficient in the techniques and *CEC* requirements described.

PREFACE

Electrical Wiring: Residential will prove a valuable resource to instructors and students alike. It includes 2009 *Canadian Electrical Code, Part I* (*CEC*) references and wiring techniques. Each chapter is a complete lesson ending with review questions to summarize the material covered. The chapters are sequenced to introduce the student to basic principles and wiring practices, and progress to more advanced areas of residential electrical wiring.

This text assumes no prior knowledge of residential electrical wiring, but the student will need a reasonable level of mechanical aptitude and skill to be successful in the practical application of the techniques discussed.

The text guides the student through the working drawings for a residential electrical installation, the proper wiring of receptacles, and the minimum required number of lighting and power branch circuits. Voltage drop calculations based on the *CEC* are shown. Grounding, bonding, and ground-fault circuit interrupters are discussed together with the lighting branch circuits for all the rooms in the house, garage, workshop, and exterior. Special-purpose outlets for ranges, dryers, air conditioners, water heaters, and water pumps are explained, as are electric, oil, and gas heating systems; heat detectors, smoke detectors, and smoke alarms; swimming pools, spas, and hot tubs; and low-voltage-signal systems, security systems, and low-voltage remote-control systems, or Smart House Wiring.

The text details service-entrance equipment and the installation of the electrical service, including the calculations for the sizing of the service-entrance conductors, conduit, switch, grounding conductor, and panel.

FEATURES

- New! All text material fully updated to reference the 2009 edition of the *CEC*.
- Sections on wireless data networks and home automation.
- Measures converted to metric.
- Complete descriptions of Canadian practice and applications.
- Numerous diagrams of CSA-approved equipment found in everyday wiring installations.
- Step-by-step explanations of wiring a typical residence.
- Complete set of blueprints to help apply *CEC* theory to real working drawings.

ACKNOWLEDGMENTS

The authors wish to acknowledge the valuable assistance of Bill Wright in the preparation of the original architectural drawings for this book. Much gratitude is extended to Miroslav Lukic for his thorough technical check of the blueprints against the text and for revising those blueprints for this edition. Great appreciation is given to David Bird for the complete technical check of the text against the blueprints. Thank you as well to the people from Nelson Education Ltd. for their encouragement and professional advice.

The authors are also indebted to reviewers past and present for their comments and suggestions on how to improve this text, edition after edition, including:

Frank Bowick, Algonquin College
Rodney Deline, Durham College
Mike Mankulich, Fanshawe College
Dave Murray, Northern Alberta Institute of Technology
Orest Staneckyj, Cambrian College

Their critical analyses and recommendations made this text what it is today.
We wish to thank the following for contributing data, illustrations, and technical information:

AFC/A Nortek Co.
Anchor Electric Division, Sola Basic Industries
Appleton Electric Co.
Arrow-Hart, Inc.
Brian Cameron
BRK Electronics, A Division of Pittway Corporation
Brazos Technologies, Inc.
Broan-NuTone
Bussmann Division, Cooper Industries
Carlon
Chromalox
Commander
Electri-Flex Co.
Emerson Electric
General Electric Co.
Gould Shawmut
Halo Lighting Division, Cooper Industries
Heyco Molded Products, Inc.
Honeywell
Hubbell Incorporated, Wiring Devices Division
International Association of Electrical Inspectors
IPEX

Juno
KEPTEL
The Kohler Co.
Leviton Manufacturing Co., Inc.
Lightolier Canada
Midwest Electric Products, Inc.
Milbank Manufacturing Co.
Moe Light Division, Thomas Industries
Noma Division, Danbel Industries Inc.
Pass & Seymour, Inc.
Progress Lighting/Hubbell
Rheem Manufacturing Co.
Seatek Co. Inc.
Siemens Energy and Automation
Sierra Electric Division, Sola Basic Industries
SMART HOUSE, L.P.
SMC
Square D Co.
Superior Electric Co.
THERM-O-DISC
Thomas & Betts Incorporated
Underwriters Laboratories of Canada (ULC is accredited by the Standards Council
     of Canada as a Certification Organization, a Testing Organization, a Registration
     Organization and Standards Organization under the National Standards System
     of Canada.)
Vitabath, DM Industries
Winegard Company
Wiremold Co.
Wolberg Electrical Supply Co., Inc.
Woodhead Industries, Inc.

Material is reproduced with the permission of the Canadian Standards Association
from *C22.1-09, Canadian Electrical Code, Part I* (21st Edition), *Safety Standard for
Electrical Installations*, which is copyrighted by the Canadian Standards Association,
5060 Spectrum Way, Suite 100, Mississauga, ON L4W 5N6. While use of this material has been authorized, the CSA shall not be responsible for the manner in which the
information is presented, nor for any interpretations thereof. For more information on
the CSA or to purchase standards, please visit the CSA website at www.csa.ca or call
1-800-463-6727.

CSA logo on page 4 used courtesy of Canadian Standards Association
SCC logo on page 4 used courtesy of Standards Council of Canada
ULC logos on page 4 used courtesy of Underwriters Laboratories of Canada

For any further information or to give feedback on this text, you can contact the
authors by e-mail at

bill.todd@humber.ca   craig.trineer@humber.ca
sandy.gerolimon@humber.ca   tony.branch@humber.ca

UNIT 1

General Information for Electrical Installations

OBJECTIVES

After studying this unit, the student will be able to

- identify the purpose of codes and standards related to electrical systems
- describe the importance of the *Canadian Electrical Code (CEC)*
- describe the layout of the *CEC*
- identify approved equipment
- explain the importance of electrical inspection
- identify links to websites relating to the electrical trade
- list the agencies that are responsible for establishing electrical standards and ensuring that materials meet the standards
- discuss systems of measurement used on construction drawings
- begin to refer to the *CEC*

SAFETY

Electrical installations are required to be safe. Because of the ever-present danger of electric shock and/or fire due to the failure of an electrical system, electricians and electrical contractors must use approved materials and methods. Therefore, codes and standards are used to ensure that the exposure of users, installers, and maintenance personnel to the shock and fire hazards associated with electrical systems is minimized.

CODES AND STANDARDS

Codes are standards that deal with life safety issues and are enforceable by law. Residential electrical wiring systems are required to meet building codes and electrical codes. Building codes dictate such

things as which rooms and areas (such as stairwells) require luminaires (lighting fixtures) and how they are to be controlled. Electrical codes deal with the actual installation of the luminaires and controls. Municipal, provincial, and federal governments can pass laws regarding the installation of electrical equipment.

The *Canadian Electrical Code (CEC)* is the standard that governs electrical work across Canada. It is adopted by the provinces, is enacted in legislation, and is the basis of a bill (with provincial amendments) that passes through the provincial legislature, becoming law and enforced by municipalities or provincial electrical safety authorities.

The first edition of the *CEC* was published in 1927. It is revised every three years to take into account changes in equipment, wiring systems, and new technologies. The standards steering committee for the *CEC* solicits comments and proposals from the electrical industry and others interested in electrical safety. Information about the structure and operation of the committee, how to propose an amendment, or how to request an interpretation can be found in *CEC Appendix C*. Copies of the *CEC* can be ordered from the Canadian Standards Association.

The object of the *CEC* is to provide a minimum safety standard for the installation and maintenance of electrical systems to safeguard people from the hazards of fire and electric shock arising from the use of electrical energy. Meeting the requirements of the *CEC* ensures an essentially safe installation.

The *CEC* does not take economics into consideration. For instance, a house built using wood frame construction may be wired with non-metallic sheathed cable or armoured cable. Both meet the requirement of the CEC, but one costs several times as much as the other.

The *CEC* is divided up into evenly numbered sections, each one covering a main division of the work. Each section has a scope that outlines what will be covered by the section. Ten general sections cover basic installation requirements. The rest are amending or supplementary sections, which change the general requirements of the code for specific types of equipment or installation. Rules in one amending section cannot be applied to another amending section unless they specifically state that they may.

The sections are divided into sub-sections and numbered rules with captions. For example, Section 12–500 is Non-metallic sheathed cable rules, where 12 identifies the section, 500 is the rule number in the section. 12–500 is the first rule of the sub-section that deals with the installation of non-metallic sheathed cable. The 12–500 series of rules (12–500 to 12–526) describe how non-metallic sheathed cables are to be installed.

Tables

At the end of the main body of the *CEC* are a number of tables. An index of the tables precedes the tables. At the top of each table is a list of rules that refer you to the table, which can be helpful in looking up rules. The tables cover such things as the ampacity of wires, number of conductors in raceways, number of conductors in boxes, and depth of cover over conductors installed underground.

Diagrams

After the tables are a number of diagrams. An index of the diagrams precedes the diagrams. Some of the diagrams included show pin configurations for both locking and non-locking receptacles.

Appendices

Of the 11 appendices, the two important to you are *Appendix B* and *Appendix D*. *Appendix B* is Notes on Rules. These notes give supplemental information about rules. Whenever a rule references *Appendix B*, you must read that information. *Appendix B* uses the same numbering system as the main body of the code. Simply look up the rule number from the main body of the code in *Appendix B*.

Example:

Rule 6–112 Support for the attachment of overhead supply or consumer's service conductors (See *Appendix B*)

Section	Title
0	Object, scope, and definitions
2	General Rules
4	Conductors
6	Services and service equipment
8	Circuit loading and demand factors
10	Grounding and bonding
12	Wiring methods
14	Protection and control
16	Class 1 and class 2 circuits
26	Installation of electrical equipment

Table 1–1   General sections of the *CEC*.

Section	Title
28	Motors and generators
30	Installation of lighting equipment
32	Fire alarm systems and fire pumps
62	Fixed electric space and surface heating systems
68	Pools, tubs, and spas
76	Temporary wiring

Table 1–2   Some supplementary or amending sections of the *CEC*.

6–112 in *Appendix B* outlines the components that may be used for an overhead mast service and methods of installation.

Appendix D is Tabulated general information. This text will reference several tables, including *Tables D1, D3,* and *D6.*

Table of Contents and Index

The *CEC* provides an index and a table of contents to help you look up information. If you know exactly what you want look up, use the index in the back of the book. If you just have a general idea of what you are looking for, use the table of contents at the front of the book.

Code Terminology

Shall: indicates a mandatory requirement.

- shall be: compulsory, mandatory, a requirement
- shall have: the same as *shall be*
- shall not: not permitted, not allowed, must not be
- shall be permitted: is acceptable, is allowed, is permitted

May: indicates permitted or allowable

Acceptable: means acceptable to the authority having jurisdiction

Notwithstanding: in spite of

Practicable: feasible or possible

This text is designed to be used in conjunction with the *CEC*. Throughout the text, reference will be made to the rules, tables, and appendices found in the 21st (2009) edition of the *CEC*. To distinguish these from the tables and appendices in this text, all references to them are in italics.

Standards

Standards describe minimum performance levels for equipment and systems. The *Canadian Electrical Code* Part II found in *Appendix A* consists of Safety Standards for Electrical Equipment and is a mandatory part of the *CEC*. Another standard used in the electrical field is TIA-570A, which is used when installing communication cabling.

PERSONNEL SAFETY

When it comes to personnel safety, you have to make it your responsibility.

When you ask most people if they work safely, they answer that they do. The question is, do you know enough about safe working practices to work safely? See if you can answer the five questions below. If you can't, you require additional safety training.

- What are the proper procedures for locking out and tagging electrical equipment?
- When is a worker required to wear a safety belt or safety harness on a construction site?
- What are five pieces of personal protective equipment?
- What are the eight classes of products identified by WHMIS symbols?
- What precautions should be taken when using ladders for electrical installations?

You can obtain information about safe working practices from the provincial government, construction safety associations, and construction unions. The *Construction Health & Safety Manual* produced by the Construction Safety Association of Ontario (www.csao.org) is an excellent example.

TESTING AND ACCREDITATION

All electrical equipment sold or installed in Canada is required to be approved for its intended use. Approved equipment has met specified safety standards of federal and provincial governments. Currently several companies are accredited to approve electrical equipment. Equipment that has been approved by an approval agency will have an identifying label such as CSA or ULC. If you are unsure about an approval label other than CSA or ULC, look for C at the eight o'clock position on the label. For equipment that has not been approved, arrangements may be made for field approval by contacting the inspection authority in your province. This authority should also be able to provide you with a complete list of companies that can approve electrical equipment in your province.

STANDARDS COUNCIL OF CANADA (SCC)

The objective of the Standards Council of Canada is to promote efficient and effective voluntary standardization as a means of advancing the national economy and benefiting the health, safety, and welfare of the public.

The council fulfills its mandate by accrediting organizations engaged in standard development, certification, testing and quality, and environmental registration. It is also responsible for approving National Standards of Canada (NSC).

The council represents Canada on international standards bodies or agencies and accredits Canadian technical experts to international standards committees. It also offers a comprehensive family of information, products, and services dealing with standards, regulations, and conformity assessment procedures. Through these activities, the council facilitates Canadian and international trade, protects consumers, and promotes international standards cooperation.

For more information, write to:

Standards Council of Canada
270 Albert St.
Suite 200
Ottawa, ON
K1P 6N7
Tel: 613 238-3222
http://www.scc.ca

CANADIAN STANDARDS ASSOCIATION (CSA)

The Canadian Standards Association is a nonprofit organization that was formed in 1919 to develop standards for a variety of organizations and disciplines. These standards are the cornerstone for product certification, and are subsequently approved by provincial and federal government authorities to become recognized throughout the country.

Thousands of manufacturers work directly with the CSA on the certification process, and the CSA mark appears on many new products every year.

For more information, write to:

Canadian Standards Association
Customer Service Department
178 Rexdale Boulevard
Etobicoke, ON   M9W 1R3
Tel: 416 747-4000
      1-800-463-6727
Fax: 416 747-4149
http://www.csa.ca

UNDERWRITERS LABORATORIES OF CANADA (ULC)

Underwriters Laboratories of Canada is a nonprofit organization that maintains and operates laboratories and the certification service for testing and classification of devices, constructions, materials, and systems to determine their relation to life, fire, and property hazards.

ULC also publishes standards, specifications, and classifications for products having a bearing on fire, accident, or property hazards.

For more information, write to:

Underwriters Laboratories of Canada
7 Underwriters Rd.
Toronto, ON
M1R 3B4
Tel: 1-866-937-3852 or 416 757-3611
Fax: 416 757-8727
http://www.ulc.ca
E-mail: customerservice@ulc.ca

Electrical Inspection

The *CEC* requires that a permit be obtained for the inspection of all work with respect to the installation, alteration, or repair of electrical equipment. It is the responsibility of the electrical contractor or person doing the work to obtain the permit. When the installation has passed inspection, the inspection authority will issue a connection permit (also called authorization for connection or current permit) to the supply authority. The supply authority will then make connection to the installation.

Since local codes vary, the electrician should also check with the local inspection authority that enforces the *CEC*. Electrical utility companies can supply additional information on local regulations. The electrician should never assume that the *CEC* is the recognized standard everywhere. Any city or province may adopt electrical installation and licensing laws. In many cases these laws are more stringent than the *CEC*.

Property	SI unit	Symbol
Length	metre	m
Area	square metre	m²
Volume	cubic metre	m³
Time	second	s
Speed (velocity)	metre/second	m/s
Acceleration	metres/second/second	m/s²
Temperature	degrees celsius	˚C
Force	newton	N
Energy	joule	J
Work	joule	J
Power	watt	W
Mass	kilogram	kg
Weight	kilogram	kg
Pressure	pascal	Pa
Flow	litres/min	l/m
Electric charge	coulomb	C
Electric current	ampere	A
Electric potential difference	volt	V
Luminous flux	lumen	lm
Illuminance	lux	lx

Table 1–3   Metric units of measure.

UNITS OF MEASURE

Two common systems of measurement are used in Canada: the System International (SI), or metric, and the American standard unit (ASU), which uses inches and pounds. Although SI is the official system of measurement in Canada, the ASU system is widely used. Refer to Tables 1–1 to 1–5.

Length

The basic SI unit of measure for length is the metre. The SI system has a base of 10 (the next larger unit is 10 times the size of the smaller unit) and uses prefixes to identify the different units of length.

Prefix	Value
Mega (M)	1 000 000
Kilo	1000
Hecto	100
Deka (Deca)	10
Base unit (Metre)	1
Deci	0.1
Centi (c)	0.01
Milli (m)	0.001
Micro (μ)	0.000 001
Nano (n)	0.000 000 001

Table 1–4   SI prefixes.

Construction drawings normally give SI dimensions in millimetres or metres to prevent confusion.

The ASU commonly uses inches, feet, yards, rods, and miles to measure length.

Unit	Value
Inch	$\frac{1}{36}$ yard
Foot	12 inches
Yard	3 feet
Rod	5$\frac{1}{2}$ yards, 16$\frac{1}{2}$ feet
Mile	320 rods, 1760 yards, 5280 feet

Table 1–5   ASU units.

Construction drawings may use feet ('), inches ("), and fractions of an inch as units of measure. A standard steel tape measure that uses ASU units is divided into feet, inches, and fractions of an inch. The first 12 inches of the tape normally has divisions of $\frac{1}{32}$ of an inch. The remainder of the tape is divided into 16ths of an inch. Fig. 1–1 shows how to read a tape measure using ASU units. An architect's scale is used for working with construction

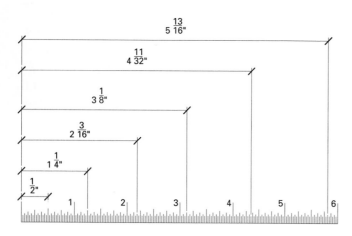

Fig. 1–1   Reading tape measure divisions.

are used to designate the trade size of the raceway. Table 1–7 lists the equivalent metric and American designator units.

Outlet boxes are still referenced in inches. *Table 23* of the *CEC* gives the trade sizes of boxes in inches. The volume of the boxes is given in both cubic inches and mL (cc).

American wire gauge sizes (AWG) are still used in the 2009 *CEC*. Metric wire sizes are not identical to AWG sizes. Standard metric sizes generally fall between American wire gauge sizes. Table 1–8 gives standard AWG sizes and their circular mil area and metric wire sizes and their equivalent circular mil size.

drawings that use ASU units. Scales are covered in Unit 2. Table 1–6 gives useful conversions from ASU to metric units.

Common Conversions
Trade Sizes

Raceways such as rigid conduit and electrical metallic tubing (EMT) are designated by trade sizes. This is because the actual inside diameter of a conduit or tubing is not its stated size. For example, $1/2$" (trade size) rigid steel conduit has an inside diameter of 0.632 inches, $1/2$" EMT has an internal diameter of 0.622 inches, and $1/2$" ENT (non-metallic tubing) has an internal diameter of 0.560 inches. All are considered to be $1/2$" (or 16 mm) trade size.

The 2009 *CEC* designates raceways in metric units only: The actual size of the raceway has not changed, only the fact that metric units of measure

Metric Designator	American Designator
12	$3/8$
16	$1/2$
21	$3/4$
27	1
35	$1 1/4$
41	$1 1/2$
53	2
63	$2 1/2$
78	3
91	$3 1/2$
103	4
129	5
155	6

Table 1–7   Metric and American designators for common trade sizes of raceways.

Property	ASU unit	Multiplication factor for conversion to SI	SI unit
Length	Inch	× 25.4	Millimetre
	Foot	× 304.8	Millimetre
	Yard	× 914.4	Millimetre
	Rod	× 5.0292	Metre
	Furlong	× 201.17	Metre
	Mile	× 1.609347	Kilometre
Area	Sq. in.	× 645.16	Sq. mm
	Sq. ft.	× 92903	Sq. mm
	Sq. yd.	× 0.8354	Sq. m
Volume	Cu. In.	× 16.39	cc or mL
	Cu. ft.	× 0.02832	Cubic metres
	Cu. yd.	× 0.7646	Cubic metres
Liquid volume	Fl. oz	× 29.6	Millilitres
	Gallons	× 3.78	Litres

Table 1–6   Converting ASU to SI, or metric, units.

American Wire Size AWG	Circular Mils or Equivalent Circular Mils	Metric Wire Size mm²
	937	0.50
20	1020	
18	1620	
	1974	1.0
16	2580	
	2960	1.5
14	4110	
	4934	2.5
12	6530	
	7894	4.0
10	10380	
	11840	6.0
8	16510	
	19740	10.0
6	26240	
	31580	16
4	41740	
	49340	25
3	52620	
2	66360	
	69070	
1	83690	
	98680	50
1/0	105600	
2/0	133100	
	138100	70
3/0	167800	
	187500	95
4/0	211600	

Source: Courtesy of Thomas & Betts Incorporated

Table 1–8   Standard AWG wire sizes and metric (mm²).

REVIEW

1. What is the difference between a code and a standard?

   _____

   _____

2. What code sets standards for the installation of electrical equipment? _____

   _____

3. What authority enforces the standards set by the *CEC*? _____

   _____

4. Does the *CEC* provide minimum or maximum standards? _____

   _____

5. What do the letters CSA signify? _____

   _____

6. Does compliance with the *CEC* always result in an electrical installation that is adequate, safe, and efficient? Why?

    _____

    _____

    _____

    _____

    _____

7. What are the general sections of the *CEC*?

    _____

    _____

8. Is the section of the *CEC* that deals with wiring methods a general section or an amending section?

    _____

9. When is an electrical installation required to be inspected?

    _____

    _____

10. What should you look for when trying to determine if a piece of electrical equipment is approved for use in Canada?

    _____

    _____

11. If a piece of electrical equipment is not approved for use in Canada, what should you do?

    _____

    _____

12. When the words "shall be" appear in a code reference, they mean that it (must) (may) (does not have to) be done. (Underline the correct answer.)

UNIT 2

Drawings and Specifications

OBJECTIVES

After studying this unit, the student will be able to

- identify the line types used in construction drawings
- visualize building views
- read common scales
- explain how electrical wiring information is conveyed to the electrician at the construction or installation site
- demonstrate how the specifications are used in estimating cost and in making electrical installations
- identify and explain the application of the various common line types used on drawings
- explain why symbols and notations are used on electrical drawings
- identify symbols used on architectural, mechanical, and electrical drawings

TECHNICAL DRAWINGS

Technical drawings are used to show the size and shape of an object to the viewer. The basic elements of a technical drawing are lines, symbols, dimensions, and notations.

Lines

The lines that make up working drawings are sometimes referred to as the alphabet of lines since each one is unique and conveys a special meaning. Fig. 2–1 shows types of lines commonly found on construction drawings.

Visible lines Visible lines are continuous lines that show the visible edges of an object. An example of a visible line would be the outside edge of the foundation wall shown on sheet 5 of the drawings provided with this book.

Hidden lines Hidden lines are short dashed lines. They show the edges of an object that are hidden from view by surfaces that are closer to the viewer. Hidden lines are used to show the building footings on sheet 1 of the drawing set.

— — — — — — — — — — — — — — —

Centre lines Centre lines show the centre of an arc or circle. They consist of alternating long and short dashes. The exact centre of a circle is normally indicated by the crossing of the two short dashes.

——— — ——— — ———

Section lines A sectional view (also called a section or cross section) is a view of a building in which we imagine a portion of the building has been sliced off to reveal the internal construction details. Section lines show the solid parts (e.g., walls, floors) of the

Border Line	**Centre Line**
Visible Object Line	**Cutting Plane Line**
Hidden Line	////////////////// **Section Lines**
Construction Line	**Extension Lines**
Projection Line	**Dimension Lines**
Phantom Line	**Arrowheads**
Break Line	**Ticks**
Contour Line	**Leader**

Fig. 2–1   Line types.

portion of the building shown in the sectional view. Section lines are also called crosshatching. Different types of crosshatching represent different types of materials used in the building. In section A-A on sheet 7 of the drawings, different types of cross-hatching indicate concrete used for the basement floor, cement blocks used for the basement walls, and bricks used on the outside walls above grade.

Cutting plane lines   Cutting plane lines consist of a long dash followed by two short dashes. They indicate where a building is being sectioned. The basement floor plan, sheet 7 of the drawings, uses a cutting plane line to show where the building is being sectioned. The arrows at each end of a cutting plane line indicate the direction you are looking when you are viewing the section. Refer to section A-A on sheet 7.

Phantom lines   Phantom lines are short dashed lines (about 1.5 times as long as the dashes for hidden lines) that show an alternate position. The diagram on page 80 shows two single-pole, double-throw (SPDT) switches. Closed switch positions are shown by a phantom line.

Break lines   Break lines are used where only a part of the drawing needs to be shown. For example, a break line might be used in a connection diagram where it is necessary to show the termination of both ends of a cable but not the cable in between.

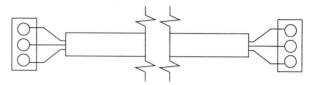

Contour lines   Contour lines are used on plot plans to show changes in elevation. Dashed contour lines indicate existing grades, and solid contour lines show finished grades.

Lineweight refers to the thickness of a line used on a drawing. Object lines have a heavier lineweight (are thicker) than other lines such as hidden lines or dimension lines. Items of other trades that appear on electrical drawings normally have a lighter weight (are thinner) than the electrical items.

VISUALIZING A BUILDING

Pictorial drawings are three-dimensional drawings that show two or more surfaces of an object in one view. Three types of pictorial drawings are shown in Fig. 2–2. Fig. 2–3 shows a pictorial drawing of the wiring of a house.

Unfortunately, pictorial drawings distort shapes such as angles, arcs, and circles and are hard to dimension. Since it is important that construc-

Fig. 2–2   Pictorial drawings.

Fig. 2–3   Three-dimensional view of house wiring.

tion drawings show the exact size and shape of an object, most construction drawings are two-dimensional drawings.

Construction drawings are made using orthographic projection, which shows the true size and shape of an object through a number of views. Each view shows two dimensions. To visualize an object (such as a house), two or more views will often be required. Fig. 2–4 shows an example of orthographic projection.

Construction drawings use the terms *plan* for a top view and *elevation* for a side view.

BUILDING VIEWS
Plan Views

A plan view shows how a building will look when viewed from directly above it. Plan views are used to show lengths and widths. Floor plans and site plans are examples of plan views. Fig. 2–5 shows the first-floor plan of the house used in this book.

Fig. 2–5   Floor plan.

Elevations

An elevation is a side view of the building. Elevations are used to show heights and widths. Elevations are commonly used to show outside walls, interior walls, and the placement of such things as cupboards and equipment. Fig. 2–6 shows the front (south) elevation of the house.

Sections

A sectional view of the building (after a portion of the building has been removed) is used to show the interior construction details of the building.

Fig. 2–4   Orthographic projection.

Fig. 2–6   Front elevation.

Fig. 2–7 shows a portion of section A-A on drawing sheet 7.

Fig. 2–7   Typical wall section.

SYMBOLS AND NOTATIONS

Graphical symbols are used on construction drawings to represent equipment and components. Symbols are used to show the size, location, and ratings of equipment. The Standards Council of Canada (SCC) approves national standards for symbols used on construction drawings and accredits organizations such as CSA that are engaged in standards development. CSA standard Z99.3 -1979 (1989) includes symbols for electrical drawings. They relate closely to the symbols published by the Institute of Electrical and Electronics Engineers (IEEE) and the American National Standards Institute (ANSI), which publish similar standards for the United States. Fig. 2–8 is a portion of a floor plan showing the symbols for two three-way switches controlling an overhead light and two duplex receptacles. Fig. 2–9 (a–e) shows symbols commonly used on construction drawings.

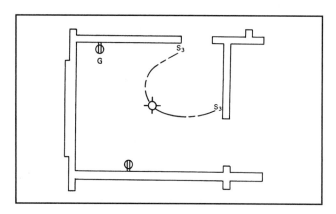

Fig. 2–8   Electrical symbols on a floor plan.

Notes to symbols:

- When symbols are used to indicate future equipment, they are shown with dashed lines and will have a note to indicate such use.
- The orientation of a symbol on a drawing does not alter the meaning of the symbol.
- Lighting fixtures are generally drawn in a manner that will indicate the size, shape, and orientation of the fixture.
- Symbols found on electrical plans indicate only the approximate location of electrical equipment. Where exact locations are required, check the drawings for dimensions and confirm all measurements with the owner or owner's representative.

A notation will generally be found on the plans (blueprints) next to a specific symbol calling attention to a variation, type, size, quantity, or other necessary information. In reality, a symbol might be considered to be a notation because, according to the dictionary, symbols "represent words, phrases, numbers, and quantities."

Another method of using notations to avoid cluttering up a blueprint is to provide a system of symbols that refer to a specific table. For example, the written sentences on plans could be included in a table referred to by a notation. Fig. 5–4 on page 73 is an example of how this could be done. The special symbols that refer to the table would have been shown on the actual plan.

If the word "typical" appears as a notation, it means that construction details are uniform throughout the building. An example would be "typical two-bedroom suite." This indicates all the two-bedroom suites are identical.

DIMENSIONS

Dimensions are used to show the size of an object. Metric dimensions are normally expressed in millimetres or metres.

When the American system of units is used on building drawings such as floor plans, measurements are expressed in feet, inches, and fractions of an inch. This is referred to as architectural practice. There are 12 inches in 1 foot. A ' mark after a number indicates a measurement in feet.

Fig. 2–9 (a)   Architectural drafting symbols.

Fig. 2–9 (b) Architectural drafting symbols (continued).

PIPING	
Piping, in general (Lettered with name of material conveyed)	
Non-intersecting pipes	
Steam	
Condensate	
Cold water	
Hot water	
Air	
Vacuum	
Gas	
Refrigerant	
Oil	

PIPE FITTINGS		
For welded or soldered fittings, use joint indication shown below	Screwed	Bell and spigot
Joint		
Elbow - 90 deg		
Elbow - 45 deg		
Elbow - turned up		
Elbow - turned down		
Elbow - long radius		
Side outlet elbow - outlet down		
Side outlet elbow - outlet up		
Base elbow		
Double branch elbow		
Single sweep tee		
Double sweep tee		
Reducing elbow		
Tee		
Tee - outlet up		
Tee - outlet down		
Side outlet tee outlet up		
Side outlet tee outlet down		
Cross		
Reducer		
Eccentric reducer		

PIPE FITTINGS (continued)		
For welded or soldered fittings, use joint indication shown below	Screwed	Bell and spigot
Lateral		
Expansion joint flanged		

VALVES		
For welded or soldered fittings, use joint indication shown below	Screwed	Bell and spigot
Gate valve		
Globe valve		
Angle globe valve		
Angle gate valve		
Check valve		
Angle check valve		
Stop cock		
Safety valve		
Quick opening valve		
Float opening valve		
Motor operated gate valve		

PLUMBING	
Corner bath	
Recessed bath	
Roll rim bath	
Sitz bath	
Foot bath	
Bidet	
Shower stall	
Shower head	(Plan) (Elev)
Overhead gang shower	(Plan) (Elev)
Pedestal lavatory	PL
Wall lavatory	WL
Corner lavatory	Lav
Manicure lavatory Medical lavatory	ML
Dental lavatory	Dental lav

PLUMBING (continued)	
Plain kitchen sink	S
Kitchen sink, R & L drain board	
Kitchen sink, L H drain board	
Combination sink & dishwasher	
Combination sink & laundry tray	S & T
Service sink	SS
Wash sink (Wall type)	
Wash sink	
Laundry tray	LT
Water closet (Low tank)	
Water closet (No tank)	
Urinal (Pedestal type)	
Urinal (Wall type)	
Urinal (Corner type)	
Urinal (Stall type)	
Urinal (Trough type)	TU
Drinking fountain (Pedestal type)	DF
Drinking fountain (Wall type)	DF
Drinking fountain (Trough type)	DF
Hot water tank	HWT
Water heater	WH
Meter	M
Hose rack	HR
Hose bibb	HB
Gas outlet	G
Vacuum outlet	
Drain	D
Grease separator	
Oil separator	
Cleanout	
Garage drain	
Floor drain with backwater valve	
Roof sump	

Types of joints

Flanged   Screwed   Bell & spigot   Welded   Soldered

Fig. 2–9 (c)   Standard symbols for plumbing, piping, and valves.

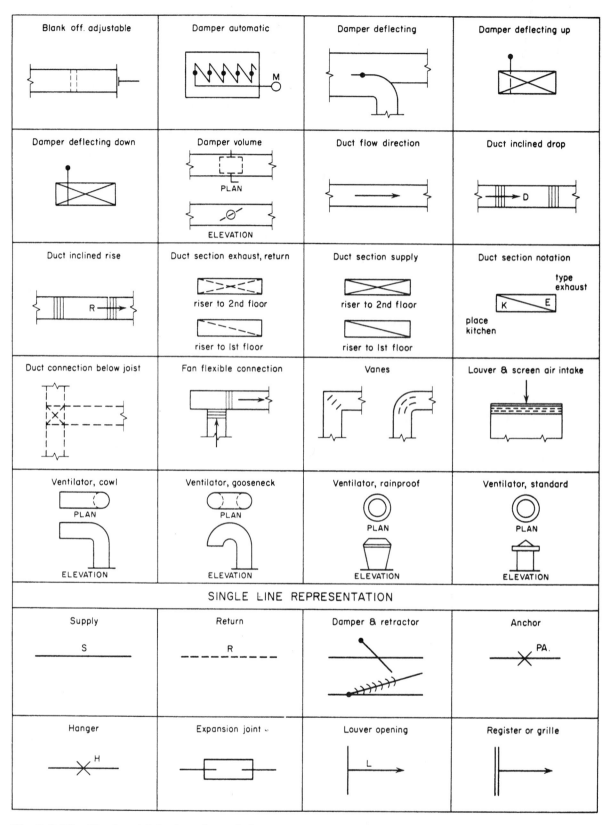

Fig. 2–9 (d)   Sheet metal duct work symbols.

ELECTRICAL FLOOR PLAN SYMBOLS

Symbol	Description
$ OR S	SINGLE POLE SWITCH
$_3 OR S_3	3 WAY SWITCH
$_4 OR S_4	4 WAY SWITCH
$_D OR S_D	DOOR SWITCH
$_{DIM} OR S_{DIM}	DIMMER SWITCH
$_{RCM} OR S_{RCM}	REMOTE CONTROL MASTER
$_{RC} OR S_{RC}	REMOTE CONTROL SWITCH
$_M OR S_M	MOTOR STARTER SWITCH
$_a OR S_a	LOWER CASE LETTER DESIGNATES SWITCHING ARRANGEMENT

SINGLE RECEPTACLE 5-15R

DUPLEX RECEPTACLE 5-15R

DUPLEX REC. 5-15R FLOOR MOUNTED

DUPLEX RECEPTACLE 5-20 RA

SPLIT SWITCHED DUPLEX RECEPTACLE 5-15 R

3 CONDUCTOR SPLIT RECEPTACLE 5-15R

DRYER RECEPTACLE

RANGE RECEPTACLE

SPECIAL PURPOSE OUTLET

MULTI-OUTLET ASSEMBLY

TRACK LIGHTING LENGTH AND NUMBER OF FIXTURES AS NOTED

CEILING OUTLET

WALL OUTLET

FLUORESCENT LUMINAIRE
UPPER CASE LETTER DENOTES STYLE
LOWER CASE LETTER DENOTES
SWITCHING ARRANGEMENT

RECESSED FLUORESCENT LUMINAIRE

FLUORESCENT STRIP LUMINAIRE

CLOCK OUTLET

EXIT LIGHT

FAN OUTLET

IONIZATION TYPE SMOKE DETECTOR

PHOTOELECTRIC TYPE SMOKE DETECTOR

THERMOSTAT

TRANSFORMER SIZE AND TYPE AS NOTED.

PUSHBUTTON

BUZZER

BELL

DOOR OPENER

DOOR CHIME

TV OUTLET

WATER HEATER

DATA OUTLET

TELEPHONE OUTLET

DATA / TELEPHONE OUTLET

PADDLE FAN

PADDLE FAN WITH LAMP

ELECTRICAL PANEL FLUSH MOUNTED

ELECTRICAL PANEL SURFACE MOUNTED
DP DISTRIBUTION PANEL
LP LIGHTING PANEL
PP POWER PANEL

BRANCH CIRCUIT WIRING. SHORT LINES INDICATE UNGROUNDED CONDUCTORS. THE LONG LINE INDICATES AN IDENTIFIED OR NEUTRAL CONDUCTOR. A BACKWARDS SLASH INDICATES A BONDING CONDUCTOR. THE ARROW INDICATES A HOME RUN.

Fig. 2–9 (e)   Electrical symbols for floor plans.

If the number is followed by ", it represents a dimension in inches. Architectural dimensions may be expressed in feet and inches or in inches; for example a dimension of 3'6" may be expressed as 42". The standard on construction drawings is to use feet and inches.

Building dimensions are normally shown on the architectural drawings from which all measurements should be taken. Be sure to confirm dimensions with the owner or architect when exact positioning of equipment is required.

SCALE

A drawing of an object that is the same size as the object is said to be drawn full scale or at a scale of 1:1.

Construction drawings reduce the size of a building down to fit on a piece of paper by drawing everything proportionally smaller than it actually is. This is called reduced scale. A drawing that uses a scale of 1:50 indicates that the building is 50 times larger than the drawing. A drawing with a scale of $\frac{1}{4}$ in = 1 ft has a scale of 1:48.

Enlarged scale would produce a drawing larger than the actual object. Printed circuit boards are drawn using enlarged scale since many of the components used in the manufacture of printed circuit boards are very small. An example of enlarged scale is 2:1, which indicates the drawing is twice as large as the component.

Types of Scales (Measuring Instruments)

Three types of scales are commonly used on construction drawings. When the drawing is in feet and inches, an architectural scale is used. Fig. 2–10 shows a portion of an architect's triangular scale showing the $\frac{1}{8}$" and $\frac{1}{4}$" scales.

For metric drawings, a metric scale is used. For working with civil drawings (e.g., roads, dams, bridges), an engineering scale is used. Table 2–1 lists common construction drawing scales.

Fig. 2–11 gives examples of reading scales.

Fig. 2–10   Architect's scale.

Metric	Architectural	Engineering
1:20	1/2" = 1'0"	1" = 10
1:50	1/4" = 1'0"	1" = 20
1:100	1/8" = 1'0"	1" = 30
1:200	1/16" = 1'0"	1" = 40
1:500	1/32" = 1'0"	1" = 50

Table 2–1   Common construction drawing scales.

METRIC SCALE 1:100

METRIC SCALE 1:50

ARCHITECTURAL SCALE 1/8"=1FT

ARCHITECTURAL SCALE 1/4"=1FT

ENGINEERING SCALE 1"=10FT

ENGINEERING SCALE 1"=50FT

Fig. 2–11   Reading scales.

THE WORKING DRAWINGS

The architect or engineer uses a set of working drawings or plans to make the necessary instructions available to the skilled trades that are to build the structure. They show sizes, quantities, and locations of the materials required and the construction features of the structural members. Each skilled construction trade—masons, carpenters, electricians, and others—must study and interpret these details of construction before the actual work is started.

The electrician must be able to convert the two-dimensional plans into an actual electrical installation and visualize the many views of the plans and coordinate them into a three-dimensional picture, as shown in Fig. 2–3.

The ability to visualize an accurate three-dimensional picture requires a thorough knowledge of blueprint reading. Since all of the skilled trades use a common set of plans, the electrician must be able to interpret the lines and symbols that refer to the electrical installation and also to interpret those used by the other construction trades. The electrician must know the structural makeup of the building and the construction materials to be used.

The Drawing Set

A set of drawings is made up of several numbered sheets. If there are many sheets in a set, it will be broken down into subsets of architectural, electrical, mechanical, etc. The electrical drawings will be numbered E-1, E-2, E-3, etc.

A typical set of construction drawings includes the following.

A Site (Plot) Plan

The site plan shows:
- the dimensions of the lot
- a north-pointing arrow
- the location of utilities and their connection points
- any outdoor lighting
- easements (rights to the property by someone other than the owner)
- contour lines to show changes in elevation
- accessory buildings
- trees to be saved

Floor Plans

Depending on the complexity of the building, a set of drawings for a single-family dwelling may have only a single floor plan for each storey of the building.

As the complexity of the building increases there will be a number of floor plans for each storey (one for each trade). A logical breakdown would be:

architectural
electrical
mechanical
plumbing
HVAC (heating, ventilating, and air
     conditioning)
structural

On very complex installations, each electrical system such as power, lighting, voice/data, fire alarm, and security/access control may have its own plan for each storey.

Use architectural floor plans to find room measurements, wall thickness, the sizes of doors and windows, and the swing of doors. The floor plans of other trades will show the locations of the equipment associated with that trade.

Elevations

Elevations show the vertical faces of buildings, structures, and equipment. Elevations are used to show both interior and exterior vertical faces. Sheet 9, details 2 to 5, shows elevations of the kitchen cupboards. On sheet 1 you will find the elevation (south, north, east, and west) of the building.

Details and Sections

Details are large-scale drawings that are used when floor plans and elevations cannot provide sufficient detail. Sheet 6 provides details of the requirements for the installation of a swimming pool. Sheet 9 shows details of the kitchen cupboards and bathroom vanities. Sections show internal details of construction. See section A-A on sheet 7.

Legend of Symbols

A legend of symbols provides the contractor or electrician with the information to interpret the symbols on the drawing.

Schedules

Schedules are lists of equipment. The door schedule on sheet 9 of the drawing set describes the size and types of doors required for the house.

Electrical schedules are made for such equipment as lighting fixtures (luminaires), panelboards, raceways and cables, motors, and mechanical equipment connections.

Diagrams

One-line and riser diagrams are provided to quickly show the flow of power throughout the building. Riser diagrams are used to show vertical details while using a one-line format to show power, telephone, or fire alarm systems.

Shop Drawings

Shop drawings are used by equipment manufacturers to show details of custom-made equipment for a specific job. Installation diagrams that come with equipment could be considered as shop drawings.

SPECIFICATIONS

Working drawings are usually complex because of the amount of information they must include. To prevent confusing detail, it is standard practice to include with each set of plans a set of detailed written specifications prepared by the architect.

Specifications are written descriptions of materials and methods of construction. They provide general information for all trades involved in the construction of the building, as well as specialized information for the individual trades. Specifications include information on the sizes, type, and desired quality of the parts to be used in the structure.

Typical specifications include a section on "General Clauses and Conditions" which is applicable to all trades involved in the construction. This section is followed by detailed requirements for the various trades—excavating, masonry, carpentry, plumbing, heating, electrical, painting, and others.

The plan drawings for the residence used as an example for this text are included in the back of the text. The specifications for the electrical work indicated on the plans are given in Appendix A.

In the electrical specifications, the list of standard electrical parts and supplies frequently includes manufacturers' names and catalogue numbers of the specified items. This ensures that these items will be the correct size, type, and electrical rating, and that the quality will meet a certain standard. To allow for the possibility that the contractor may be unable to obtain the specified item, the words "or equivalent" are usually added after the manufacturer's name and catalogue number.

The specifications are also useful to the electrical contractor as all of the items needed for a specific job are grouped together and the type or size of each item is indicated. This information allows the contractor to prepare an accurate cost estimate without having to find all of the data in the plans.

REVIEW

Note: Refer to the *CEC* or the plans where necessary.

1. Identify three line types shown on drawing 3.

   _____

   _____

   _____

2. Determine the length of the lines below.

   Scale   Length

   1:50   _____   _____

   1:75   _____   _____

Scale   Length

1:100   _____   _____

$1/8$" = 1FT   _____   _____

$1/4$" = 1FT   _____   _____

3. What is the purpose of specifications? _____

   _____

4. In what additional way are the specifications particularly useful to the electrical

   contractor? _____

   _____

5. What is done to prevent a plan from becoming confusing because of too much detail?

   _____

   _____

6. Name three requirements contained in the specifications regarding material (Appendix A).

   a. _____

   b. _____

   c. _____

7. What are the two main hazards that the *CEC* is designed to prevent?

   a. _____   b. _____

8. What phrase is used when a substitution is permitted for a specific item in a specification?

   _____

   _____

9. What is the purpose of an electrical symbol? _____

   _____

10. What is a notation? _____

   _____

11. List 5 electrical notations found on the plans for this residence. Refer to the plans
    at the back of the text.

    1 _____

    2 _____

    3 _____

    4 _____

    5 _____

PART 2—STRUCTURAL FEATURES

1. To what scale is the basement plan drawn? _____

2. What is the size of the footing for the steel support columns in the basement?

   _____

3. To what kind of material will the front porch lighting bracket fixture be attached?

   _____

4. Give the size, spacing, and direction of the ceiling joists in the workshop.

   _____

5. What is the size of the lot on which this residence is located? _____

6. In what compass direction is the front of the house facing? _____

7. How far is the front garage wall from the curb? _____

8. How far is the side garage wall from the property lot line? _____

9. How many steel support columns are in the basement, and what size are they?

   _____

10. What is the purpose of the I-beams that rest on top of the steel support columns?

    _____

11. Is the entire garage area to have a concrete floor? _____

12. Where is access to the attic provided? _____

13. Give the thickness of the outer basement walls. _____

14. What material is indicated for the foundation walls? _____

    _____

15. Where are the smoke detectors located in the basement? _____

    _____

16. What is the ceiling height in the basement workshop from bottom of joists to floor?

    _____

17. Give the size and type of the front door. _____

    _____

18. What is the stud size for the partitions between the bathrooms in the bedroom area
    where substantial plumbing is to be installed? _____

    _____

    _____

19. Who is to furnish the range hood? _____

20. Who is to install the range hood? _____

UNIT 3

Service Entrance Calculations

OBJECTIVES

After this studying this unit, the student will be able to

- determine the total calculated load of the residence
- calculate the size of the service entrance conductors, including the size of the neutral conductors
- understand the *CEC* requirements for services, *Rule 8–200*
- fully understand special *CEC* rules for single-family dwelling service entrance conductor sizing
- perform an optional calculation for computing the required size of service entrance conductor for the residence

---

Before calculating the ampacity of service and service entrance conductors required for the residence, you must take an inventory of all loads in excess of 1500 watts. The calculations are based on the *CEC*. Always consult local electrical codes for any variations in requirements that may take precedence over the *CEC*.

SIZE OF SERVICE ENTRANCE CONDUCTORS AND SERVICE DISCONNECTING MEANS

Service entrance conductors (*Rule 8–200*) and the disconnecting means shall not be smaller than:

1. 60-ampere, three-wire, for a single-family dwelling where the floor area, excluding the basement floor area, is less than 80 m², *Rule 8–200(1)(b)(ii)*.
2. 100-ampere, three-wire, for a single-family dwelling when the floor area, excluding the basement floor area, is more than 80 m², *Rule 8–200(1)(b)(i)*.

The load on a residential service is a noncontinuous load, *Rule 8–200(3)*. This means that the service conductors, the service switch, and the panel are sized based upon the demand amperes. To correctly calculate the demand amperes for the service, you must follow the steps listed in *Rule 8–200(1)(a)*.

Step 1. Calculate the basic load based on the living area of the house, *Rule 8–200(1)(a)(i)*.

Step 2. Calculate the electric space-heating load and the air-conditioning load. Use the larger of these loads in calculating the demand if you know that they cannot be used at the same time, *Rule 8–106(4)*. If it is possible for the electric heat and the air conditioning to operate at the same time, you must include both loads in calculating the demand. If baseboard heating is used with individual thermostats in each room or area, refer to *Rule 62–116(2)* for demand watts. If the residence has an electric furnace, the load must be included at 100% of its rating, *Rule 62–116(1)*.

Step 3. Calculate the load for the electric range or the combined load for a separate cooktop and oven.

Step 4. Add the kilowatt rating of any swimming pool heaters or hot-tub and spa heaters. In this case there should be no derating, *Rule 8–200(1)(a)(v).*

Step 5. A demand factor of 25% must be applied to any additional loads in excess of 1500 watts if an electric range is included. If there is no electric range because the residence has a gas range, the first 6 kW loads in excess of 1500 watts are added at 100%. If the load is greater than 6 kW, the remainder should be added at 25%, *Rule 8–200(1)(a)(vi).*

For clarity, these calculations have been summarized in Tables 3–1 and 3–2.

SERVICE CALCULATIONS FOR A SINGLE-FAMILY DWELLING, *RULE 8–200:* CALCULATING FLOOR AREA (GROUND FLOOR)

First Floor Area

To estimate the total basic load for a dwelling, you must calculate the occupied floor area of the dwelling. Note in the residence plans that the first-floor area has an irregular shape. In this case, the simplest method of calculating the occupied floor area is to determine the total floor area using the inside dimensions of the dwelling. Then, subtract open porches, garages, and other unfinished or unused spaces from the total area if they are not adaptable for future use, *Rule 8–110.*

SERVICE CALCULATION

| Load | Demand Watts
Line 1, Line 2 | Demand Watts
Neutral (Min.) |
|---|---|---|
| Basic | 7 000 W | 7 000 W |
| Heat | 13 000 W | — |
| Range | 6 820 W | 6 820 W |
| Other—Dryer | 1 425 W | 1 425 W |
| —Water Heater | 750 W | — |
| —Water Pump | 600 W | — |
| | 29 595 W | 15 245 W |
| Line 1-2; I = P/E | 29 595/240 = 123.3 A | |
| Neutral; I = P/E | | 15 245/240 = 63.52 A |

Table 3–1 Demand watts calculation for service and neutral conductor.

MINIMUM CODE REQUIREMENTS FOR A SERVICE

Reference	Minimum Service Calculation	Panel A for Residence
Minimum service ampacity, *Rule 8–104(1)*	123.3 A	200 A
Minimum service switch	200 A	200 A
Minimum fuse or breaker, *Table 13*	125 A	200 A
Minimum ground conductor, *Table 17*	No. 6 copper	No. 6 copper
Minimum bonding jumper, *Table 41*	No. 6 copper	No. 6 copper
Minimum AWG L1, L2, *Table 2*	No. 1 RW90	No. 3/0 RW90
Min AWG neutral, *Table 2* and *Table 16*	No. 6 RW90	No. 3/0 RW90
Minimum conduit size, *Tables 8, 9, 10*	35-mm	41-mm
Minimum panel ampacity	125 A	200 A
Minimum number of circuits, *Rule 8–108(1)(c)*	30	30

Table 3–2 Service requirements.

Many open porches, terraces, patios, and similar areas are commonly used as recreation and entertainment areas. Therefore, adequate lighting and receptacle outlets must be provided.

To simplify the calculation, the author has chosen to round off dimensions for those areas (garage, porch, and portions of the inset at the front of the house) not to be included in the computation of the general lighting load.

Fig. 3–1 shows the procedure for calculating the watts needed for the first floor of this residence.

For practical purposes, the outside dimensions of this building could be taken as being 17 m and 11 m.

Basement Area

The basement area of the home contains the furnace, the hot water tank, and other service equipment. According to *Rule 8–110,* only 75% of the inside dimensions are to be used in calculating the living area of the basement because 25% houses the above equipment.

For our calculations, we make these assumptions:

a. Basement walls are 305 mm thick.
b. Garage is unexcavated so that additional support for the weight of vehicles is not required.

On this basis our calculations can be made as follows:

Main Floor Area

$16.765 - (0.1 + 0.1)$ m $= 16.565$ m
$10.670 - (0.1 + 0.1)$ m $= 10.470$ m

16.765 m $\times 10.470$ m $\quad = 175.5$ m^2

Less area not included:
2.135 m $\times 3.675$ m $= -\;7.85$ m^2
3.255 m $\times 3.740$ m $= -12.17$ m^2
3.255 m $\times 4.775$ m $= \underline{-15.54}$ m^2

Total main floor area $\qquad 139.9$ m^2

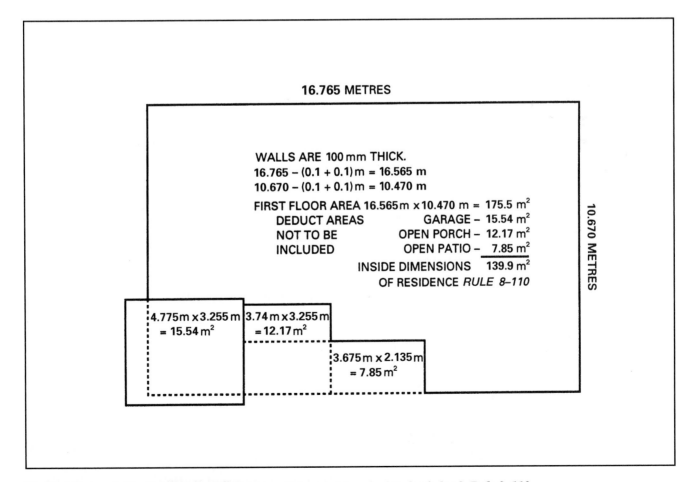

Fig. 3–1 Determining the first floor living area to be used to calculate basic load, *Rule 8–110.*

Basement Area

$$17.045 - (0.305 + 0.305) \text{ m} = 16.435 \text{ m}$$
$$10.945 - (0.305 + 0.305) \text{ m} = 10.34 \text{ m}$$

$$16.435 \text{ m} \times 10.34 \text{ m} = 169.94 \text{ m}^2$$

Less area not included:

2.1 m × 3.75 m	=	− 7.88 m²
3.45 m × 3.2 m	=	−11.04 m²
3.2 m × 4.71 m	=	−15.07 m²

Total basement area   135.95 m²

Basic Load Calculations, *Rule 8–200(1)(a)(i,ii)*

For this calculation, *Rule 8–200(1)(a)(i,ii)* requires that the load be based on 100% of the living area on the main floor plus 75% of the living area in the basement.
Area to be considered:

100% of first floor (139.9 m²)	139.9 m²
75% of the basement area (135.95 m²)	101.96 m²
Total area	241.86 m²

$$\text{Total area} = 241.86 \text{ m}^2$$
$$\text{1st } 90 \text{ m}^2 = 5000 \text{ W}$$

This leaves 151.86 m² (241.86 m² − 90 m²)

An additional 1000 W must be added for each 90 m² *or portion thereof* in excess of 90 m². Since 151.86/90 = 1.7, use 2 times.

$$2 \times 1000 \text{ W} = 2000 \text{ W}$$

A. Basic load = 5000 W + 2000 W = 7000 W

Electric furnace and air conditioning, *Rule 8–200(1)(a)(iii)*

Since you know that the air conditioning and heating will not be used simultaneously, use whichever is the greater load in calculating the demand, *Rule 8–106(4)*. Therefore, if the air conditioner load is less than the heating load, it need not be included in the calculations. Air conditioner load is:

$$30 \text{ amps} \times 240 \text{ volts} = 7200 \text{ W}$$

This is less than the heating load.

The electric furnace for this home is applied at a demand of 100%, based on *Rule 62–116(1)*. If this home had been heated with baseboard heaters, we would have been able to derate the amount over 10 kW at a demand factor of 75%, *Rule 62–116(2)(a)*.

B. Electric furnace = 13 000 W

Electric range or wall-mounted oven and counter-mounted cooking unit combined, *Rule 8–200(1)(a)(iv)*

Wall-mounted oven = 6600 W
Counter-mounted cooking unit = 7450 W
Total = 14 050 W (14.05 kW)
For rating of 12 kW or less load = 6000 W
Add 2.05 kW × 40% = 820 W

C. Range load = 6820 W
D. Dryer 5700 W × 25% = 1425 W
E. Water heater 3000 W × 25% = 750 W
F. Pump 2400 W × 25% = 600 W
A + B + C + D + E + F = Total demand load
= 29 595 W

Summary of Calculations

Step 1: Basic Load
241.86 m²
−90 m²   1st 90 m² = 5 000 W
151.86 ÷ 90
= 1.7 or 2 × 1 000 = 2 000 W
Demand watts = 7 000 W   **7 000 W**

Step 2: Heat   13 000 W **13 000 W**
(electric furnace @ 100%)

Step 3: Range 1st 12 kW = 6 000 W
2.05 kW × 40% = 820 W
Demand watts = 6 820 W **6 820 W**

Step 4: Other Loads:
Dryer × 25% = 1 425 W
Water Heater × 25% = 750 W
Pump × 25% = 600 W
Demand watts = 2 775 W **2 775 W**

Total load (demand watts)   **29 595 W**

Service Entrance Conductor Size

The ungrounded conductors are sized from *CEC Table 2* for copper wires and *Table 4* for aluminum. It is common to have aluminum conductors installed up to the line side of the meter base and from the load side of the meter base into the panel.

Take care in this type of installation to ensure that the aluminum conductors are installed in lugs rated and approved for the purpose, *Rule 12–118(3)*.

An appropriate joint compound, e.g., Pentrox™, must be used to penetrate the oxide film and prevent it from reforming when terminating or splicing all sizes of stranded aluminum conductors, *Rule 12–118(2)*.

It is important to follow the requirements of *Rule 2–130*. All conductors that are exposed to direct sunlight must be approved for the purpose. Conductors that are approved for exposure to sunlight will be surface-marked appropriately.

Neutral Conductor

According to *Rule 4–022(1)*, the neutral conductor is sized according to the unbalanced load on the service. This wire shall be the larger of the bonding conductor, sized from *CEC* as specified in *Rule 10–204(2)*, or the calculated ampacity as determined from the demand watts calculation shown in Table 3–1 on page 24. In this example, both methods require a minimum #6 AWG conductor.

This conductor may be run bare provided it is made of copper and is run in a raceway, *Rule 6–308(a)*. Verify with your local inspection authority before proceeding.

The neutral conductor is sized by the maximum unbalanced load between line 1, line 2, and the neutral.

This means that 240-volt loads are not included in calculating the demand watts on the neutral. These may include:

- central air conditioner
- electric furnace
- water pump
- water heater

Since some utilities do not allow a reduction in the size of the neutral conductor for services less than 200 amperes, check with provincial codes or the local inspection authority before proceeding with a reduced neutral installation.

Consumer Service Ampacity Rating

The minimum rating of a consumer service is based on a comparison between the overcurrent device and the ampacity of the line conductors.

Whichever is the lower value determines the ampacity rating of the service. This would also apply for branch circuits or feeders, *Rule 8–104(1)*.

Main Service Switch

The main disconnect for a service is described in the *CEC* as the service box. This could be a separate switch containing fuses or a circuit breaker. In reality it is more commonly installed as a combination panel comprising the service entrance portion and the distribution portion.

The standard available sizes of disconnecting means are 30 amperes, 60 amperes, 100 amperes, 200 amperes, and 400 amperes.

Optional available sizes are 125-ampere and 225-ampere panels. However, most residential applications are now either 100 amperes or 200 amperes. Using the demand watts calculation in Table 3–1, it is determined that a 125-ampere service would be sufficient for this residence. This would, however, leave little room for future growth, such as an outbuilding for a workshop or a hot tub. For this reason, it has been decided that a 200-ampere service will be installed.

Main Service Fuse or Circuit Breaker (*Table 13*)

The size of the main fuse or circuit breaker is based on the ampacity of the line conductors, as found in *Table 2*. This ampacity is used to determine the size of the fuse or circuit breaker from *Table 13, Rule 14–104(a)*.

Main Service Ground (*Table 17*)

The main service ground is sized from *Table 17* (see Table 3–3), and is based on the ampacity of the line 1 and line 2 conductors, *Rule 10–206(1)*.

Table 17
Minimum size of grounding conductor
(See *Rules 10–206*, *10–700*, and *10–812*.)

Ampacity of largest service conductor or equivalent for multiple conductors not exceeding—A	Size of copper grounding conductor, AWG
100	8
200	6
400	3
600	1
800	0
Over 800	00

Note: *The ampacity of the largest service conductor, or equivalent if multiple conductors are used, is to be determined from the appropriate table in the CEC, taking into consideration the number of conductors in the raceway or cable and the type of insulation.*

Table 3–3   *Table 17, CEC:* **Minimum size of grounding conductor for systems or common grounding conductor.**

Main Service Bonding Conductor (*Table 41*)

The bonding jumper for metallic service raceways is sized from *Table 41* and is based on the ampacity of the largest service conductor, *Rule 10–614(1) (a,b)*. This jumper connects the service raceway by means of a ground bushing to the metal enclosure. Accordingly, it provides a low impedance path for adequate bonding to prevent dangerous conditions that may result from equipment failure or lightning strikes.

Main Service Conduit Size (*Tables 8, 9, and 10*)

The minimum size conduit for the service entrance is based on 40% conduit fill (*Table 8*). The total area of the conductors is calculated using *Table 10*. Then the 40% column in *Table 9* is used to find the minimum size of the raceway. If all the conductors are the same type and size, *Table 6* may be used to size the raceway. For this service, a 41-mm conduit is adequate for the three No. 3/0 RW90 conductors.

Note that the service mast is usually 63-mm rigid metal, but the conduit from the meter base into the panel must be sized using the method described.

Main Service Panel Ampacity and Number of Circuits

The ampacity of the service panel is based on the ampacity of the main service switch, for example, a 100-ampere switch and a 100-ampere panel.

The minimum number of circuits for the panel is listed in *Rule 8–108(1)*. In this case, the 200-ampere service requires a minimum of 30 circuits, *Rule 8–108(1)(c)(ii)*. If the house had been heated with baseboard heaters, it would have required 40 circuits. The baseboard heaters would have required a number of two-pole breakers (30 amperes maximum), whereas a central furnace could be fed by a 70-ampere two-pole breaker, taking up only two positions.

Rule 8–108(2) states that two additional spaces must be left for future overcurrent devices. This is sometimes interpreted to mean that two circuits must be added to the required number in *Rule 8–108(1)*, but for the purposes of this calculation *Rule 8–108(2)* is satisfied by leaving two spaces available in the 30-circuit panel. These two spaces could be used to feed a future subpanel.

According to *Table 2*, No. 1 AWG RW90 will be the minimum size of conductor for the 29 595 watt (123.3 ampere) load.

For the service to be 200 amperes, No. 3/0 AWG RW90 (200-ampere) will be required.

According to *Table 2*, the minimum size neutral is No. 6 (RW90, 65 amperes).

The 240-volt loads are not to be included in the calculation for neutral conductors.

Calculation for Subpanel B

The calculation for the subpanel is based on the load supplied. *Rule 6–102(1)* allows for only one service to be installed to the residence. We cannot apply *Rule 8–200* to a subpanel because that rule must be applied to the whole house.

We must provide enough power at the second panel to handle the known loads. Based on the circuit schedule for Panel B in Fig. 4–13 on page 44, a 60-ampere panel would be adequate for the load. However, if the dryer, oven, and range were operated at the same time, the subpanel feeder breaker could be overloaded and trip. Therefore, the recommended feeder to Panel B is 100 amperes.

To install a 100-ampere panel, we need No. 3 T90. From *Table 6*, three No. 3 T90 require 27-mm conduit.

Checking the specifications and also Fig. 4–2 on page 35, we find that three No. 3 T90/TWN75 conductors supply Panel B.

Main Disconnect

The main disconnect in the residence is a 200-ampere combination panel. Many electric utilities state that the conductors feeding this type of disconnect must have the same ampere rating as the disconnect. Some utilities also state that the neutral conductor cannot be reduced in size because it is not always possible to foresee the type of load that may be connected to the panel. As a result, it is sometimes difficult to conform to all the rules of the electrical utility as well as the local electrical code. By installing three No. 3/0 T90 conductors for the line and neutral conductors, you will satisfy most local and provincial electrical code requirements.

Grounding Electrode Conductor

The grounding electrode conductor connects the main service equipment neutral bar to the grounding electrode. The grounding electrode might be the underground metallic water piping system, a driven ground rod, or a concrete-encased ground. All of this is discussed in Unit 4.

Grounding electrode conductors are sized according to *Table 17*.

For example, a residence is supplied by No. 3/0 AWG copper service-entrance conductors. Checking *Table 17*, we find that the minimum grounding electrode conductor must be a No. 6 AWG copper conductor.

SERVICE CALCULATIONS FOR APARTMENTS

Often, single-family homes are constructed or renovated to divide the space into two or more units. This type of building is considered to be an apartment building and requires that the services be calculated according to *Rule 8–202*.

For this example, we assume that our building is constructed with a completely independent apartment in the basement. The living area has not changed.

The heat will be changed to electric baseboard heaters with 7.5 kW located in the main floor unit and 5.5 kW in the basement. The floor plans will be modified so that all loads previously allocated to the service will be connected to the main-floor unit service. The basement unit will have a 3 kW water heater and a 12 kW range. Laundry facilities will not be available in the basement unit.

The service calculation for an apartment unit is almost identical to that of a single dwelling. The only difference is in the basic load calculation.

Main Floor Unit

Basic load based on a floor area of 139.9 m²

First 45 m²	3500 watts	
Next 45 m²	1500 watts	
Next 90 m² or portion	1000 watts	
Total	6000 watts	6 000 watts
Range First 12 kW	6000 watts	
Remaining 2.05 kW × 40%	820 watts	
Total	6820 watts	6 820 watts

Other loads

Dryer × 25%	1425 watts	
Water Heater × 25%	750 watts	
Pump × 25%	600 watts	
Total	2775 watts	2 775 watts
Subtotal		15 595 watts
Electric Heat		7 500 watts
Grand Total		23 095 watts

Basement Unit

Basic load based on a floor area of 180.98 m²

First 45 m²	3500 watts	
Next 45 m²	1500 watts	
Next 90 m² or portion	1000 watts	
Total	6000 watts	6 000 watts
Range		
First 12 kW	6000 watts	6 000 watts
Other loads		
Water heater × 25%	750 watts	750 watts
Subtotal		12 750 watts
Electric Heat × 100%		5 500 watts
Grand Total		18 250 watts

To determine the ampacity of the service for the main floor:

$$\frac{\text{Demand Watts}}{\text{System Voltage}} = \frac{23\ 095\ \text{watts}}{240\ \text{volts}} = 96.2\ \text{amps}$$

Therefore, a 100-ampere service would be acceptable. However, the installer may wish to increase the service to 125 amperes to allow for future growth.

To determine the ampacity of the basement unit service:

$$\frac{\text{Demand Watts}}{\text{System Voltage}} = \frac{18\ 250\ \text{watts}}{240\ \text{volts}} = 76\ \text{amps}$$

Therefore a 100-ampere service would be installed.

This type of service installation usually is in the form of a single consumer service, divided into several parts in a multigang meter socket and terminating in separate service boxes. The ampacity of the consumer service conductors is not the sum of the individual services, but a calculation based on the individual demand loads and *Rule 8–202(3)*. To perform this calculation, we must begin with the demand of each unit, without any electric heat or air-conditioning loads.

The calculation is:

100% of the largest unit = 15 595 × 100% =	15 595 watts
65% of the next largest unit = 12 750 × 65% =	8 287.5 watts
Subtotal	23 882.5 watts

Add the electric heat as outlined in *Rule 62–116(2)*

Total installed electric heat =	7.5 kW + 5.5 kW = 13 kW
100% of the first 10 kW = 10 kW × 100% =	10 000 watts
75% of the remaining 3 kW = 3 kW × 75% =	2 250 watts
Subtotal	12 250 watts
Grand Total	36 132.5 watts

Minimum Service Ampacity

$$\frac{\text{Calculated Demand}}{\text{System Voltage}} = \frac{36\ 132.5\ \text{watts}}{240\ \text{volts}} = 150.6\ \text{amps}$$

The minimum consumer service ampacity is considerably less than the sum of the individual services. For this application, the installer would normally install a 200-ampere service to allow for future growth.

REVIEW

Note: Refer to the *CEC* or the plans where necessary.

1. When a service entrance calculation results in a value of 15 kW or more, what is the minimum size service required by the *CEC*? _____

   _____

2. a. What is the demand load per 90 m² for the general lighting load of a residence?

      _____

   b. What are the demand factors for the baseboard heating load in dwellings?

      _____

      _____

      _____

      _____

3. a. What is the ampere rating of the circuits that are provided for the lighting and power loads? _____

   b. What is the maximum number of outlets permitted on a 15-ampere branch circuit by the *CEC*? _____

   c. How many 15-ampere circuits are included in this residence? _____

4. Why is the air-conditioning load for this residence omitted in the service calculations? What is the rule number? _____

   _____

   _____

   _____

   _____

5. What demand factor may be applied to loads greater than 1500 watts, such as a clothes dryer and water heater, in addition to an electric range, *Rule 8–200(1)(a)(vi)*.

   _____

6. What load may be used for an electric range rated at not over 12 kW?

   _____

7. What is the load for an electric range rated at 16 kW, *Rule 8–200(1)(a)(v)*? Show calculations.

8. What is the computed load when fixed electric heating is used in a residence? (*Rule 62–116*) _____

9. On what basis is the neutral conductor of a service entrance determined? _____

    _____

    _____

    _____

10. Why is it permissible to omit an electric space heater, water heater, and certain other 240-volt equipment when calculating the size of the neutral service entrance conductor for a residence? _____

    _____

    _____

    _____

11. Calculate the minimum size of service entrance conductors required for a residence containing the following: floor area 8 m × 12 m; 12-kW electric range; 5-kW dryer 120/240-volt; 2200-watt sauna heater, 120-volt; 12 kW of baseboard heaters with individual thermostat control, 240-volts; two 3-kW, 240-volt air conditioners; 3-kW, 240-volt water heater.

    Determine the sizes of the ungrounded conductors and the neutral conductor. Use type RW90 copper conductors.

    Two No. _____ RW90 ungrounded conductors

    One No. _____ RW90 neutral (or bare neutral if permitted)

    One No. _____ AWG system grounding conductor to water meter

    STUDENT CALCULATIONS
    Minimum service ampacity

    Minimum switch

    Minimum fuse or circuit breaker

    Minimum bonding jumper

    Minimum conduit size

    Minimum panel ampacity

    Minimum number of circuits in panel

UNIT 4

Service Entrance Equipment

OBJECTIVES

After studying this unit, the student will be able to

- define electrical service, supply service, consumer's service, overhead service, and underground service
- list the various *CEC* rules covering the installation of a mast-type overhead service and an underground service
- discuss the *CEC* requirements for disconnecting the electric service using a main panel and load centres
- discuss the grounding of interior alternating current systems and the bonding of all service entrance equipment
- describe the various types of fuses
- select the proper fuse for a particular installation
- explain the operation of fuses and circuit breakers
- explain the term *interrupting rating*
- determine available short-circuit current using a simple formula
- perform cost-of-energy calculations
- understand how to read a watt-hour meter

An electric service is required for all buildings containing an electrical system and receiving electrical energy from a utility company. The *CEC* describes the term *service* in two ways: *consumer's service* and *supply service*.

A consumer's service is considered to be all of the wiring from the service box to the point where the utility makes connection. A service box is a main switch and fuses or the main breaker in a combination panel.

Section 0 defines a supply service as the single set of conductors run by the utility from its mains to the consumer's service. On an overhead service, this is usually from the transformer on the pole

to the point of attachment on the service mast. With an underground service, this may be from the transformer to the line side of the meter base. If the electric utility defines the point of connection as the transformer, the wires that are run from the transformer to the meter base form part of the consumer's service. Always check with your local utility for the specific bylaws and requirements that pertain to services.

In general, watt-hour meters are located on the exterior of a building. Local codes may permit the watt-hour meter to be mounted inside the building. In some cases, the entire service entrance equipment may be mounted outside the building. This includes the watt-hour meter and the disconnecting means, *Rule 6–408*.

The electric utility generally must be contacted to determine the location of the meter, *Rule 6–408*. This will usually be within 1 m from the front of the house and 1.8 m above grade to the centre of the socket.

OVERHEAD SERVICE

An overhead supply service consists of the overhead service conductors (including any splices) that are connected from the mains of the supply authority to the service entrance conductors at the mast for the consumer's service.

A combination panel has two sections, the service box (service entrance portion) and the panelboard (the distribution portion), Fig. 4–1.

In Fig. 4–1 there is a dashed line between the two portions of the combination panel. The panelboard (distribution portion of the combination panel) is not considered to be part of the service entrance wiring. Therefore, it is very important that all of the service entrance wiring enters only the service entrance portion of the panel.

Fig. 4–2 illustrates the *CEC* terms for the various components of a service entrance. Fig. 4–1 shows how a pad-mounted transformer feeds the service in the residence, which in turn feeds Panel B.

Figure 4–1   Service entrance, main panel, subpanel, and grounding for the service in this residence. The metal water pipe from a public water main provides the best ground for a service, *Rule 10–700(1)(a)*.

THESE CONDUCTORS
FURNISHED AND INSTALLED
BY UTILITY IN MOST
LOCALITIES. CHECK WITH
LOCAL ELECTRIC UTILITY.

3 3/0
RW 90 XLPE
IN 41 mm PVC CONDUIT

GROUND BUSHING
IF METAL CONDUIT,
RULE 10–606(1)

BONDING JUMPER
AS PER *TABLE 41*

SYSTEM GROUNDING
CONDUCTOR AS PER
TABLE 17,
RULE 10–812

LB
FITTING

TRANSFORMER

200 AMP
MAIN

SERVICE-
ENTRANCE
PORTION OF
PANEL

PANEL A
200A 120/240V

DISTRIBUTION
PORTION
OF PANEL

EQUIPMENT
BONDING CONDUCTOR
BUS

NEUTRAL BUS

SEE TABLE 53 FOR
COVER REQUIREMENTS.

100 A

27 mm CONDUIT
WITH 3 #3 T90/TWN75

100A 2-POLE BREAKER

WATER METER

REMOVE BRASS
BONDING SCREW

EQUIPMENT BONDING
CONDUCTOR BUS

PANEL B
100A 120/240V
24 CCTS

NEUTRAL BUS

BONDING JUMPER AS PER
TABLE 17, RULE 10-902

MAIN
WATER
SHUT-OFF

UNDERGROUND
METALLIC WATER
PIPE TO MAIN,
RULE 10–700(4)

Figure 4–2   *CEC* **terms for the services. See Fig. 4–6 for grounding requirements.**

MAST-TYPE SERVICE ENTRANCE

The mast service (Figs. 4–3, 4–4, and 4–5) is a commonly used method of installing a service entrance. The mast service is often used on buildings with low roofs, such as ranch-style dwellings, to ensure adequate clearance between the ground and the lowest service conductor.

The mast must be run through the roof, as shown in Fig. 4–3, and must be fastened to comply with *CEC* requirements. The specific requirements for installing this type of service are set out in *Rule 6–112* and *Appendix B.*

Clearance Requirements for Mast Installations

When installing an overhead service, you must maintain certain clearances between the conductors and other surfaces such as the ground and nearby parts of the building. The point of attachment of the conductors to the building must not be higher than 9 m above grade, and minimum clearances from the conductors to grade should not be less than:

5.5 m for highways, lanes, and alleys
5.0 m commercial and industrial driveways

Figure 4–3   Clearances for a typical service entrance installation, *Rule 6–112(2)*.

4.0 m for residential driveways
3.5 m for sidewalks

The service conductors must not be less than 1 m from a window, door, or porch, *Rule 6–112(3)*. Normally there must be a clearance of 915 mm between the point of attachment and the roof. However, this may be reduced if a minimum of 600 mm is maintained between the roof and the bottom of the drip loops on the supply conductors, Fig. 4–4.

Service Mast as Support (*Rule 6–112*)

The bending force on the conduit increases with an increase in the distance between the roof support and the point where the supply service conductors are attached. The pulling force of service conductors on a mast service conduit increases as the length of the supply service conductors increases. As the length of the service conductors decreases, the pulling force on the mast service conduit decreases.

Figure 4–4   Clearance requirements for consumer's service conductors passing over residential roofs, *Rule 6–112* and *Appendix B*, where voltage between conductors does not exceed 300 volts.

If extra support is not provided, the mast must be an approved tubular support member (most common) or 63-mm rigid steel conduit, *Rule 6–112(4,5)*. This size prevents the conduit from bending due to the strain of the supply service conductors. If extra support is provided, it should be in the form of a guy wire that attaches to the roof rafters of the house and the service mast by means of approved fittings. This provides support for long runs of supply service conductors and for the extra weight that ice storms can add to the conductors. If the mast is more than 1.5 m above the uppermost support, the mast must have a guy wire attached to the roof rafters by an approved fitting, *Rule 6–112(4,7,8)* and *Appendix B*. Consult the utility company and electrical inspection authority for information relating to their specific requirements for clearances and support for service masts.

The *CEC* rules for installation and clearances apply to the supply service and consumer service conductors. For example, the service conductors must be insulated, except where the provisions of *Rule 6–308* are met and approved for exposure to direct sunlight as required by *Rule 2–130*. In this case, the grounded neutral conductors are not required to have insulation.

Rule 6–112 and *Appendix B* give clearance allowances for the supply service passing over the roof of a dwelling, Fig. 4–4.

The installation requirements for a typical service entrance are shown in Figs. 4–5 and 4–6. Fig. 4–3 shows the required clearances above the ground. The wiring connections, grounding requirements, and *CEC* references are given in Figs. 4–5 and 4–6.

UNDERGROUND SERVICE

The underground service is the cable installed underground from the customer's meter socket to the utility's supply. This cable is usually, but not always, supplied by the utility company.

New residential subdivisions often include underground installations of the electrical system. The transformer and primary high-voltage conductors are installed by, and are the responsibility of, the utility company.

The conductors installed between the pad-mounted transformers and the meter are called con-sumer service conductors. The utility runs the services to the lot line, and the electrical contractor completes the installation. Otherwise, the electrical contractor supplies and installs the consumer service conductors, but the local utility may require a specific type of cable that it may supply to the contractor at a reasonable cost. Always check with the local utility in addition to the inspection authority before commencing work. Fig. 4–7 shows a typical underground installation.

The wiring from the external meter to the main service equipment is the same as the wiring for a service connected from overhead lines, Fig. 4–6. Some local codes may require conduit to be installed underground from the pole to the service entrance equipment. Requirements for underground services are given in *Rule 6–300*. The underground conductors must be suitable for direct burial in the earth.

If the electric utility installs the underground service conductors, the work must comply with the rules established by the utility, *Section 0, Scope*. These rules may not be the same as *Rule 6–300*.

When the underground conductors are installed by the electrician, *Rule 6–300* applies. This rule deals with the protection of conductors against damage and sealing of underground conduits where they enter a building.

MAIN SERVICE DISCONNECT LOCATION

The main service disconnect means shall be installed at a readily accessible location so that the service entrance conductors within the building are as short as possible. See Fig. 4–8.

The reason for this rule is that the service entrance conductors do not have overcurrent protection other than that provided by the utility's transformer fuses. **Should a fault occur on these service entrance conductors within the building (at the bushing where the conduit enters the main switch, for instance), the arcing could result in a fire.**

The electrical inspector (authority having jurisdiction) must make a judgment as to what is considered to be a readily accessible location nearest the point of entrance of the service entrance conductors, *Rule 6–206(1)*.

Rule 2–100(3)(a) states that all panels must have a legible circuit directory indicating what the

Figure 4–5 A method of connecting conduit to meter base on a 200-ampere overhead service entrance installation.

POINT OF ATTACHMENT TO
BUILDINGS SHALL BE A
MAXIMUM OF 9 M ABOVE
GRADE, *RULE 6–112(2)*

SYSTEM GROUNDING
CONDUCTOR

PVC CONDUIT

SERVICE HEAD
TO BE 150 TO 300 mm
ABOVE POINT OF
ATTACHMENT,
RULE 6–116.

DRIP LOOP

750 mm MIN.
LENGTH,
RULE 6–302(3).

RACK AND
SPOOL
ASSEMBLY

THREE TWO-HOLE
CLAMPS OR EYE-
BOLT MAST
SUPPORT

CONDUIT SIZE,
RULE 6–112(4,5).

RECOMMENDED
HEIGHT OF
METER SOCKET
1.8 m. VERIFY
WITH UTILITY,
RULE 6–408.

NEUTRAL
BONDED TO
METER
ENCLOSURE

METER
BASE HUB

LB FITTING

N

200 A MAIN BREAKER

L₁

L₂

TO
GROUND
ELECTRODE

BRASS MAIN
BONDING
SCREW,
RULE 10–204.

NEUTRAL
BAR BONDED
TO GROUND

NOT TO
SCALE

BONDING
BUS

NEUTRAL
BUS

MAIN SWITCH
AND OVERCURRENT
PROTECTION,
RULE 14–010.

100 A 2-POLE
BREAKER

15 A 1-POLE
BREAKER

15 A
BRANCH
CIRCUIT

ELECTRICAL EQUIPMENT REQUIRED
TO INTERRUPT FAULT CURRENTS
SHALL HAVE RATINGS SUFFICIENT
FOR THE VOLTAGE EMPLOYED AND
FOR THE FAULT CURRENT THAT IS
AVAILABLE AT THE TERMINALS,
RULE 14–012.

FUSES OR CIRCUIT
BREAKERS MUST HAVE
INTERRUPTING RATING
SUFFICIENT FOR VOLTAGE
EMPLOYED AND CURRENT
THAT MUST BE INTERRUPTED,
RULE 14–012.

GRADE

DISCONNECT MEANS
LOCATED AT READILY
ACCESSIBLE LOCATION NEAREST
POINT OF ENTRANCE OF SERVICE
CONDUCTORS, *RULE 6–206(1).*

Figure 4–6   The wiring of a typical overhead service entrance installation.

Figure 4–7 Underground service.

Figure 4–8 Main service disconnect location, *Rule 6–206(1)(d).*

circuits are for. You can see the circuit directory on the inside of the door of the panel or on the side of the panel front cover. The main service Panel A circuit directory is shown in Fig. 4–9. The Panel B circuit directory is shown in Fig. 4–13.

Disconnect Means (Panel A)

The requirements for disconnecting the electrical services are covered in *Rule 6–200. Rule 6–200(1)* requires that each consumer's service be provided with a single service box.

However, *Subrule (2)* sets out the circumstances in which more than one service box is permitted.

- The subdivision is made at the meter base.
- Maximum rating is 600 A and 150 V to ground.
- The meter base is located outdoors.

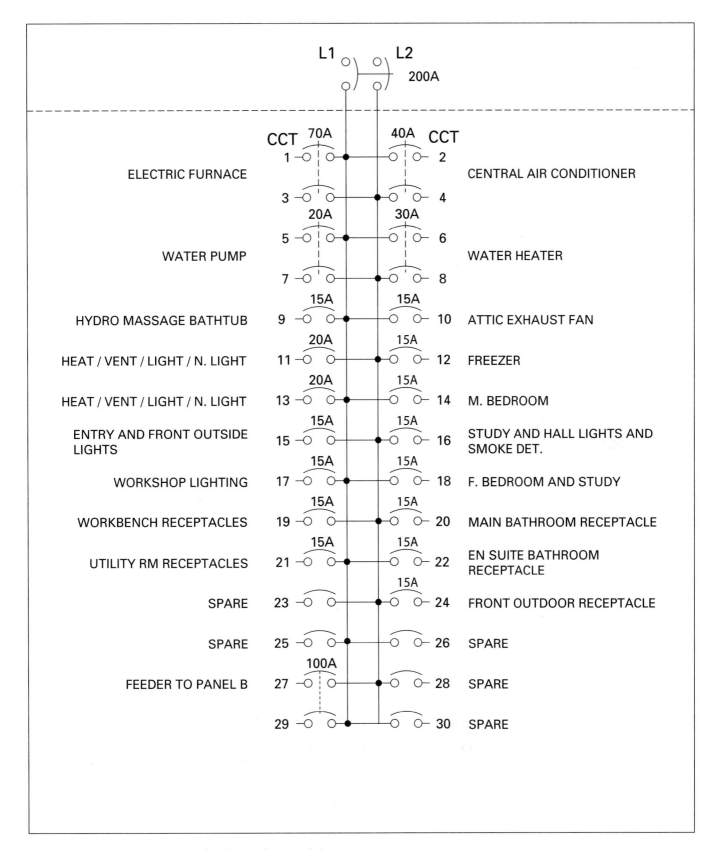

Figure 4–9   Circuit schedule of main service panel A.

A panelboard must have the minimum number of overcurrent devices or positions, as required by *Rule 8–108(1, 2)*.

The *minimum* rating for a single-family residential service is 100-ampere, three-wire when:

- the area excluding the basement is 80 m² or more, *Rule 8–200(1)(b)(i);*

or

- the initial computed load is more than 14.4 kW.

The minimum rating for a single-family residential service is 60-ampere, three-wire when

- the area is 80 m² or less, *Rule 8–200(1)(b)(ii);*

and

- the initial computed load is less than 14.4 kW.

For this residence, Panel A provides a 200-ampere circuit breaker for the main disconnect. Panel A also has a number of branch-circuit overcurrent devices to protect the many circuits originating from this panel, Figs. 4–9 and 4–10. This type of panel is listed by the CSA as a load centre and as service equipment. The panel meets *CEC*

Figure 4–10   Typical main panel with 200-ampere main breaker suitable for use as service entrance equipment, *Rule 6–200(1).* **Courtesy of Siemens Energy and Automation.**

requirements and is designed to accommodate full- and half-width breakers. This panel may be surface or flush mounted.

Panel A is located in the workshop. The placement of the main disconnect is determined by the location of the meter. The disconnect is mounted at an accessible point as close as possible to the place where the service conductors enter the building. The local utility company generally decides where the supply service, or the consumer's service in the case of an underground service, is to be located, and also provides the electrician with a meter location. The consumer's service is the responsibility of the electrician, and the supply service is the responsibility of the utility.

Rule 10–624(1) states that a grounded circuit conductor shall not be used for bonding noncurrent-carrying equipment on the load side of the service disconnecting means.

The panelboard must have an approved means for attaching the bonding conductors when cable is used, Fig. 4–11.

Bonding: At the main service entrance equipment, the grounded neutral conductor must be bonded to the metal enclosure. For most residential-type panels, this main bonding jumper is a bonding screw furnished with the panel. This bonding screw is inserted through the neutral bar into a threaded hole in the back of the panel itself. The bonding screw will be brass or green in colour and must be clearly visible after it is in place.

As previously mentioned, *Rule 10–624(1)* prohibits bonding the grounded neutral of a system to any equipment on the load side of the main service disconnect. Therefore, the brass or green main-bonding jumper screw furnished with residential-type panels will be inserted at the main service panel *only*. This screw must be removed at subpanels, such as Panel B in the recreation room of the residence, *Rule 10–204(1)(c)*. See Fig. 4–11.

Load Centre (Panel B)

When the main service panel is located some distance from areas having many circuits and/or a heavy load concentration such as the kitchen or laundry of this residence, it is recommended that load centres be installed near these concentrations

GROUND
BUSHING

BONDING
JUMPER

MAIN SYSTEM
GROUND

3030
CONNECTOR

BRASS BONDING
SCREW (MAY BE
COLOURED GREEN)

SERVICE
ENTRANCE
PORTION

MAIN SERVICE

NEUTRAL

NEUTRAL
BLOCK

NEUTRAL
BAR

EQUIPMENT BONDING
TERMINAL BLOCK

BRANCH-CIRCUIT
BONDING CONDUCTORS

DISTRIBUTION
PORTION

10–32 SCREW
CONNECTING BUS TO
PANEL ENCLOSURE.

TERMINAL FOR
CONNECTION
OF EQUIPMENT
BONDING CONDUCTORS (BARE)
NOT MORE THAN THREE EQUIPMENT GROUNDING
CONDUCTORS PER SET-SCREW. TERMINAL BAR
MUST BE BONDED TO PANEL ENCLOSURE. THE
EQUIPMENT BONDING BUS SHALL NOT BE
CONNECTED TO THE NEUTRAL BAR.

BRANCH CIRCUIT
NEUTRALS. MUST BE
WHITE OR GREY INSULATED.
ONLY ONE NEUTRAL
CONDUCTOR (IDENTIFIED
CONDUCTOR) PER SET-
SCREW. MUST BE
INSULATED FROM
EXCLOSURE EXCEPT AT
SERVICE ENTRANCE
EQUIPMENT,
RULE 4–020 & 4–028.

Figure 4–11   Connections of service neutral, branch circuit neutral, and equipment bonding conductors at main panel boards, *Rule 10–210(1)(b).*

of load. The individual branch-circuit conductors are run to the load centre, not back to the main panel. Thus, the branch-circuit runs are short, and line losses (voltage and wattage) are less than if the circuits had been run all the way back to the main panel.

To determine if there is a cost or benefit and if the lower line losses justify installing an extra load centre, the cost of material and labour to install the extra load centre should be compared to the cost of many branch circuits run back to the main panel.

A typical load centre is shown in Fig. 4–12.

The circuit schedule for Panel B, the load centre of the residence, is shown in Fig. 4–13.

Figure 4–12   Typical load centre of the type installed for Panel B in this residence. The bonding screw will be removed from the neutral block, *Rule 10–204(1)(c).* **Courtesy of Siemens Energy and Automation.**

Figure 4–13 Circuit schedule of Panel B.

Panel B is located in the recreation room. It is fed by three No. 3 T90/TWN75 conductors run in a 27-mm conduit originating from Panel A. The conductors are protected by a 100-ampere, 240-volt, two-pole overcurrent device in Panel A. The overcurrent protection can be fuses or circuit breakers. This is discussed later in this unit.

SERVICE ENTRANCE CONDUIT SIZING

To determine the proper size of conduit, the conduit fill is calculated using the same method as outlined in Unit 3.

Look up the necessary data in *Tables 8, 9,* and *10* (see Table 4–1). If all the wires in the conduit are the same size, *Table 6* may be used for sizing the conduit.

PANEL A CONDUCTOR SIZE BASED ON AMPACITY PER *TABLE 2*		PANEL B CONDUCTOR SIZE BASED ON AMPACITY PER *TABLE 2*	
Three No. 3/0 RW90 XLPE (600 V) Conduit Size	170.6 mm² × 3 = 511.8 mm² 41 mm	Three No. 3 T90/TWN75 (600 V) Conduit Size	61.99 mm² × 3 = 185.97 mm² 27 mm

Table 4–1 Examples of conduit fill for the service entrance conduit and subpanel feeder conduit in this residence. Obtain conductor area from *Table 10*. Using *Table 8*, conduit fill is 40%. Use *Table 9*'s 40% column for correct conduit size.

METER

Some electric utility companies offer lower rates for water heaters and electric heating units connected to separate meters. The residence in the plans has only one meter. Thus, all lighting, heating, cooking, and water heater loads are registered on that meter.

For a typical overhead service, a meter socket (Fig. 4–14) is mounted at eye level on the outside wall of the house, Fig. 4–7. The service conduit is connected to the socket at a threaded boss (hub) on the top of the socket.

If non-metallic raceways are used for the service conduit, the method of fastening to the top of the meter socket will be different. If a threaded hub is used at the meter socket, a short threaded conduit nipple will be required along with a non-metallic female adapter. Non-metallic raceways must never be threaded directly into a threaded boss

(Rule 12–1112(2)). Alternatively, a non-metallic hub may be used in place of the threaded hub, with the raceway fastened with solvent cement.

This conduit would be carried down into the basement, connecting to the main service equipment as shown in Figs. 4–7, 4–8, and 4–15. Proper fittings must be used, and the conduit must be sealed where it penetrates the wall, Fig. 4–15.

This is required by *Rule 6–312(1)* and is achieved by placing duct sealing compound around the conductors in the raceway where they pass through the wall. *Rule 6–312* also requires that the service raceway be suitably drained, either inside or outside the building. However, if the service raceway terminates in the top of the service box, the drain must be outdoors. This is usually accomplished by drilling a small hole in the bottom of the "L" fitting on the outside of the building. The

THREADED BOSS

LINE

LOAD

Figure 4–14 Typical meter sockets.

RULE 6–312(1) REQUIRES THAT WHEN RACEWAYS
PASS THROUGH AREAS HAVING GREAT TEMPERATURE
DIFFERENCES, SOME MEANS MUST BE PROVIDED TO
PREVENT PASSAGE OF AIR BACK AND FORTH THROUGH THE
RACEWAY. NOTE THAT OUTSIDE AIR IS DRAWN IN THROUGH
THE CONDUIT WHENEVER A DOOR OPENS. COLD OUTSIDE
AIR MEETING WARM INSIDE AIR CAUSES THE
CONDENSATION, WHICH CAN CAUSE
RUSTING AND CORROSION OF VITAL ELECTRICAL
COMPONENTS. EQUIPMENT HAVING MOVING PARTS, SUCH
AS CIRCUIT BREAKERS, SWITCHES, AND CONTROLLERS, IS
ESPECIALLY AFFECTED BY MOISTURE. SLUGGISH
ACTION OF THE MOVING PARTS IN THIS EQUIPMENT IS
UNDESIRABLE.

INSULATION OR OTHER TYPE OF SEALING COMPOUND CAN
BE INSERTED AS SHOWN TO PREVENT THE PASSAGE OF AIR.

THE RACEWAY MUST ALSO BE DRAINED WHERE IT ENTERS THE
BUILDING. THIS IS USUALLY ACCOMPLISHED BY DRILLING A
SMALL HOLE IN THE BOTTOM OF THE L FITTING.

INSIDE

OUTSIDE

INSULATION OR OTHER
SEALING COMPOUND

Figure 4–15   Installation of conduit through a basement wall.

electrician's responsibility is to install the consumer's service conductors from the load side of the meter to the line side of the main service disconnect switch.

Fig. 4–16 shows one type of underground meter base. Note the following:

1. The line side is designated. This is important because if the meter is installed with the load at the top it will run backwards.

2. These are stud-type connections and require the line conductors to have a crimped-on lug.

3. Line-side insulated jaw shrouds help prevent shocks and faults when installing the cover on the meter base.

4. Insulated support blocks also provide a location for a flat-rate water heater lug, which can be bolted on at the three o'clock position.

5. Load-side lugs are angled for ease of connection without too much cable bending.

6. Two-position ground lug. This is not normally used because the meter base is bonded through the grounded service conductor (neutral), *Rule 10–624(2)*.

7. The enclosure is 450 × 300 × 112.5 mm, which gives plenty of room for pulling in conductors and terminating the lugs.

Figure 4–16   An underground type of meter base with studs for utility connection of line conductors. Courtesy of Thomas & Betts Incorporated.

8. The knockouts are provided for the conduit from the transformer to the meter base and for the conduit to the main disconnect.

GROUNDING—WHY GROUND?

Electrical systems and their conductors are grounded to minimize voltage spikes when lightning strikes, or when other line surges occur. Grounding stabilizes the normal voltage to ground.

EMT (electrical metallic tubing) and equipment are grounded so that the equipment voltage to ground is kept to a low value. In this way, the shock hazard is reduced.

Proper grounding means that overcurrent devices can operate faster when responding to ground faults. Effective grounding occurs when a low-impedance (opposition to AC current flow) ground path is provided. A low-impedance ground path means that there is a high value of ground-fault current. As the ground-fault current increases, there is an increase in the speed with which a fuse will open or a circuit breaker will trip. Thus, as the overcurrent device senses and opens the circuit faster, less equipment damage results. This *inverse time characteristic* means that the higher the value of current, the less time it will take to operate the overcurrent device.

Rule 10–500 and *Appendix B* provide the purpose and requirements for effective grounding in an electrical system. The arcing damage to electrical equipment and conductor insulation is closely related to the value of I^2t, where:

I = current flowing from phase to ground, or from phase to phase, in amperes

t = time needed by the overcurrent device to open the circuit, in seconds

I^2t is measured in ampere-squared-seconds.

This expression shows that there will be less equipment and/or conductor damage when the fault current is kept to a low value and when the time that the fault current is allowed to flow is kept to a minimum.

Grounding electrode conductors and equipment grounding conductors carry an insignificant amount of current under normal conditions. However, when a ground fault occurs, these grounding conductors must be capable of carrying whatever value of fault current might flow for the time it takes the overcurrent protective device to clear the fault.

GROUNDING ELECTRODE SYSTEMS

In the grounding electrode system, rather than grounding a single item such as the neutral conductor, the electrician must be concerned with grounding and bonding together an entire system. The term *system* means the service neutral conductor, the grounding electrode, cold water pipes, gas pipes, service entrance equipment, and jumpers installed around meters. If any of these system parts become disconnected or open, the integrity of the grounding system is maintained through other paths. This means that all parts of the system must be tied (bonded) together.

Fig. 4–17 and the following steps illustrate what can happen if an entire system is not grounded.

1. A live wire contacts the gas pipe. The bonding jumper (A) is not installed originally.

2. The gas pipe now has 120 volts with respect to ground. The pipe is hot.

3. The insulating joint in the gas pipe results in a poor path to ground; assume the resistance is 8 ohms.

4. The 20-ampere overcurrent device does not open:

$$I = \frac{E}{R} = \frac{120}{8} \text{ amperes} = 15 \text{ amperes}$$

5. If a person touches the hot gas pipe and the water pipe at the same time, current flows through the person's body. If the body resistance is 12 000 ohms, the current is:

$$I = \frac{E}{R} = \frac{120}{12\,000} \text{ amperes} = 0.01 \text{ amperes}$$

This value of current passing through a human body can cause death.

6. The overcurrent device is "seeing" (15 + 0.01) amperes = 15.01 amperes; however, it still does not open.

Figure 4–17   System grounding. See text for explanation.

7. If the *system grounding* concept had been used, bonding jumper A would have kept the voltage difference between the water pipe and the gas pipe at zero, *Rule 10–406(2)(a)* and *(4)*. Thus, the overcurrent device would open. If 3.05 m of No. 6 AWG copper wire is used as the jumper, the resistance of the jumper is 0.003 95 ohms. The current is

$$I = \frac{E}{R} = \frac{120}{0.003\ 95}\ \text{amperes} = 30\ 380\ \text{amperes}$$

(In an actual system, the impedance of all parts of the circuit is much higher, resulting in a much lower current. The current, however, would be high enough to cause the overcurrent device to open.)

Advantages of System Grounding

To appreciate the concepts of proper grounding, let us review some important *CEC* rules.

Rule 10–002 sets out the objectives of grounding and bonding, summarized as follows:

- to protect life from the danger of electric shock
- to limit the voltage on a circuit
- to facilitate the operation of electrical apparatus and systems
- to limit voltage on a circuit when exposed to lightning
- to limit AC circuit voltages to ground to 150 volts or less on circuits supplying interior wiring systems.

Section 0 provides definitions for the terms associated with grounding and bonding electrical systems.

Bonding: a low impedance path obtained by permanently joining all non-current-carrying metal parts to ensure electrical continuity. Any current likely to be imposed must be conducted safely.

Grounding: a permanent path to earth with sufficient ampacity to carry any fault current liable to be imposed on it.

Bonding Conductor: a conductor that connects the non-current-carrying parts of electrical equipment to the service equipment or grounding conductor.

Grounding Conductor: the conductor used to connect the service equipment to the grounding electrode.

- The potential voltage differences between the parts of the system are minimized, reducing the shock hazard.
- The impedance of the ground path is minimized. This results in a higher current flow in the event of a ground fault: the lower the impedance, the higher the current flow. This means that the overcurrent device will open faster under fault conditions.

Methods of Grounding

Fig. 4–6 shows that the metal cold water pipes, the service raceways, the metal enclosures, the service switch, and the neutral conductor are bonded together to form a *grounded system, Rules 10–406(2)* and *10–700.*

Regarding the residence in this text, the grounding electrode will be the service water pipe from the public water main. This is considered an "in-situ" grounding electrode, as outlined in *Rule 10–700(4).*

Where a driven ground rod, as approved by *Rule 10–700(2)(a),* is the grounding electrode, the size of the grounding electrode conductor that is the sole connection to the driven rod shall not be smaller than the requirements of *Rule 10–812* and *Table 17.*

If the driven ground rod is not selected as the grounding electrode, any one of these items may be used: at least 6 m of bare copper conductor not smaller than No. 4 AWG (*Table 43*) or an approved manufactured ground plate, Fig. 4–18. Both of these may be encased in concrete at least 50 mm thick and in direct contact with the earth, such as near the bottom of a foundation or footing.

Either a bare copper conductor or a ground plate may be used when buried directly in the ground to a minimum depth of 600 mm below finished grade.

There is little doubt that this concept of a grounded system gives rise to many interpretations of the *CEC.* The electrician must check with the local inspection authority to determine the local interpretation.

Refer to Fig. 4–19.

GROUNDING THE SERVICE WHEN NON-METALLIC WATER PIPE IS USED

It is quite common to find the main water supply to a residence installed with plastic (PVC) piping. Even interior water piping systems may be non-

metallic piping. Some local building codes prohibit the use of non-metallic piping. Regardless, it is important that the electrical system be properly grounded and bonded.

When non-metallic water systems are used, one of the other methods of system grounding as identified in *Rule 10–700* must be used. One of the most common methods is the ground plate, Fig. 4–18.

If the water piping inside the building is non-metallic, and only short isolated sections of metal water piping are used, such as at water heaters, these short sections may not need to be bonded. Check with your local inspection authority to be sure. Sometimes, the water system inside the dwelling is metallic, even though the water service to the street is non-metallic. With this large amount of metal water piping present, it must now be bonded to ground.

SUMMARY—SERVICE ENTRANCE EQUIPMENT GROUNDING

When grounding service entrance equipment, Fig. 4–20, you must observe these *CEC* rules:

- The system must be grounded when the maximum voltage to ground does not exceed 150 volts, *Rule 10–106(1)(a).*

- The system must be grounded if it incorporates a neutral conductor, *Rule 10–106(1)(b).*

- All grounding schemes shall be installed so that no objectionable currents will flow over the grounding conductors and other grounding paths, *Rule 10–200(1,2).*

- The ground electrode conductor must be connected to the supply side of the service-disconnecting means. It must not be connected to any grounded circuit conductor on the load side of the service disconnect, *Rule 10–204(1)(b)(d).*

- The neutral conductor must be grounded, *Rule 10–210(1).*

- All electrical equipment must be bonded together. See *Rule 10–400.*

- The grounding electrode conductor used to connect the grounded neutral conductor to the grounding electrode must not be spliced, except with thermit welding or compression connectors. *Rule 10–806(1).*

Figure 4–18   Ground plate.

Figure 4–19 *Rule 10–700* lists many options for a system grounding electrode. Here a concrete-encased No. 3 AWG bare copper conductor is laid near the bottom of the footing, encased by at least 50 mm of concrete. The *minimum* length of the conductor is 6 m. See *Rule 10–700(3)(a)* and *Table 43*. *Note:* The *CEC* requires that if more than one ground electrode connection exists, they must be bonded together, *Rule 10–702*.

- The grounding electrode conductor must be sized according to *Table 17, Rule 10–812(1)*.
- The minimum conductor size for concrete-encased electrodes is found in *Table 43, Rule 10–700*.
- The metal water supply system shall be bonded to the grounding conductor of the electrical system, *Rule 10–406(2)* and *Appendix B*.
- When more than one grounding means exist, they must be bonded together, *Rule 10–702*.

- The grounding electrode conductor must be a minimum of No. 6 AWG copper if the run is exposed and without metal covering or protection. This wire must be free from exposure to mechanical injury, *Rule 10–806(2)*.
- The grounding electrode conductor shall be copper, and may be insulated or bare, *Rule 10–802*.
- Bonding shall be provided around all insulating joints or sections of the metal water

Figure 4–20   One method that may be used to provide proper grounding and bonding of service entrance equipment for a typical single-family residence.

piping system that may be disconnected, *Rule 10–902(3)*.

- The connection to the grounding electrode must be accessible, *Rule 10–902(2)*.
- The grounding conductor must be connected tightly using the proper lugs, connectors, clamps, or other approved means, *Rule 10–908(1)*.

Fig. 4–21 illustrates three common types of ground electrodes. These are connected to the neutral in the main service disconnect.

This residence is supplied by three No. 3/0 RW75 XLPE service entrance conductors. According to *Table 17*, a No. 6 AWG grounding electrode conductor is required. This conductor may be run in conduit or cable armour, or it may be run exposed if it is not subjected to severe physical damage.

BONDING

Bonding must be done at service entrance equipment. *Rule 10–606* lists the methods approved for bonding this equipment. Grounding bushings,

Figure 4–21 Three common types of grounding electrodes: (1) a field-assembled grounding electrode, (2) two rod electrodes, and (3) a plate electrode. In many instances, the grounding electrode is better than the service water pipe (in-situ) ground electrode, particularly since the advent of non-metallic (PVC) water piping systems.

Fig. 4–22, and bonding jumpers are installed on metallic service conduits at service entrance equipment to ensure a low-impedance path to ground if a fault occurs on any of the service entrance conductors, *Rule 10–606(d)*. Service entrance conductors are not fused at the service head. Thus, the short-circuit cur-

rent on these conductors is limited only by the capacity of the transformer or transformers supplying the service equipment, and the distance between the service equipment and the transformers. The short-circuit current can easily reach 20 000 amperes or more in dwellings. Fault currents can easily

Figure 4–22   Insulated ground bushing with bonding lug.

Figure 4–23   Insulated plastic bushings.

reach 40 000 to 50 000 amperes or more in apartments, condominiums, and similar dwellings. These installations are usually served by a large-capacity transformer located close to the service entrance equipment and the metering. **This extremely high fault current produces severe arcing, which is a fire hazard.** The use of proper bonding reduces this hazard to some extent.

Fault current calculations are presented later in this unit. *Electrical Wiring: Commercial* covers these calculations in much greater detail. *Rule 10–614(1)(b)* states that the main bonding jumpers must have an ampacity not less than the corresponding bonding conductor. The grounding lugs on grounding bushings are sized by the trade size of the bushing. The lugs become larger as the size of the bushing increases.

The conductor may be sized according to *Table 41* if used in conjunction with two locknuts and a ground bushing.

Rule 12–906(2) states that if No. 8 AWG or larger conductors are installed in a raceway, an insulating bushing or equivalent must be used, Fig. 4–23. This bushing protects the wire from shorting or grounding itself as it passes through the metal bushing. Combination bushings can be used. These bushings are metallic (for mechanical strength) and have plastic insulation. When conductors are installed in EMT, they can be protected at the fittings by the use of connectors with insulated throats.

If the conduit bushing is made of insulating material only, as in Fig. 4–23, two locknuts must be used, Fig. 4–24, *Rule 10–610(b)*.

Figs. 4–25 to 4–28 show various types of ground clamps. These clamps and their actual connection to

Figure 4–24   The use of locknuts. See *Rule 10–610(b)*.

the grounding electrode must conform to *Rule 10–908*.

MULTIPLE METER INSTALLATIONS

If multiple meter sockets are required, as in a multifamily dwelling or a water heater metered at a different rate, a multigang meter socket is used. This is a single enclosure that is capable of holding more than one meter. The meter socket is rated in two ways: the maximum ampacity of the consumer service conductors feeding the enclosures and the maximum ampacity that may be taken off each meter position. A commonly used two-position

Figure 4–25   Typical ground clamps used in residential systems.

NO. 6 AWG COPPER BONDING JUMPER FOR WATER METER

WATER PIPE

Figure 4–27   Ground clamp of the type used to bond (jumper) around water meter.

Figure 4–26   Armoured grounding conductor connected with ground clamp to water pipe.

Figure 4–28   Ground clamp of the type used to attach ground wire to well casings. Courtesy of Thomas & Betts Incorporated.

meter socket may have a maximum ampacity of 200 amperes on the consumer service with a maximum ampacity of 100 amperes per meter position.

Since the system grounding conductor is connected in the service box of each service, the grounding of the multiple services is the same as for a single service, that is, based on the individual service ampacity and *Table 17*. However, *Rule 10–204(1)(b)* states that the system grounding connection may be made in other service equipment located on the supply side of the service disconnects. The service can therefore be grounded at the multigang meter socket by connecting the neutral conductor to the grounding electrode using a conductor sized according to *Table 17,* based on the ampacity of the consumer service conductors feeding the meter socket.

If the raceway system from the meter socket is metallic, the requirements for bonding the raceway

to the service enclosure, as previously outlined in this section, will apply. If the service raceways are non-metallic, a separate bonding conductor is required, sized according to *Table 16,* based on the ampacity of the individual service overcurrent device ratings. Because the neutral conductor is already bonded to the case of the meter enclosure and connected to the grounding electrode from this location, you must apply *Rule 10–204(1)(c)*. This rule requires that the neutral not be connected to the enclosure in the service box of the individual services. Therefore, the brass or green bonding screw must be removed from the neutral bar in the individual services.

The neutral must only be bonded to non-current-carrying components in one location, that location being at the point at which the neutral is connected to the system-grounding electrode.

BRANCH CIRCUIT OVERCURRENT PROTECTION

The overcurrent devices commonly used to protect branch circuits in dwellings are fuses and circuit breakers. *Section 14, CEC* discusses overcurrent protection.

Plug Fuses, Fuseholders, and Fuse Rejecters

Fuses are a reliable and economical form of overcurrent protection. *Rules 14–200* through *14–212* give the requirements for plug fuses, fuseholders, and fuse rejecters, including:

- Plug fuses shall not be used in circuits exceeding 125 volts between conductors. An exception to this rule is for a system having a grounded neutral where no conductor is more than 150 volts to ground. (This is the case for the 120/240-volt system used in the residence in the plans.) See *Rule 14–202*.

- Plug fuses shall have ratings between 0 and 30 amperes, *Rule 14–208(1)*.

- Plug fuses with low–melting point characteristics shall be marked "P," and fuses with time delay shall be marked "D," *Rule 14–200*.

- The screw shell of the fuseholder must be connected to the load side of the circuit.

- Where there is an alteration to a fuse panel, the fuse panel is required to be equipped with fuse rejecters that are designed to limit the size of fuse that can be accommodated by the fuseholder, *Rule 14–204(2)*.

- All new installations require fuse rejecters.

- Fuse rejecters are classified at 0 to 15 amperes, 16 to 20 amperes, and 21 to 30 amperes. The reason for this classification is given in the following paragraph.

When the electrician installs fusible equipment, the ampere rating of the various circuits must be determined. Based on this rating, a fuse rejecter of the proper size is inserted into the fuseholder. The proper Type P or D fuse is then placed in the fuseholder. Because of the rejecter, the fuse is non-interchangeable, as required by *Rule 14–204(1)*. For example, assume that a 15-ampere rejecter is inserted for the 15-ampere branch circuits in the residence. It is impossible to substitute a fuse with a larger rating without removing the 15-ampere rejecter. If any alterations are made to an existing panel, these rejecters must be installed, *Rule 14–204(2)*. Fig. 4–29 shows typical fuse rejecters.

Often the term *Dual-Element* is used when referring to time-delay fuses. *Dual-Element* is a trade name of fuses made by Bussmann Division, Cooper Industries. The term is used in much the same manner as *Romex* is used for type NMD90 cable and *BX* is used for armoured cable.

CSA has classified fuses according to their size and operating characteristics. Table 4–2 provides information on common low-voltage fuses.

Dual-Element Fuse (Time-Delay)

Fig. 4–30 shows the most common type of Dual-Element cartridge fuse. These fuses are available in 250-volt and 600-volt sizes with ratings from 0 to

Figure 4–29 **Type C fuses and rejection washers are designed to prevent overfusing in panelboards that use plug fuses. Rejection washers can be added to any existing panelboard that uses plug fuses.**

FUSE CLASS	VOLTAGE	CURRENT	COMMENTS
Plug Fuses			
C	125	0 – 30	Has rejection feature when used with rejection washers; low-melting point plug fuses are Type P nontime-delay and Type D time-delay
S	125	0 – 30	May be time-delay or nontime-delay
Cartridge Fuses			
H	250 600	0 – 600	Standard fuse used in circuits where interrupting capacity of 10 000 A or less is required
J	250 600	0 – 600	UL class J: 200 000 A interrupting capacity, smaller physical dimensions than Class H fuses
K	250 600	0 – 600	UL class K: will be found on equipment imported from USA
L	250 600	601 – 6000	UL class L: high temperature fuses
R	250 600	0 – 600	UL class R: rejection fuse may be placed in a standard fuseholder, but standard fuses will not fit in a Class R fuseholder
T	250 600	0 – 600	UL class T: has smaller physical dimensions than either Class H or Class J

Table 4–2   Characteristics of common low-voltage fuses.

600 amperes, and have an interrupting rating of 200 000 amperes.

A Dual-Element fuse has two fusible elements connected in series: the overload element and the short-circuit element. When an excessive current flows, one of the elements opens. The amount of excess current determines which element opens.

The electrical characteristics of these two elements are very different. Thus, a Dual-Element fuse has a greater range of protection than the single-element fuse. The overload element opens circuits on current in the low overload range (up to 500% of the fuse rating). The short-circuit element handles only short circuits and heavy overload currents (above 500% of the fuse rating).

Overload Element: The overload element opens when excessive heat is developed in the element. This heat may be the result of a loose connection or poor contact in the fuseholder or excessive current. When the temperature reaches 138°C, a fusible alloy melts and the element opens. Any excessive current produces heat in the element. However, the mass of the element absorbs a great deal of heat before the cutout opens. A small excess of current will cause this element to open if it continues for a long period of time. This characteristic gives the thermal cutout element a large time lag on low overloads (up to 10 seconds at a current of 500% of the

fuse rating). In addition, it provides very accurate protection for prolonged overloads.

Short-Circuit Element: The capacity of the short-circuit element is high enough to prevent it from opening on low overloads. This element is designed to clear the circuit quickly of short circuits or heavy overloads above 500% of the fuse ratings.

(A)

SHORT-CIRCUIT ELEMENTS

OVERLOAD ELEMENT

(B)

Figure 4–30   Cartridge-type Dual-Element fuse (A) is a 250-volt, 100-ampere fuse. The cutaway view in (B) shows the internal parts of the fuse. Courtesy of Bussmann Division, Cooper Industries.

Dual-element fuses are used on motor and appliance circuits where the time-lag characteristic of the fuse is required. Single-element fuses do not have this time lag. Such fuses blow as soon as an overcurrent condition occurs. The homeowner may be tempted to install a fuse with a higher rating. The fuseholder adapter, however, prevents the use of such a fuse. Thus, the dual-element fuse is recommended for this type of situation.

The standard plug fuse can be used only to replace blown fuses on existing installations. A Type P or D fuse is required on all new installations. The cartridge fuse is a dual-element fuse that is available in both ferrule and knife blade styles.

Cartridge Fuses

Fig. 4–31 shows another style of small-dimension cartridge fuse.

A type of cartridge fuse that is also in use is rated at not over 60 amperes for circuits of 300 volts or less to ground. The physical size of this fuse is smaller than that of standard cartridge fuses. The time-delay characteristics of this fuse mean that it can handle harmless current surges or momentary overloads without blowing. However, these fuses open very rapidly under short-circuit conditions.

These fuses prevent over-fusing since they are size limiting for their ampere ratings. For example, a fuseholder designed to accept a 15-ampere fuse will not accept a 20-ampere cartridge fuse. In the same manner, a fuseholder designed to accept a 20-ampere fuse will not accept a 30-ampere cartridge fuse.

Figure 4–31 Cartridge-type fuses.

Class J Fuses

Another type of physically small fuse is the Class J fuse, Fig. 4–32. This has a high-interrupting rating in sizes from 0 to 600 amperes in both 300-volt and 600-volt ratings. Such fuses are used as the main fuses in a panel having circuit-breaker branches, Fig. 4–33. In this case, the Class J fuses protect the low-interrupting capacity breakers against high-level short-circuit currents.

Most manufacturers list service equipment, metering equipment, and disconnect switches that make use of Class J fuses. This allows the safe installation of this equipment where fault currents exceed 10 000 amperes, such as on large services, and any service equipment located close to pad-mounted transformers, such as are found in apartments, condominiums, shopping centres, and similar locations.

The standard ratings for fuses range from 1 ampere to 6000 amperes.

Circuit Breakers

Installations in dwellings normally use thermal-magnetic circuit breakers. On a continuous overload, a bimetallic element in such a breaker moves

Figure 4–32   Class J fuses rated at 30, 60, 100, 200, 400, and 600 amperes, 300 volts. Courtesy of Bussmann Division, Cooper Industries.

Figure 4–33   Fused main/fused branch circuits.

until it unlatches the inner tripping mechanism of the breaker. Momentary small overloads do not cause the element to trip the breaker. If the overload is heavy or if there is a short circuit, a magnetic coil in the breaker causes it to interrupt the branch circuit instantly.

Rules 14–300 to *14–308* give the requirements for circuit breakers, including:

- Circuit breakers shall be trip free so that the internal mechanism will trip to the off position even if the handle is held in the on position (which can be achieved using a lock-on device), *Rule 14–300(1).*

- Breakers shall indicate clearly whether they are on or off, *Rule 14–300(2).*

- A breaker shall be non-tamperable so that it cannot be readjusted (trip point changed) without dismantling the breaker or breaking the seal.

- The rating shall be durably marked on the breaker. For small breakers rated at 100 amperes or less and 600 volts or less, the rating must be moulded, stamped, or etched on the handle (or on another part of the breaker that will be visible after the cover of the panel is installed).

- Every breaker with an interrupting rating other than 5000 amperes shall have this rating marked on the breaker. Most have a 10 000 ampere mimimum rating.

- Circuit breakers rated at 120 volts and 347 volts and used for fluorescent loads shall

not be used as switches unless marked "SWD," *Section 30–710(3)(b).*

Most circuit breakers are ambient temperature compensated. This means that the tripping point of the breaker is not affected by an increase in the surrounding temperature. An ambient-compensated breaker has two elements. One element heats up due to the current passing through it and the heat in the surrounding area. The other element heats up because of the surrounding air only. The actions of these elements oppose each other. Thus, as the tripping element tends to lower its tripping point because of external heat, the second element opposes the tripping element and stabilizes the tripping point. As a result, the current through the tripping element is the only factor that causes the element to open the circuit. It is a good practice to turn the breaker off and on periodically to "exercise" its moving parts.

Table 13 gives the standard ampere ratings of fuses and circuit breakers up to 600 amperes.

INTERRUPTING RATINGS FOR FUSES AND CIRCUIT BREAKERS

Rule 14–012(a) states that all fuses, circuit breakers, and other electrical devices that break current shall have an interrupting capacity sufficient for the voltage employed and for the current that must be interrupted.

According to *Rule 14–012(a)* and *Appendix B,* all overcurrent devices, the total circuit impedance, and the withstand capability of all circuit components (wires, contractors, switches) must be selected so that minimal damage will result in the event of a fault, either line-to-line or line-to-ground.

The overcurrent protective device must be able to interrupt the current that may flow under any condition (overload or short circuit). Such interruption must be made with complete safety to personnel and without damage to the panel or switch in which the overcurrent device is installed.

Overcurrent devices with inadequate interrupting ratings are, in effect, bombs waiting for a short circuit to trigger them into an explosion. Personal injury may result, and serious damage will be done to the electrical equipment.

Cartridge-type fuses have a maximum interrupting rating of 10 000 amperes root mean square (RMS) symmetrical. Plug fuses can interrupt no more than 10 000 amperes RMS symmetrical. Cartridge dual-element fuses (Fig. 4–30) and Class J fuses (Fig. 4–32) have interrupting ratings of 200 000 amperes RMS symmetrical. These interrupting ratings are described in *CSA Standard C22.2 No. 59.1-M1987*. The interrupting rating of a fuse is marked on its label when the rating is other than 10 000 amperes.

Short-Circuit Currents

This text does not cover in detail the methods of calculating short-circuit currents (see *Electrical Wiring: Commercial*). The ratings required to determine the maximum available short-circuit current delivered by a transformer are the kilovolt-ampere (kVA) and impedance values of the transformer. The size and length of wire installed between the transformer and the overcurrent device must be considered as well.

The transformers used in modern electrical installations are efficient and have very low impedance values. A low-impedance transformer having a given kVA rating delivers more short-circuit current than a transformer with the same kVA rating and a higher impedance. When an electrical service is connected to a low-impedance transformer, the problem of available short-circuit current is very serious. The possible situations are shown in Fig. 4–33.

The examples given in Fig. 4–33 show that the available short-circuit current at the transformer secondary terminal is 20 800 amperes.

Determining Short-Circuit Current

The local power utility and the electrical inspector are good sources of information for determining short-circuit current.

You can use this simplified method to determine the approximate available short-circuit current at the terminals of a transformer.

Step 1. Determine the normal full-load secondary current delivered by the transformer.

For single-phase transformers:

$$I = \frac{kVA \times 1000}{E}$$

For three-phase transformers:

$$I = \frac{kVA \times 1000}{E \times 1.73}$$

I = current, in amperes
kVA = kilovolt-amperes (transformer nameplate rating)
E = secondary line-to-line voltage (transformer nameplate rating)

Step 2. Using the impedance value given on the transformer nameplate, find the multiplier to determine the short-circuit current.

$$Multiplier = \frac{100}{percent\ impedance}$$

Step 3. The short-circuit current = normal full-load secondary current × multiplier.

EXAMPLE: A transformer is rated at 100 kVA and 120/240 volts. It is a single-phase transformer with an impedance of 1% (from the transformer nameplate). Find the short-circuit current.

For a single-phase transformer:

$$I = \frac{kVA \times 1000}{E}$$

$$= \frac{100 \times 1000}{240} \text{ amperes}$$

$$= 417 \text{ amperes, full-load current}$$

For a transformer impedance of 1% is:

$$\text{multiplier} = \frac{100}{\text{percent impedance}}$$

$$= \frac{100}{1}$$

$$= 100$$

The short-circuit current $= I \times$ multiplier

$$= 416 \text{ amperes} \times 100$$

$$= 41\,600 \text{ amperes}$$

Thus, the available short-circuit current at the terminals of the transformer is 41 600 amperes.

This value decreases as the distance from the transformer increases. If the transformer impedance is 1.5%,

$$\text{multiplier} = \frac{100}{1.5} = 66.6$$

The short-circuit current $= 416 \times 66.6$ amperes

$$= 27\,706 \text{ amperes}$$

If the transformer impedance is 2%:

$$\text{multiplier} = \frac{100}{2} = 50$$

The short-circuit current $= 416 \times 50$ amperes

$$= 20\,800 \text{ amperes}$$

Note: A short-circuit current of 20 800 amperes is used in Figs. 4–33, 34, 35, and 36.

CAUTION: The line-to-neutral short-circuit current at the secondary of a single-phase transformer is about 1.5 times greater than the line-to-line short-circuit current. For the previous example,

Line-to-line short-circuit current = 20 800 amperes

Line-to-neutral short-circuit current
$= 20\,800 \times 1.5$ amperes $= 31\,200$ amperes

Note that *CSA Standard C22.2 No.47* allows the marked impedance on the nameplate of a transformer to vary plus or minus ($\pm$) 10% from the actual impedance value of the transformer. Thus, for a transformer marked "2%Z," the actual impedance could be as low as 1.8%Z or as much as 2.2%Z. Therefore, the available short-circuit current at the

secondary of a transformer as calculated above should be increased by 10% if you wish to be on the safe side when considering the interrupting rating of the fuses and breakers to be installed.

An excellent point-to-point method of calculating fault currents at various distances from the secondary of a transformer that uses different size conductors is covered in detail in *Electrical Wiring: Commercial*. It is simple and accurate. The point-to-point method can be used to compute both single-phase and three-phase faults.

Calculating fault currents is just as important as calculating load currents. Overloads will cause conductors and equipment to run hot, will shorten their life, and will eventually destroy them. Fault currents that exceed the interrupting rating of the overcurrent protective devices can cause violent electrical explosions of equipment with the potential of serious injury to personnel standing near it. Fire hazard is also present.

PANELS AND LOAD CENTRES
Fused Main/Fused Branches (Fig. 4–33)

- Fused main/fused branch circuits must be listed as suitable for use as service equipment.
- They must be sized to satisfy the required ampacity as determined by service entrance calculations and/or local codes.
- All fuses must have interrupting rating adequate for the available fault current. Panel must have short-circuit rating adequate for the available fault current.
- For overload conditions on a branch, only the branch-circuit fuse(s) will open. All other circuits remain energized. System is selective.*
- For short circuits (line-to-line or line-to-neutral) and for ground faults (line-to-ground), only the fuse protecting the faulted circuit opens. All other circuits remain energized. System is selective.*

  * See asterisked note on page 62.

Fused Main/Breaker Branches (Fig. 4–34)

- Fused main/breaker branches must be listed as suitable for use as service equipment.

- They must be sized to satisfy the required ampacity as determined by service entrance calculations and/or local codes.

- Main fuses must have interrupting rating adequate for the available fault circuit.

- Panel must have short-circuit rating adequate for the available fault current.

- Main fuses must be current limiting, such as Class J fuses, so that if a fault exceeding the branch-circuit breaker's interrupting rating (A.I.R.) occurs, the fuse will limit the let-through fault current to a value less than the maximum interrupting rating of the branch-circuit breaker. Branch-circuit breakers generally have 10 000 amperes interrupting rating.

- For branch-circuit faults (L-N or L-G) exceeding the branch-circuit breaker's interrupting rating, one fuse opens, de-energizing one-half of the panel. The system is nonselective.*

- For branch-circuit faults (L-L) exceeding the branch-circuit breaker's interrupting rating, both main fuses open, de-energizing the entire panel. The system is nonselective.*

- Check time-current curves of fuses and breakers and check the unlatching time of breakers to determine at what point in "time" and "current" the breakers are slower than the current-limiting fuses.

- For normal overloads and low-level faults, this system is selective.

* See note on page 62.

Breaker Main/Breaker Branches (Fig. 4–35) (All Standard Plastic-Case Breakers)

- These breaker main/breaker branches must be listed as suitable for use as service equipment.

- They must be sized to satisfy the required ampacity as determined by service entrance calculations and/or local codes.

- Main and branch breakers must have interrupting rating adequate for the available fault current.

- Panel must have short-circuit rating adequate for the available fault current.

- For overload conditions and low-level faults on a branch circuit, only the branch-circuit breaker will trip off. All other circuits remain energized. The system is selective.*

- For branch-circuit faults (short circuits or ground faults) that exceed the instant trip setting of the main breaker (usually five times the breaker's ampere rating), both the branch-circuit breaker and main breaker trip off. For instance, for a branch fault of 700 amperes, a 100-ampere main breaker also trips because the main will trip instantly for faults of 500 (100 × 5) amperes or more. Entire panel is de-energized. System is nonselective.*

- Check manufacturer's data for time-current characteristic curves and unlatching times of the breakers.

* See note on page 62.

AVAILABLE FAULT CURRENT 20 800 AMPERES

SERVICE EQUIPMENT MUST HAVE SHORT-CIRCUIT RATING EQUAL TO OR GREATER THAN THE AVAILABLE FAULT CURRENT AT SUPPLY TERMINALS.

10 000 INTERRUPTING RATING BREAKERS

CURRENT-LIMITING FUSES MUST LIMIT SHORT-CIRCUIT CURRENT TO PROTECT THE BREAKERS. THE FUSES MUST HAVE AT LEAST 20 800-AMPERE INTERRUPTING RATING.

MEETING REQUIREMENTS OF *RULE 14–012* FOR FUSED MAIN/BREAKER BRANCH CIRCUITS.

Figure 4–34 Fused main/breaker branch circuits.

Figure 4–35   Breaker main/breaker branch circuits.

Breaker Main/Breaker Branches (Fig. 4–36) (Series Connected— Sometimes Called Series Rated, *Rule 14–014*)

- These breaker main/breaker branches must be listed as suitable for use as service equipment.

- They must be sized to satisfy the required ampacity as determined by service entrance calculations and/or local codes.

- They must be listed as series connected.

* The terms and theory behind selective and nonselective systems are covered in detail in *Electrical Wiring: Commercial.* In all instances, you should check the time-current curves of fuses and circuit breakers to be installed. Also check the circuit breaker manufacturer's unlatching time data, as these will help determine at what values of current the system will be selective or nonselective.

Fault current calculations are covered in *Electrical Wiring: Commercial.*

- Main breaker will have interrupting rating higher than the branch breaker. For example: main breaker has a 22 000-ampere interrupting rating, whereas branch breaker has a 10 000-ampere interrupting rating.

- Panel must be listed and marked with its maximum fault current rating.

- For branch-circuit faults (L-L, L-N, L-G) above the instant trip factory setting (i.e.,

200-ampere main breaker with 5 × setting: $200 \times 5 = 1000$ amperes), both branch and main breakers will trip. The system is nonselective.

- For overload conditions and low-level faults on a branch, only the branch breaker will trip. The system is selective.

- Obtain time-current curves and unlatching time data from the manufacturer.

- The panel's integral rating is attained because when a heavy fault occurs, both main and branch breakers open. The two arcs in series add impedance; thus the fault current is reduced to a level less than the 10 000 A.I.R. rating of the branch-circuit breaker.

- For series-rated panelboards where the available fault current exceeds the interrupting rating of the branch-circuit breakers, the equipment must be marked by the manufacturer as having a series combination interrupting rating at least equal to the maximum available fault current, *Rule 14–014(d)*.

- When panels containing low interrupting rating circuit breakers are separated from the equipment in which the higher interrupting rating circuit breakers protecting the panel are installed, the system becomes a field-installed *series-rated system*. This panel must be field-marked (by the installer) and this marking must be readily visible, *Rule 14–014(e)*. This marking calls attention to the fact that the actual available fault current

Figure 4–36 Breaker main/breaker branch circuits (series-connected plastic-case type).

exceeds the interrupting rating of the breakers in the panel, and that the panel is protected by properly selected circuit breakers back at the main equipment. The label will appear on panelboards and enclosed panelboards and may appear as follows:

When protected by ___ A maximum HRC ___ fuse or (manufacturer's name and type designation) circuit breaker rated not more than ___ A, this panelboard is suitable for use on a circuit capable of delivering not more than ___ A RMS symmetrical, ___ V maximum.

CAUTION

SERIES-RATED SYSTEM

- CAUTION: Closely check the manufacturer's catalogue numbers to be sure that the circuit breaker combination you are intending to install *in series* is recognized by the CSA for that purpose. This is extremely important because many circuit breakers have the same physical size and dimensions, yet have different interrupting and voltage ratings, and have not been tested as *series-rated* devices. To install these as a series combination constitutes a violation of *Rules 14–012* and *14–014*, which could be dangerous.

READING THE METER

Fig. 4–37 shows a typical single-phase watt-hour meter with five dials. From left to right, the dials represent tens of thousands, thousands, hundreds, tens, and units of kilowatt-hours.

Starting with the first dial (on the left), record the last number the pointer has passed. Continue doing this with each dial until the full reading is obtained. The reading on the five-dial meter in Fig. 4–38 is 18 672 kilowatt-hours.

If the meter reads 18 975 one month later, Fig. 4–39, by subtracting the previous reading of 18 672 it is found that 303 kilowatt-hours were used during the month.

The number is multiplied by the rate per kilowatt-hour, and the power company bills the consumer for the energy used. The utility may also add a fuel adjustment charge.

COST OF USING ELECTRICAL ENERGY

A watt-hour meter is always connected into some part of the service entrance equipment. In residential metering, a watt-hour meter is normally installed as part of the service entrance.

All electrical energy consumed is metered so that the utility can bill the customer on a monthly or bimonthly basis.

The kilowatt (kW) is a convenient unit of electrical power. One thousand watts (W) is equal to one kilowatt (kW). The watt-hour meter measures

Figure 4–37   Typical single-phase watt-hour meter.

both wattage and time. As the dials of the meter turn, the kilowatt-hour (kWh) consumption is continually recorded.

Utility rates are based upon "so many cents per kilowatt-hour."

EXAMPLE: One kWh will light a 100-watt light bulb for ten hours. One kWh will operate a 1000-watt electric heater for one hour. Therefore, if the electric rate is 8 cents per kilowatt-hour, the use of a 100-watt bulb for ten hours would cost 8 cents. A 1000-watt electric heater could be used for one hour at a cost of 8 cents.

$$kWh = \frac{watts \times hours}{1000} = \frac{100 \times 10}{1000} \, kWh = 1 \, kWh$$

The cost of the energy used by an appliance is:

$$cost = \frac{watts \times hours \, used \times cost \, per \, kWh}{1000}$$

PROBLEM: Find the cost of operating a colour television for eight hours. The set is rated at 175 watts. The electric rate is 9.6 cents per kilowatt-hour.

SOLUTION:

$$Cost = \frac{175 \times 8 \times 9.6}{1000} \, cents = 13.4 \, cents$$

Assume that a meter now reads 18 672 kWh. The previous meter reading was 17 895 kWh. The difference is 777 kWh.

Here is how a typical bill might look:

GENERIC ELECTRIC COMPANY			
Days of Service	from	to	Due Date
33	01–28–09	03–01–09	03–25–09
Present reading			18 672
Last reading			17 895
Kilowatt-hours			777
Rate/kWh			0.08
Amount			$62.16
Total			$62.16 + applicable taxes

Some utilities apply a fuel adjustment charge that may increase or decrease the electric bill. These charges, on a "per kWh" basis, enable the electric utility to recover from the consumer extra expenses it might incur for fuel costs used in generating electricity. These charges can vary each time the utility prepares the bill, without it having to apply to the regulatory agency for a rate change.

Some utilities increase their rates during the summer months when people turn on their air conditioners. This air-conditioning load taxes the utility's generating capabilities during the peak summer months. The higher rate structure also gives peo-

Figure 4–38   The reading of this five-dial meter is 18 672 kilowatt-hours.

Figure 4–39 One month later, the meter reads 18 975 kilowatt-hours, indicating that 303 kilowatt-hours were used during the month.

ple an incentive for not setting their thermostats too low, thus saving energy.

To encourage customers to conserve energy, a utility company might offer the following rate structure:

First 500 kWh @ 6.12 cents per kWh
Over 500 kWh @ 10.64 cents per kWh
(summer rate)
Over 500 kWh @ 8.64 cents per kWh
(winter rate)

REVIEW

Note: Refer to the *CEC* or the plans where necessary.

1. Where does a consumer's service start and end? _____

_____

_____

_____

_____

2. What are supply service conductors? _____

_____

_____

_____

3. Who is responsible for determining the service location? _____

_____

4. a. The service head must be located (above) (below) the point where the supply service conductors are spliced to the consumer's service conductors. Circle one.

   b. What *CEC* rule provides the answer to part (a)? _____

5. a. What size and type of conductors are installed for this service? _____

   _____

   b. What size conduit is installed? _____

   _____

   c. What size grounding electrode conductor (system grounding conductor) is installed? (not neutral) _____

    d. Is the grounding electrode conductor insulated, armoured, or bare? _____

    _____

6. How and where is the grounding electrode conductor attached to the water pipe? _____

    _____

    _____

7. When a service mast is extended through a roof, must it be guyed? _____

    _____

8. What are the minimum distances or clearances for the following?

    a. Supply service conductors, clearance over private driveway _____

    b. Supply service conductors, clearance over private sidewalks _____

    c. Supply service conductors, clearance over alleys _____

    d. Supply service conductors, clearance over a roof having a roof pitch that may be walked on. (Voltage between conductors does not exceed 300 volts.) _____

    e. Supply service conductors, horizontal clearance from a porch _____

    f. Supply service conductors, clearance from a window or door _____

9. What size ungrounded conductors are installed for each of the following residential services? (Use type RW90 XLPE copper conductors.)

    a. 60-ampere service   No. _____ RW90 XLPE copper

    b. 100-ampere service   No. _____ RW90 XLPE copper

    c. 200-ampere service   No. _____ RW90 XLPE copper

10. What size grounding electrode conductors are installed for the services listed in Question 9? (See *Table 17*.)

    a. 60-ampere service   No. _____ AWG grounding conductor

    b. 100-ampere service   No. _____ AWG grounding conductor

    c. 200-ampere service   No. _____ AWG grounding conductor

11. What is the recommended height of a meter socket from the ground?

    _____

12. a. What is the minimum length of conductor that must be left extending out of the service head? _____

       _____

    b. What *CEC* rule covers this? _____

13. What are the restrictions on the use of bare neutral conductors? _____

    _____

    _____

    _____

14. How far must mechanical protection be provided when underground service conductors are carried up a pole? _____

_____

15. What is the maximum number of consumer's services allowed without special permission? Quote the *CEC* rule. _____

_____

16. Complete the following table by filling in the columns with the appropriate information.

	CIRCUIT NUMBER	AMPERE RATING	POLES	VOLTS	WIRE SIZE
A. LIVING ROOM RECEPTACLE OUTLETS					
B. WORKBENCH RECEPTACLE OUTLETS					
C. WATER PUMP					
D. ATTIC EXHAUST FAN					
E. KITCHEN LIGHTING					
F. HYDROMASSAGE TUB					
G. ATTIC LIGHTING					
H. COUNTER-MOUNTED COOKING UNIT					
I. ELECTRIC FURNACE					

17. a. What size conductors supply Panel B? _____

b. What size conduit? _____

c. Is this raceway run in the form of EMT or rigid conduit? _____

d. What size overcurrent device protects the feeders to Panel B? _____

_____

18. How many electric meters are provided for this residence? _____

19. According to the *CEC*, is it permissible to ground rural service entrance systems and equipment to driven ground rods when a metallic water system is not available? _____

_____

20. What table of the *CEC* lists the sizes of grounding electrode conductors to be used for service entrances of various sizes similar to the type found in this residence?

_____

_____

21. Do the following conductors require mechanical protection?

    a. No. 8 grounding conductor _____

    b. No. 6 grounding conductor _____

    c. No. 4 grounding conductor _____

22. Why is bonding service entrance equipment necessary? _____

    _____

    _____

    _____

    _____

    _____

    _____

    _____

23. What special types of bushings are required on service entrances when metallic raceways are used? _____

    _____

24. When No. 8 AWG conductors or larger are installed, what additional provision is required on the conduit ends? _____

    _____

    _____

    _____

    _____

25. What minimum size copper bonding jumpers must be installed to properly bond the electrical service for the residence discussed in this text? _____

    _____

    _____

    _____

    _____

    _____

26. a. What is a Type D fuse? _____

    _____

    _____

    b. Where could a Type D fuse be installed? _____

    _____

27. a. What is the maximum voltage permitted between conductors when using plug fuses?

    _____

    _____

b. May plug fuses be installed in a switch that disconnects a 120/240-volt clothes dryer?

_____

c. Give a reason for the answer to (b). _____

_____

28. What part of a circuit breaker causes the breaker to trip

   a. on an overload? _____   b. on a short circuit? _____

29. List the standard sizes of circuit breakers up to and including 100 amperes. _____

_____

30. Using the method shown in this unit, what is the approximate short-circuit current available at the terminals of a 50-kVA single-phase transformer rated 120/240 volts? The transformer impedance is 1%.

   a. line-to-line? _____

   b. line-to-neutral? _____

31. Where is the service for this residence located? _____

32. a. On what type of wall is Panel A fastened? _____

   b. On what type of wall is Panel B fastened? _____

_____

_____

_____

_____

_____

_____

33. When conduits pass through the wall from outside to inside, the conduit must be _____ to prevent air circulation through the conduit.

34. Briefly explain why electrical systems and equipment are grounded. _____

_____

_____

_____

_____

35. What *CEC* rule states that all overcurrent devices must have adequate interrupting ratings for the current to be interrupted? _____

36. All electrical components have some sort of "withstand rating," which indicates the ability of the component to withstand fault currents for the time required by the overcurrent device to open the circuit. What *CEC* rule refers to withstand ratings with reference to fault current protection? _____

37. Arcing fault damage is closely related to the value of _____.

38. In general, systems are grounded so that the maximum voltage to ground does not exceed (circle one)

   a. 120 volts   b. 150 volts   c. 300 volts

39. An electric clothes dryer is rated at 5700 watts. The electric rate is 10.091 cents per kilowatt-hour. The dryer is used continuously for three hours. Find the cost of operation, assuming the heating element is on continuously. _____

   _____

   _____

   _____

   _____

   _____

40. *Rule 14–012* requires that the service equipment (breakers, fuses, and the panel itself) be rated (circle one)

   a. equal to or greater than the current it must interrupt

   b. equal to or greater than the available voltage and the fault current that is available at the terminals

41. Read the meter below. Last month's reading was 22 796. How many kilowatt-hours of electricity were used for the current month? _____

UNIT 5

Electrical Outlets

OBJECTIVES

After studying this unit, the student will be able to

- identify and explain the electrical outlet symbols used in the plans of the single-family dwelling
- discuss the types of outlets, boxes, fixtures, and switches used in the residence
- explain the methods of mounting the various electrical devices used in the residence
- understand the meaning of the terms *receptacle outlet* and *lighting outlet*
- understand the preferred way to position receptacles in wall boxes
- position wall boxes in relation to finished wall surfaces
- make surface extensions from concealed wiring
- determine the number of wires permitted in a given size box

ELECTRICAL OUTLETS

The *CEC* defines an *outlet* as a point on a wiring system where current is taken to supply utilization equipment.

The type and location of electrical outlets used for an installation are shown on electrical floor plans (Fig. 5–1).

A receptacle outlet is an outlet where one or more receptacles are installed (Fig. 5–2).

A lighting outlet is "an outlet intended for the direct connection of a lampholder, a lighting fixture, or a pendant cord terminating in a lampholder" (Fig. 5–3).

A toggle switch is *not* an outlet; however, the term *outlet* is used broadly by electricians to include noncurrent-consuming switches and similar control devices in a wiring system when estimating the cost of the installation. Each type of outlet is represented on the plans as a symbol.

Fig 5–1 shows the outlets by the symbols ⊖ and -○-. Standard electrical symbols are shown in Fig. 2–9e on page 17.

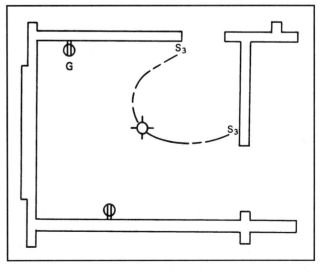

Fig. 5–1   Use of electrical symbols and notations on a floor plan.

ALWAYS ALLOW AT LEAST 150 mm
OF FREE CONDUCTOR AT ALL OUTLETS
AND JUNCTIONS SO YOU CAN
WORK EASILY WITH THE WIRING
DEVICES TO BE INSTALLED.

Fig. 5–2   When a receptacle is connected to the wires, the outlet is called a *receptacle outlet*. For ease in working with wiring devices, the *CEC Rule 12–3000(5)* requires that at least 150 mm of free conductor be provided.

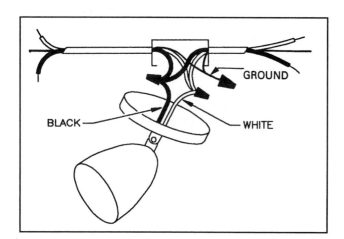

GROUND

BLACK

WHITE

Fig. 5–3   When a lighting fixture is connected to the wires, the outlet is called a *lighting outlet*. The *CEC* requires that at least 150 mm of free conductor be provided.

The dashed lines in Fig. 5–1 run from the outlet to the switch or switches that control the outlet. These lines are usually curved so that they cannot be mistaken for hidden edge lines. Outlets shown on the plan without curved dashed lines are independent and have no switch control.

When preparing electrical plans, most architects, designers, and electrical engineers use symbols approved by the Standards Council of Canada (SCC) whenever possible. However, plans may contain symbols that are not found in these standards. When unlisted (non-standard) symbols are used, the electrician must refer to a legend that interprets these symbols. The legend may be included on the plans or in the specifications. In many instances, a notation on the plan will clarify the meaning of the symbol. See Fig. 5–4.

Figs. 5–6A, B, and C show electrical symbols for outlets and their meanings. Although several symbols have the same shape, differences in the interior presentation indicate their different meanings. For example, different meanings are shown in Fig. 5–5 for the outlet symbols. A good practice in studying symbols is to learn the basic forms first and then add the supplemental information to obtain different meanings.

LIGHTING FIXTURES AND OUTLETS

Architects often include in the specifications an amount of money to purchase electrical fixtures. The electrical contractor includes this amount in

SYMBOL	NOTATION
①	PLUGMOLD ENTIRE LENGTH OF WORKBENCH. OUTLETS 90 cm O.C. INSTALL 122 cm TO CENTRE FROM FLOOR. GFCI PROTECTED.
②	TRACK LIGHTING. PROVIDE 5 LAMPHOLDERS.
③	TWO 40-WATT RAPID START FLUORESCENT LAMPS IN VALANCE. CONTROL WITH DIMMER SWITCH.

Fig. 5–4 How notations might be added to a symbol to explain its meaning. The architect or engineer has a choice of fully explaining the meaning directly on the plan if there is sufficient room; if there is insufficient room, a notation could be used.

SPLIT-SWITCHED RECEPTACLE OUTLET   GFCI RECEPTACLE OUTLET   WEATHERPROOF RECEPTACLE OUTLET   SPLIT-WIRED RECEPTACLE OUTLET (THREE-WIRE CCT)

Fig. 5–5 Variations in outlet symbols.

the bid, and the choice of fixtures is then left to the homeowner. If the owner selects fixtures whose total cost exceeds the fixture allowance, the owner is expected to pay the difference between the actual cost and the specification allowance. If the fixtures are not selected before the roughing-in stage of wiring of the house, the electrician usually installs outlet boxes having standard fixture mounting studs.

Most modern surface-mount lighting fixtures can be fastened to an octagon box (Fig. 5–7A and B) or an outlet box with a plaster ring using appropriate No. 8–32 metal screws and mounting strap furnished with the fixture.

A box must be installed at each outlet or switch location, *Rule 12–3000(1)*. There are exceptions to this rule, but in general they relate to special manufactured wiring systems where the "box" is an

integral part of the system. For standard wiring methods, such as cable or conduit, a box is usually required.

Figs. 5–8 and 5–9 illustrate types of non-metallic and metallic boxes.

Outlet boxes must be accessible, *Rule 12–3014(1)*.

Be careful when roughing in boxes for fixtures and also when hanging ceiling fixtures to:

- non-metallic device (switch) boxes
- non-metallic device (switch) plaster rings
- metallic device (switch) boxes
- metallic device (switch) plaster rings
- any non-metallic box, unless specifically marked on the box or carton for use as a fixture support, or to support other equipment, or to accommodate heat-producing equipment.

Therefore, unless the box, device plaster ring, or carton is marked to indicate that it has been listed by CSA for the support of fixtures, do not use where fixtures are to be hung. Refer to *Rule 12–1110*.

Be careful when installing a ceiling outlet box for the purpose of supporting a fan. See Unit 11 for detailed discussion of paddle fans and their installation.

If the owner selects fixtures before construction, the architect can specify these fixtures in the plans and/or specifications, giving the electrician advance information on any special framing, recessing, or mounting requirements. This information must be provided in the case of recessed fixtures, which require a specific wall or ceiling opening.

Many types of lighting fixtures are available. Fig. 5–10 shows several typical lighting fixtures that may be found in a dwelling unit. It also shows the electrical symbols used on plans to designate these fixtures and the type of outlet boxes or switch boxes on which the lighting fixtures can be mounted. A standard receptacle outlet is shown as well. These switch boxes are made of steel; they may also be made of plastic, as shown in Fig. 5–8.

When mounting an outlet box in a wall or ceiling that is to be insulated and covered with a vapour barrier, you must make special provision to

⊘ OR ⊖ OR ◯	CEILING MOUNTED INCANDESCENT
⊘— OR ⊖— OR ◯—	WALL MOUNTED INCANDESCENT
[⊘]	RECESSED CEILING MOUNTED INCANDESCENT
▽ ▽◯▽ ▽	TRACK LIGHTING LENGTH AND NUMBER OF FIXTURES AS NOTED
A ◯ d	FLUORESCENT LUMINAIRE UPPER CASE LETTER DENOTES STYLE LOWER CASE LETTER DENOTES SWITCHING ARRANGEMENT
[◯] OR [◯ R]	RECESSED FLUORESCENT LUMINAIRE
[◯ ◯ ◯]	CONTINUOUS ROW FLUORESCENT LUMINAIRES
[◯ ◯ ◯]	CONTINUOUS ROW RECESSED FLUORESCENT LUMINAIRES
	FLUORESCENT LUMINAIRES

Fig. 5–6A   Symbols for different types of lighting outlets.

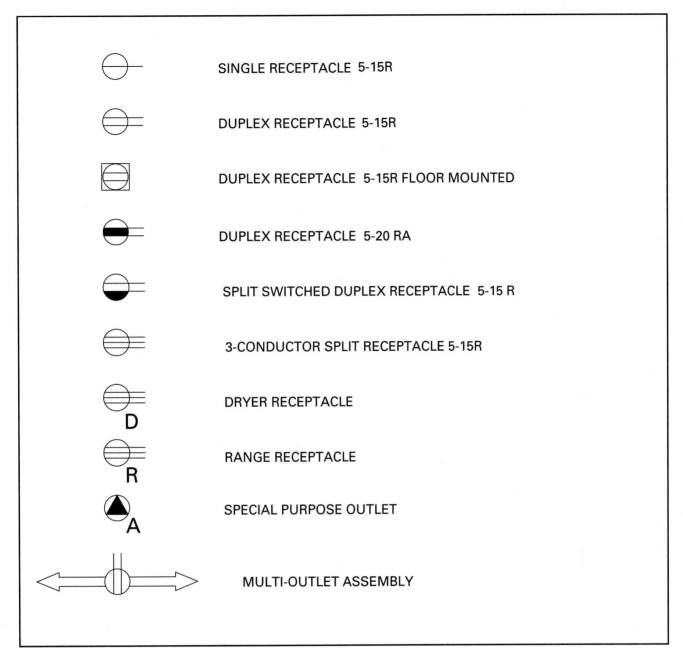

SINGLE RECEPTACLE 5-15R

DUPLEX RECEPTACLE 5-15R

DUPLEX RECEPTACLE 5-15R FLOOR MOUNTED

DUPLEX RECEPTACLE 5-20 RA

SPLIT SWITCHED DUPLEX RECEPTACLE 5-15 R

3-CONDUCTOR SPLIT RECEPTACLE 5-15R

DRYER RECEPTACLE

RANGE RECEPTACLE

SPECIAL PURPOSE OUTLET

MULTI-OUTLET ASSEMBLY

Fig. 5–6B   Symbols for receptacle outlets.

maintain the integrity of the vapour barrier. A polyethylene boot covers the outlet box and is sealed to
the 6-mm polyethylene vapour barrier.

Fig. 5–7B shows a typical polyethylene boot
installed on a 102-mm (4") octagon box.

SWITCHES

Fig. 5–11 shows some of the standard symbols for
various types of switches and typical connection
diagrams. Any sectional switch box or 102-mm

(4-in) square box with a side-mounting bracket and
raised switch cover can be used to install these
switches (Fig. 5–10).

JUNCTION BOXES AND SWITCH (DEVICE) BOXES *(RULES 12–3000 TO 12–3034)*

Junction boxes are sometimes placed in a circuit
for convenience in joining two or more cables or
conduits. All conductors entering a junction box

Fig. 5–6C   Miscellaneous symbols.

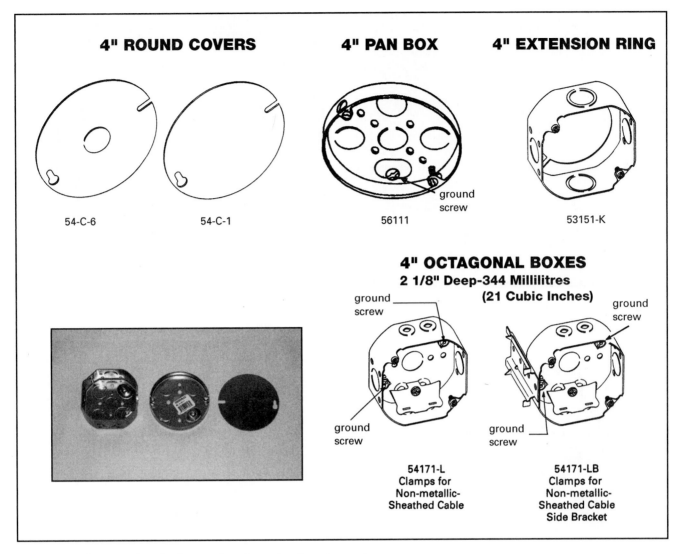

Fig. 5–7A   Octagonal outlet boxes, showing extension ring and covers.

Fig. 5–7B   Outlet box complete with vapour boot.

are joined to other conductors entering the same box to form proper connections so that the circuit will operate in its intended manner.

All electrical installations must conform to the *CEC* standards requiring that junction boxes be installed so that the wiring contained in them shall be accessible without removing any part of the building. In house wiring, this requirement limits the use of junction boxes to unimproved basements, garages, and open attic spaces because blank box covers exposed to view detract from the appearance of a room. An outlet box such as the one installed for the front hall ceiling fixture is really a junction box because it contains splices. Removing the fixture makes the box accessible, thereby meeting *CEC* requirements. Refer to Figs. 5–12 and 5–13.

Fig. 5–8A Non-metallic outlet box.

Fig. 5–8B Non-metallic outlet box.

GROUND SCREW
TAPPED HOLE

Fig. 5–9A Gang-type switch (device) box.

GROUND SCREW
TAPPED HOLE

Fig. 5–9B Octagon box.

Boxes shall be rigidly and securely mounted, *Rule 12–3010(1)*.

NON-METALLIC OUTLET AND DEVICE BOXES

The *CEC* permits non-metallic outlet and device boxes to be installed where the wiring method is non-metallic sheathed cable, *Rule 12–524*. Non-metallic boxes may be used with armoured cable or metal raceways, but only if a proper means is provided inside the boxes to bond any and all metal raceways and/or armoured cables together, *Rule 12–3000(2)*.

Rule 12–3000(1) requires that a box or fitting be installed wherever splices, switches, outlets, junction points, or pull points are required (Fig. 5–12). However, sometimes a change is made from one wiring method to another, for example, armoured cable to electrical metallic tubing (EMT). The fitting where the change is made must be accessible after installation, Fig. 5–14.

SYMBOL	TYPE OF LUMINAIRE (LIGHTING FIXTURE) OR OUTLET	BOXES THAT MAY BE USED
◯ OR ◯ OR ◯ **CEILING LIGHTING OUTLET**	**SURFACE-MOUNTED LUMINAIRE (LIGHTING FIXTURE)**	(A) ADJUSTABLE BRACKET WITH OUTLET BOX. (B) OUTLET BOX WITH BRACKET FOR JOIST MOUNTING. (C) OUTLET BOX WITH CAPTIVE NAILS/BRACKET FOR JOIST MOUNTING. BOXES ARE "LISTED" FOR LUMINAIRE (LIGHTING FIXTURE) SUPPORT
RECESSED CEILING LIGHTING OUTLET	**RECESSED LUMINAIRE (LIGHTING FIXTURE)**	CONNECTIONS TO THE LIGHT FIXTURE (LUMINAIRE) ARE MADE IN THE BOX SHOWN AT THE SIDE OF THE FIXTURE. THE CONNECTIONS ARE ACCESSIBLE THROUGH THE OPENING FOR MOUNTING THE LUMINAIRE ARE SHOWN. *CEC RULE 30–910.*
WALL LIGHTING OUTLET	**WALL-MOUNTED LUMINAIRE (LIGHTING FIXTURE) ALSO CALLED A "SCONCE"**	(A) OUTLET BOX WITH CAPTIVE NAILS/BRACKET FOR STUD MOUNTING. (B) ADAPTER RING FOR OUTLET BOX. BOX "LISTED" FOR LUMINAIRE (LIGHTING FIXTURE) SUPPORT.
DUPLEX RECEPTACLE OUTLET		(A) SINGLE-GANG SWITCH (DEVICE) BOX. (B) 102 mm SQUARE BOX WITH CAPTIVE NAILS FOR STUD MOUNTING. (C) 102 mm SQUARE SINGLE-GANG RAISED ADAPTER RING.

Fig. 5–10 Typical electrical symbols, items they represent, and boxes that could be used. Photographs courtesy of Progress Lighting/Hubbell (1, 3) and Halo/Cooper Lighting (2).

Fig. 5–11 Standard switches and symbols.

Fig. 5–12 A box (or fitting) must be installed wherever there are splices, outlets, switches, or other junction points. Refer to the points marked X. A potential *CEC* violation is shown at point XX. However, this is a violation only if the cable is not supplying equipment with an integral connection box or if the apparatus has not been approved as a connection box. *Rules 12–3000(1)* and *12–3014(1)*.

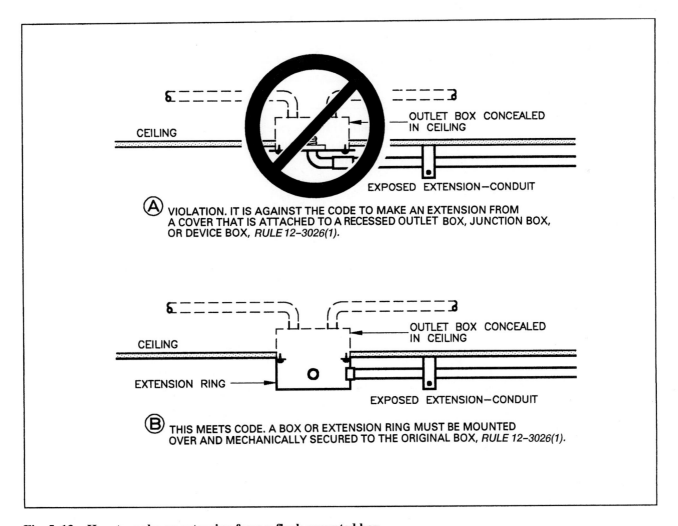

Fig. 5–13 How to make an extension from a flush-mounted box.

Fig. 5–14 A transition from one wiring method to another. In this case, the armour of the AC 90 cable is removed, allowing sufficient length of the conductors to be run through the conduit. A proper fitting as shown must be used at the transition point, and the fitting must be accessible after installation.

The house wiring system is usually formed by a number of specific circuits, each consisting of a continuous run of cable from outlet to outlet or from box to box. The residence plans show many branch circuits for general lighting, appliances, electric heating, and other requirements. The specific *CEC* rules for each of these circuits are covered in later units.

GANGED SWITCH (DEVICE) BOXES

A flush switch or receptacle outlet for residential use fits into a standard $3 \times 2 \times 2\frac{1}{2}$-in sectional switch box (sometimes called a *device box*). When two or more switches (or outlets) are located at the same point, single switch boxes may be ganged or fastened together to provide the required mounts, Fig. 5–15, or a welded (two or more) gang box may be used.

Three switch boxes can be ganged together by removing and discarding one side from both the first and third switch boxes, and both sides from the second (centre) switch box. The boxes are then joined together (Fig. 5–15B). After the switches are installed, the gang is trimmed with a gang plate having the required number of switch handle or receptacle outlet openings. These plates are called two-gang wall plates, three-gang wall plates, and so on, depending on the number of openings.

The dimensions of a standard sectional switch box, 3×2 in, are the dimensions of the opening of the box. The depth of the box may vary from $1\frac{1}{2}$ inches to $3\frac{1}{2}$ inches (38 to 89 mm), depending on the requirements of the building construction and the number of conductors and devices to be installed. *Rules 12–3000* to *12–3034* cover outlet, switch, and junction boxes. See Fig. 5–16 for a list of common box dimensions.

Rule 12–3016(1) states that boxes must be mounted so that they will be set back not more than 6 mm from the finished surface when the boxes are mounted in noncombustible walls or ceilings made of concrete, tile, or similar materials. When the wall or ceiling construction is of combustible material (wood), the box must be set flush with the surface or project from it, Fig. 5–17A. These requirements are meant to prevent the spread of fire if a short circuit occurs within the box.

Single gang switch boxes are normally mounted to a structural member of the building with nails or screws. Ganged sectional switch (device) boxes may be installed using a pair of

Fig. 5–15A Device boxes. Box on left has welded sides. Box on right has removable sides, which permit several boxes to be ganged together.

Fig. 5–15B Standard flush switches installed in ganged sectional device boxes.

metal mounting strips, *Rule 12–3010(2)*. These strips may also be used to install a switch box between wall studs, Fig. 5–18. When an outlet box is to be mounted at a specific location between joists, as for ceiling-mounted fixtures, an offset bar hanger is used (Fig. 5–10).

Older installations of switch boxes may have nails passing through the box. This is no longer permitted unless the nails are not more than 6.4 mm from the back or ends of the box. See *Rule 12–3010(5)*.

BOXES FOR CONDUIT WIRING

For some types of construction, conduit rather than cable wiring is used.

When conduit is installed in a residence, it is quite common to use 102-mm square boxes trimmed with suitable plaster covers (Fig. 5–16). There are sufficient knockouts in the top, bottom, sides, and back of the box to permit a number of conduits to run into it. Plenty of room is available for the conductors and wiring devices. Note how easily these 102-mm square outlet boxes can be mounted back-to-back by installing a small fitting between them, Fig. 5–19 (2) and (3).

You can trim 4-in (102-mm) square outlet boxes with one-gang or two-gang plaster rings where wiring devices will be installed. Where lighting fixtures will be installed, use a plaster ring having a round opening.

Any unused openings in outlet and device boxes must be closed, as per *Rule 12–3024*, using knockout fillers, Fig. 5–20.

The number of conductors allowed in outlet and device boxes is covered elsewhere in this text.

SPECIAL-PURPOSE OUTLETS

Special-purpose outlets are normally shown on floor plans. They are described by a notation and are also detailed in the specifications. The plans included in this text indicate special-purpose outlets by a triangle inside a circle with subscript letters. In some cases, a subscript number is added to the letter.

When a special-purpose outlet is indicated on the plans or in the specifications, the electrician must check for special requirements, such as a separate circuit, a special 240-volt circuit, a special grounding or polarized receptacle, a device box with extra support (Fig. 5–21), or other preparation. Special-purpose outlets include:

- central vacuum receptacle
- weatherproof receptacle
- dedicated receptacle, perhaps for a home computer
- air-conditioning receptacle
- clock receptacle

QUICK-CHECK BOX SELECTION GUIDE
FOR BOXES GENERALLY USED FOR RESIDENTIAL WIRING

DEVICE BOXES			$3 \times 2 \times 1\frac{1}{2}$	$3 \times 2 \times 2$	$3 \times 2 \times 2\frac{1}{4}$	$3 \times 2 \times 2\frac{1}{2}$	$3 \times 2 \times 3$
	BOX SIZE						
	VOLUME	Millilitres	131	163	163	204	245
	WIRE SIZE 14	14	5	6	6	8	10
	12	12	4	5	5	7	8

SQUARE BOXES			$4 \times 4 \times 1\frac{1}{2}$	$4 \times 4 \times 2\frac{1}{8}$	$4\frac{11}{16} \times 4\frac{11}{16} \times 1\frac{1}{2}$	$4\frac{11}{16} \times 4\frac{11}{16} \times 2\frac{1}{8}$
	BOX SIZE					
	VOLUME	Millilitres	344	491	491	688
	WIRE SIZE 14	14	14	20	20	28
	12	12	12	17	17	24

OCTAGON BOXES			$4 \times 1\frac{1}{2}$	$4 \times 2\frac{1}{8}$	
	BOX SIZE				
	VOLUME	Millilitres	245	344	
	WIRE SIZE 14	14	10	14	
	12	12	8	12	

UTILITY BOXES			$4 \times 2 \times 1\frac{1}{2}$	$4 \times 2\frac{1}{8} \times 1\frac{7}{8}$	$4 \times 2\frac{3}{8} \times 1\frac{7}{8}$	
	BOX SIZE					
	VOLUME	Millilitres	147	229	262	
	WIRE SIZE 14	14	6	9	10	
	12	12	5	8	9	

RAISED COVERS

WHERE RAISED COVERS ARE MARKED WITH THEIR VOLUME IN CUBIC INCHES, THAT VOLUME MAY BE ADDED TO THE BOX VOLUME TO DETERMINE MAXIMUM NUMBER OF CONDUCTORS IN THE COMBINED BOX AND RAISED COVER.

NOTE: BE SURE TO MAKE DEDUCTIONS FROM THE ABOVE MAXIMUM NUMBER OF CONDUCTORS PERMITTED FOR WIRE CONNECTORS, WIRING DEVICES, CABLE CLAMPS, FIXTURE STUDS, AND GROUNDING CONDUCTORS. THE VOLUME IS TAKEN DIRECTLY FROM *TABLE 23* OF THE *CEC*.

Fig. 5–16 Quick-check box selection guide.

Fig. 5–17A   Box position in walls and ceilings constructed of various materials, *Rule 12–3016*.

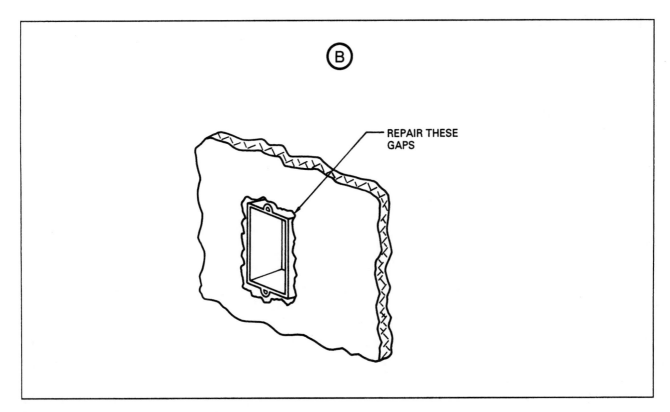

Fig. 5–17B   Repair these gaps, *Rule 26–710*.

Fig. 5–18A   Switch (device) boxes installed between studs using metal mounting strips. If wooden boards are used, they must be at least 19 mm thick and securely fastened to the stud, *Rule 12–3010(2)*.

NUMBER OF CONDUCTORS IN A BOX

Rule 12–3034 dictates that outlet boxes, switch boxes, and device boxes be large enough to provide ample room for the wires without having to jam or crowd the wires into them. The *CEC Table 23* specifies the maximum number of conductors allowed in standard outlet boxes and switch boxes, Table 5–1. A conductor running through the box is counted as one conductor. Each conductor originating outside the box and terminating inside it is counted as one conductor. Conductors that originate and terminate within the box are not counted.

When conductors are the same size, the proper box size can be selected by referring to *CEC Table 23*. When conductors of different sizes are used, refer to *CEC Table 22*, Table 5–2.

Tables 22 and *23* do not consider fittings or devices such as fixture studs, wire connectors, hickeys, switches, or receptacles that may be in the box. For assistance in this case, see Table 5–3 (p. 91) and Fig. 5–22 (p. 89). When the box contains one or more fittings (such as fixture studs or hickeys), the number of conductors must be one less than shown in the table for each type of device.

Deduct two conductors for each device mounted on a single strap (switches, receptacles). The deduc-

Fig. 5–18B   A method of mounting sectional boxes to wood studs.

tion of two conductors has become necessary because of severe crowding of conductors in the box when the switch or receptacle is mounted or when dimmer switches are installed. In most cases, dimmer switches are larger than conventional switches.

Deduct one conductor for every two-wire connector. For example, for two- or three-wire connectors deduct one wire; for four or five connectors deduct two wires. There is no deduction for the first wire connector because it will probably be used for the second ground connection. Be sure to include all conductor sizes (Nos. 14, 12, 10, 8, and 6 AWG) when determining the size of the box to be installed. See Figs. 5–23 and Fig. 5–24 for examples of this *CEC* requirement, *Rule 12–3034(1)*.

Fig. 5–19 Boxes for conduit wiring.

SELECTING A BOX WHEN ALL CONDUCTORS ARE THE SAME SIZE

EXAMPLE: A box contains one fixture stud and two cable clamps. The number of conductors permitted in the box shall be one less than shown in the table. (Deduct one conductor for the fixture stud.)

A further deduction of two conductors is made for each wiring device in the box, but if two more devices are mounted on the same strap, they count

CEC TABLE 23
NUMBER OF CONDUCTORS IN BOXES
(See *Rule 12–3034*.)

Box Dimensions Trade Size		Capacity, mL (in³)	Maximum Number of Conductors (per AWG size)				
			14	12	10	8	6
Octagonal	4 x 1-1/2	245 (15)	10	8	6	5	3
	4 x 2-1/8	344 (21)	14	12	9	7	4
Square	4 x 1-1/2	344 (21)	14	12	9	7	4
	4 x 2-1/8	491 (30)	20	17	13	10	6
	4-11/16 x 1-1/2	491 (30)	20	17	13	10	6
	4-11/16 x 2-1/8	688 (42)	28	24	18	15	9
Round	4 x 1/2	81 (5)	3	2	2	1	1
Device	3 x 2 x 1-1/2	131 (8)	5	4	3	2	1
	3 x 2 x 2	163 (10)	6	5	4	3	2
	3 x 2 x 2-1/4	163 (10)	6	5	4	3	2
	3 x 2 x 2-1/2	204 (12.5)	8	7	5	4	2
	3 x 2 x 3	245 (15)	10	8	6	5	3
	4 x 2 x 1-1/2	147 (9)	6	5	4	3	2
	4 x 2-1/8 x 1-7/8	229 (14)	9	8	6	5	3
	4 x 2-3/8 x 1-7/8	262 (16)	10	9	7	5	3
Masonry	3-3/4 x 2 x 2-1/2	229 (14)/gang	9	8	6	5	3
	3-3/4 x 2 x 3-1/2	344 (21)/gang	14	12	9	7	4
	4 x 2-1/4 x 2-3/8	331 (20.25)/gang	13	11	9	7	4
	4 x 2-1/4 x 3-3/8	364 (22.25)/gang	14	12	9	8	4
Through Box	3-3/4 x 2	3.8/mm (6/in) depth	4	3	2	2	1
Concrete Ring	4	7.7/mm (12/in) depth	8	6	5	4	2
FS	1 Gang	229 (14)	9	8	6	5	3
	1 Gang tandem	557 (34)	22	19	15	12	7
	2 Gang	426 (26)	17	14	11	9	5
	3 Gang	671 (41)	27	23	18	14	9
	4 Gang	917 (56)	37	32	24	20	12
FD	1 Gang	368 (22.5)	15	12	10	8	5
	2 Gang	671 (41)	27	23	18	14	9
	3 Gang	983 (60)	40	34	26	21	13
	4 Gang	1392 (85)	56	48	37	30	18

Table 5–1   *CEC Table 23.*

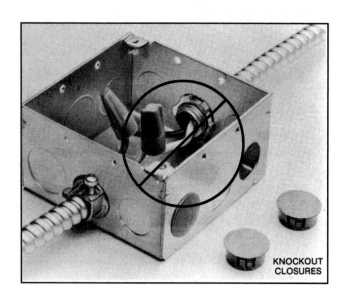

Fig. 5–20   Unused openings in boxes must be closed according to the *CEC Rule 12–3024*. This is done to contain electrical short-circuit problems inside the box or panel and to keep rodents out.

WIRES MUST BE 150 mm MINIMUM. *CEC Rule 12–3000(4)*. See Fig. 5–2.

CEC TABLE 22
SPACE FOR CONDUCTORS IN BOXES
(See Rule 12–3034.)

Size of conductor, AWG	Usable space required for each insulated conductor, mL
14	24.6
12	28.7
10	36.9
8	45.1
6	73.7

Table 5–2 *CEC Table 22.*

Fig. 5–21 Flush-mounted device box for steel studs with required additional support.

HICKEY   FIXTURE STUD

Fig. 5–22 Hickey and fixture stud.

GROUND

BLACK   WHITE

as a single device. For example, if a switch and a receptacle are mounted on a single strap, only two conductors are deducted.

SELECTING A BOX WHEN CONDUCTORS ARE DIFFERENT SIZES (*RULE 12-3034(4)*)

When the box contains different size wires, do the following:

- Determine the size of conductors to be used and the number of each (e.g., two No. 10 and two No. 12).
- Determine if the box is to contain any wiring devices (switch or receptacle), wire connectors, or fixture studs. This volume adjustment must be based on the volume of the largest conductor in the box.
- Deduct one conductor for each pair of wire connectors. For example, two or three equals one wire, four or five equals two wires. *Note:* If only one wire connector is used, it is not counted.
- Deduct two conductors for each switch or receptacle that is less than 25 mm between the mounting strap and the back of the device.
- Make the necessary adjustments for devices, wire connectors, and fixture studs.
- Size the box based upon the total volume required for conductors according to *Table 22*.

EXAMPLE: What is the minimum volume required for a box that will contain one switch, two No. 14 wires, two No. 12 wires, and three No. 33 wire connectors? The cable is armoured cable.

2 No. 14 AWG wires $\times$ 24.6 cm^3/wire = 49.2 cm^3
2 No. 12 AWG wires $\times$ 28.7 cm^3/wire = 57.4 cm^3
3 wire connectors
  each pair counts as 1 wire   = 28.7 cm^3
switch
  each switch counts as 2 wires   = 57.4 cm^3
               Total volume   = 192.7 cm^3

Fig. 5–23 The transformer leads are not counted, since no part of the conductors from the transformer leaves the box.

Therefore, select a box having a minimum volume of 192.7 mL of space. A $3 \times 2 \times 2\frac{1}{2}$-in device box has a volume of 204 mL from *Table 23*, and this would be sufficient. The cubic-inch volume may be marked on the box; otherwise refer to the column of *Table 23* titled "Capacity."

When sectional boxes are ganged together, the volume to be filled is the total volume of the assembled boxes. Fittings, such as plaster rings, raised covers, and extension rings, may be used with the sectional boxes. If these fittings are marked with their volume, or have dimensions comparable to the boxes in *Table 23*, their volume may be considered when determining the total volume (*Rule 12–3034(6)*). In some cases, the volume listed in *Table 23* is not equivalent to the length × width × depth of the box. When doing calculations using boxes listed in *Table 23*, use the volume found in the table. Otherwise multiply L × W × D to find the actual volume of the box. For example, a $12 \times 12 \times 6$-in box is not found in *Table 23*. Therefore the box volume is $12 \times 12 \times 6 = 864$ in³ (14.1 L). Figs. 5–25 and 5–26 show how a 19-mm raised cover (plaster ring) increases the wiring space of the box to which it is attached.

Electrical inspectors are very aware that GFCI receptacles, dimmers, and certain types of timers take more space than regular receptacles. Therefore, it is a good practice to install device boxes that will provide lots of room for wires, instead of pushing and crowding the wires into the box.

EXAMPLE: How many No. 12 conductors are permitted in the box and raised plaster ring shown in Fig. 5–26?

SOLUTION: See *Rule 12–3034* and *Tables 22* and *23* for volume required per conductor.
This box and cover will take:

$$\frac{418 \text{ mL}}{28.7 \text{ mL per No. 12 conductor}} =$$

14 No. 12 conductors maximum, less the deductions for devices and connectors, per *Rule 12–3034(2)*.

Fig. 5–23 shows how transformer leads No. 18 AWG or larger are treated when selecting a box.

Wire No. 16 or No. 18 AWG supplying a lighting fixture is not counted when sizing the box that the fixture is mounted on, Fig. 5–24. This is stated in *Rule 12–3034(1)(d)*.

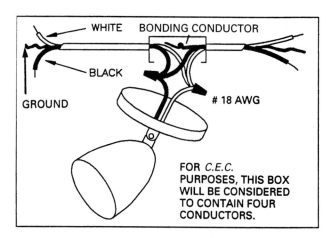

Fig. 5–24 Fixture wires are not counted in determining correct box size. The *CEC Rule 12–3034(1)(d)* specifies that No. 18 and No. 16 AWG fixture wires supplying a lighting fixture, mounted on the box containing the fixture wires, shall not be counted.

Fig. 5–16, Quick-check box selection guide, shows some of the most popular types of boxes used in residential wiring. You can refer to it as you work your way through this text.

When wiring with cable, "feeding through" a box is impossible. But, when installing the wiring using conduit, it is possible to feed straight through a box. Only one wire is deducted for a wire running straight through a box.

BOX FILL

The following method can make box fill easier to calculate when determining the proper size junction box or wall box to install:

1. Count the number of circuit wires.
2. Add one wire for fixture stud (if any).
3. Add one wire for each pair of wire connectors.
4. Add one wire for one or more isolated (insulated) grounding conductors (if any).
5. Add two wires for each wiring device.

Look up the total count in *Table 23* to find a box appropriate for the intended use that will hold the number of conductors required.

EXAMPLE: Six circuit conductors   6
One wiring device (switch)   + 2
Three wire connectors   + 1
_____
Total   9

Note: 1 mL = 0.061 cu. in.

• If box contains NO fittings, devices, fixture studs, hickeys, switches, or receptacles	• refer directly to *Table 23*.
• If box contains ONE or MORE fittings, fixture studs, or hickeys...	• deduct ONE from maximum number of conductors permitted in *Table 23* for each type.
• For each device on a single strap containing ONE or MORE devices such as switches, receptacles, or pilot lights...	• deduct ANOTHER TWO from *Table 23*.
• For ONE or MORE isolated (insulated) grounding conductors...	• deduct ANOTHER ONE from *Table 23*.
• For each PAIR of insulated wire connectors. . .	• deduct another conductor from the number permitted by *Table 23*.
• For conductors running through the box...	• count ONE conductor for each conductor running through the box.
• For conductors that originate outside box and terminate inside box...	• count ONE conductor for each conductor originating outside and terminating in box.
• If no part of the conductor leaves the box—for example, a "jumper" wire used to connect a receptacle.	• don't count the conductors

Table 5–3 Quick checklist for determining proper size of boxes.

ADDITIONAL WIRING SPACE PROVIDED BY THE RAISED COVER IS 2" X 3" X 3/4" = 4 1/2 in³ = 74 mL (cm³)

Fig. 5–25 Raised cover. Raised covers are sometimes called plaster rings.

Fig. 5–26 Area of box and plaster ring.

Therefore, select a box capable of containing nine or more conductors. See *Table 23*.

It is possible to select a smaller box when using EMT than when using the cable wiring method because it is possible to "loop" conductors through the box. These only count as one conductor for the purpose of box fill count, Fig. 5–27.

LOCATING OUTLETS

Electricians commonly consult the plans and specifications to determine the proper heights and clearances for installing electrical devices. The electrician then has these dimensions verified by the architect, electrical engineer, designer, or homeowner to avoid unnecessary and costly changes in the locations of outlets and switches as the building progresses.

There are no code rules for locating the height of outlets. A number of conditions determine the proper height for a switch box. For example, the height of the kitchen counter backsplash determines where the switches and receptacle outlets are located between the kitchen countertop and the cabinets.

The location of receptacles is covered in *Rule 26–712* of the *CEC* and Unit 6 of this text.

When locating receptacles near electric baseboard heaters, refer to the section "Location of electric baseboard heaters" in Unit 17.

POSITIONING RECEPTACLES

Although no actual *CEC* rules exist on positioning receptacles, there is a concern in the electrical industry that a *metal* wall plate could come loose and fall downward onto the blades of an attachment plug cap that is loosely plugged into the receptacle, thereby creating a potential shock and fire hazard.

Therefore, when metal wall plates are used, position the grounding hole to the top. A loose

FIGURE A. CONDUIT: SINCE TWO CONDUCTORS ARE LOOPED THROUGH THE BOX,
THE CONDUCTOR COUNT IN THIS ILLUSTRATION IS FOUR.

FIGURE B. CABLE: THE CONDUCTOR COUNT IN THIS
ILLUSTRATION IS FOUR. THE TWO EQUIPMENT
BONDING CONDUCTORS ARE NOT COUNTED.

FIGURE C. CABLE AND A BOX WITH INTERNAL CABLE CLAMPS:
THE CONDUCTOR COUNT IN THIS ILLUSTRATION IS FOUR.
THE TWO EQUIPMENT BONDING CONDUCTORS ARE NOT
COUNTED. THE TWO CABLE CLAMPS ARE NOT COUNTED.

Fig. 5–27 Conductor count for both metal conduit and cable installations.

metal plate could fall onto the grounding blade of the attachment plug cap, but no hazard would result.

Or when metal cover plates are used, position grounded neutral blades on top. A loose metal plate could fall onto these grounded neutral blades with no hazardous effect.

When metal cover plates are used, do not position the "hot" terminal on top. A loose metal plate could fall onto these live blades. If this was a split-circuit receptacle fed by a three-wire, 120/240-volt circuit, the short would be across the 240-volt line.

To ensure uniform installation and safety, and in accordance with long-established custom, standard electrical outlets are located as shown in Fig. 5–28. These dimensions usually are satisfactory. However, the electrician must check the blueprints, specifi-

SWITCHES	
	*Height above floor
Regular	1.17 m (46 in)
Between counter and kitchen cabinets (depends on backsplash)	1.12–1.17 m (44–46 in)
RECEPTACLE OUTLETS	
	*Height above floor
Regular (not permitted above electric baseboard heaters)	300 mm (12 in)
Between counter and kitchen cabinets (depends on backsplash)	1.12–1.17 m (44–46 in)
In garage	1.22 m (48 in)
WALL BRACKETS	
	*Height above floor
Outside	1.8 m (72 in)
Inside	1.52 m (60 in)
Side of medicine cabinet	1.52 m (60 in)

*Note: All dimensions given are from the finished floor to the centre of the outlet box. Verify all dimensions before roughing in.

Fig. 5–28   Outlet locations.

cations, and details for measurements that may affect the location of a particular outlet or switch.

The cabinet spacing, available space between the countertop and cabinet, and the tile height may influence the location of the outlet or switch. For example, if the top of the wall tile is exactly 1.22 m (48 in) from the finished floor line, a wall switch should not be mounted 1.22 m (48 in) to centre. This is considered poor workmanship since it prevents the wallplate from sitting flat against the wall. The switch should be located entirely within the tile area or entirely out of the tile area, Fig. 5–29. This situation requires the full cooperation of all trades involved on the construction job and a careful review of the architectural drawings before roughing in the outlets.

See Fig. 5–30 for how to locate switches next to a doorway.

Faceplates

Faceplates for switches shall be installed to completely cover the wall opening and seat against the wall surface.

WHERE DOORWAYS ARE FRAMED WITH A SINGLE STUD, INSTALL BLOCKING SHOWN BEFORE MOUNTING SWITCH BOXES. THIS WILL ENSURE THE SWITCH DOES NOT INTERFERE WITH THE DOOR TRIM.

Fig. 5–30   Locating switches next to doorways.

(A)   (B)   (C)   (D)

WHERE A WALL IS PARTIALLY TILED, A SWITCH OR CONVENIENCE OUTLET MUST BE LOCATED ENTIRELY OUT OF THE TILE AREA (A) OR ENTIRELY WITHIN THE TILE AREA (B), (C).

THE FACEPLATE IN (D) DOES NOT "HUG" THE WALL PROPERLY. THIS INSTALLATION IS CONSIDERED UNACCEPTABLE BY MOST ELECTRICIANS. *RULE 12–3002.*

Fig. 5–29   Locating an outlet on a tiled wall.

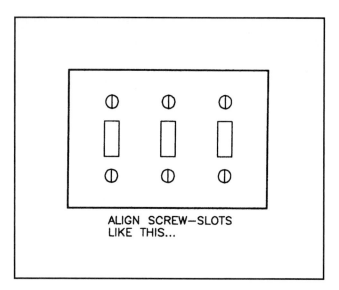

ALIGN SCREW—SLOTS
LIKE THIS...

Fig. 5–31   Aligning the slots of the faceplate mounting screws in the same direction makes the installation look neater.

Faceplates for receptacles shall be installed to completely cover the opening and seat against the mounting surface. The mounting surface may be the wall or the gasket of a weatherproof box.

Faceplates should be level. In addition, a minor detail that can improve the appearance of the installation is to align the slots in all the faceplate mounting screws in the same direction, Fig. 5–31.

The location of lighting outlets is determined by the amount and type of illumination required to provide the desired lighting effects. (It is not the intent of this text to describe how proper and adequate lighting is determined; rather, it covers the proper methods of installing the circuits for such lighting.)

BONDING

Bonding is the joining together of all the non-current-carrying metal parts of an electrical system with a conductor that has a low impedance (opposition to the flow of current). Bonding provides protection from shock hazards by maintaining all the non-current metal parts of the electrical system at the same potential (voltage). It prevents a voltage from developing between (the enclosures of) two pieces of electrical equipment due to static build-up or the failure of insulation on conductors.

Bonding is the most important part of the installation of electrical outlets. It will permit the system to "fail safe." Your electrical inspector will look closely at the workmanship involved in the installation of the bonding conductors.

It is important that the continuity of the bonding conductor be maintained. Therefore, the bonding conductor is brought into the box and attached to the bonding screw in the box before it is attached to the green bonding screw on a receptacle. With such connections, if someone cut the connections to the receptacle to remove it, it would not interfere with the bonding connection to the box.

When connecting the bonding conductor to the bonding screws in outlet boxes, ensure that the conductor does not get pinched between the head of the screw and the shoulder of the box (which prevents the conductor from slipping out from under the head of the screw as the screw is tightened down). This can damage the bonding conductor, which reduces its size (cross-sectional area) and its effectiveness.

REVIEW

Note: Refer to the *CEC* or the plans where necessary.

1. What does a plan show about electrical outlets? _____

_____

_____

2. What is an outlet? _____

_____

_____

_____

3. Match the following switch types with the proper symbol.

   a. single-pole S_p

   b. three-way S_4

   c. four-way   S

   d. single-pole with pilot light S_3

4. The plans show dashed lines running between switches and various outlets. What do these dashed lines indicate? _____

_____

_____

_____

5. Why are the dashed lines usually curved? _____

6. a. What are junction boxes used for? _____

   b. Are junction boxes normally used in wiring the first floor? _____

   _____

   _____

   c. Are junction boxes normally used to wire exposed portions of the basement?

   _____

   _____

7. How are standard sectional switch (device) boxes mounted? _____

_____

_____

_____

8. a. What is an offset bar hanger? _____

_____

_____

_____

   b. What types of boxes may be used with offset bar hangers? _____

_____

_____

_____

9. What methods may be used to mount lighting fixtures to an outlet box fastened to an offset bar hanger? _____

_____

_____

10. What is the size of the opening of a switch (device) box for a single device?

_____

_____

11. The space between a door casing and a window casing is 89 mm. Two switches are to be installed at this location. What problems could you encounter when placing the switches in this location? What would you recommend as a possible solution?

_____

_____

_____

_____

_____

_____

12. Three switches are mounted in a three-gang switch (device) box. The wall plate for this assembly is called a _____ plate.

13. For each fixture stud inside a box, (increase) (decrease) the number of conductors allowed by one. Circle the correct answer.

14. a. How high above the finished floor are the switches located in the garage of this dwelling? _____

   b. In the living room of this dwelling? _____

15. How high above the finished floor are the receptacle outlets in the garage located? _____ In the living room? _____

16. Outdoor receptacle outlets in this dwelling are located _____ mm above grade.

17. In the spaces provided, draw the correct symbol for each of these items:

   a. _____ Lighting panel   j. _____ Special-purpose outlet
   b. _____ Clock outlet   k. _____ Fan outlet
   c. _____ Duplex outlet   l. _____ Range outlet
   d. _____ Outside line telephone   m. _____ Power panel
   e. _____ Single-pole switch   n. _____ Three-way switch
   f. _____ Four-way switch   o. _____ Push button
   g. _____ Duplex outlet, split-circuit   p. _____ Thermostat
   h. _____ Lampholder with pull switch   q. _____ Electric door opener
   i. _____ Weatherproof outlet   r. _____ Multi-outlet assembly

18. The front edge of a box installed in a combustible wall must be _____ with the finished surface. Rule number? _____

19. List the maximum number of No. 12 AWG conductors permitted in a

   a. $4 \times 1^{1}/_{2}$-in (102×38-mm) octagon box. _____
   b. $4^{11}/_{16} \times 1^{1}/_{2}$-in ($119 \times 38$-mm) square box. _____
   c. $3 \times 2 \times 2^{1}/_{2}$-in ($76 \times 51 \times 64$-mm) device box. _____

20. Hanging a ceiling fixture directly from a plastic outlet box is permitted only if _____

   _____

   _____

   _____

   _____

21. It is necessary to count fixture wires when counting the permitted number of conductors in a box according to *Rule 12–3034(1)(d)*.

   (True)   (False)   Underline or circle the correct answer.

22. *CEC Table 23* allows a maximum of ten wires in a certain box. However, the box will have two wire connectors and one fixture stud in it. What is the maximum number of wires allowed in this box? _____

   _____

23. When laying out a job, the electrician will usually make a layout of the circuit, taking into consideration the best way to run the cables and/or conduits and how to make up the electrical connections. Doing this ahead of time, the electrician determines exactly how many conductors will be fed into each box. With experience, the electrician will probably select two or three sizes and types of boxes that will provide adequate space to meet code requirements. *CEC Table 23* shows the maximum number of conductors permitted in a given size box. However, the number of conductors shown in the table must be reduced:

by _____ conductor(s) for two wire connectors

by _____ conductor(s) for the fixture stud

by _____ conductor(s) for each wiring device mounted on a single strap

by _____ conductor(s) for one or more bare copper bonding conductors.

24. Define the term *bonding*. Why is the installation of bonding conductors important?

_____

_____

_____

_____

_____

_____

UNIT 6

Determining the Number and Location of Lighting and Receptacle Branch Circuits

OBJECTIVES

After studying this unit, the student will be able to

- determine the number and location of lighting outlets in a dwelling unit
- determine the minimum number of and location of receptacles required in a dwelling unit
- identify *CEC* requirements for lighting and receptacle branch circuits

In the design and planning of dwelling units, it is standard practice to permit the electrician to plan the circuits. Thus, the residence plans do not include layouts for the various branch circuits. The electrician may follow the general guidelines established by the architect; however, all wiring systems must conform to the standards of the *CEC* and the *National Building Code of Canada,* as well as local and provincial requirements.

This unit focuses on lighting branch circuits and small appliance circuits. The circuits supplying the electric range, oven, clothes dryer, and other specific circuitry not considered a lighting branch circuit or a small appliance circuit are covered later in this text.

LIGHTING

In this unit, we look at where lighting fixtures and receptacles must be located to meet the minimum requirements of the *CEC*. Lighting fixtures (luminaires), lamps, and ballasts will be covered in Unit 10.

The provisions for supplying a light fixture in a room or area and whether it is to be controlled by a wall switch are found in Part 9 of the *National Building Code of Canada*. The *CEC* has adopted these requirements, which are found in *Rules 30–500* to *510*. Table 6–1 (p. 100) outlines these requirements for lighting in dwelling units. Fig. 6–1 (p. 101) shows them graphically.

RECEPTACLES

Receptacles are required in dwelling units for the connection of large and small appliances. Receptacles installed for small appliances are 15 ampere, 125 volt (two pole, three wire) or 20 ampere T slot 125 volt (two pole, three wire).

Some residences are lighted by table and floor lamps plugged into switch-controlled wall receptacle outlets. In bedrooms and living rooms, switch-controlled wall receptacles are permitted in lieu of ceiling and/or wall lighting outlets. Only half the receptacle may be switched if the receptacle is

ROOM OR AREA	PROVIDE	REFERENCE RULE
Entrance	A luminaire controlled by a wall switch inside the building	30-500
Living room, bedroom	At least one luminaire controlled by a wall switch or a receptacle controlled by a wall switch	30-502
Kitchen, dining room, hall, bathroom, vestibule, utility room	At least one luminaire controlled by a wall switch	30-502
Stairway	All stairways must be lighted A luminaire controlled by a switch at the top and bottom of the stairway when the stairway has four or more risers. If the stairway leads to an unfinished basement that does not have an outside exit, a single switch at the top of the stairs may be used	30-504
Basement	One luminaire for each 30 m² or portion thereof of unfinished basement The luminiare closest to the stairs must be controlled by switch at the top of the stairs	30-506
Storage room	At least one luminaire in the storage room	30-508
Garage and carport	At least one luminaire controlled by a wall switch near the doorway When a wall-mounted or ceiling luminaire that is not located above a space that would be occupied by a car is used, a built-in switch is permitted on the luminaire A carport that is lighted by a luminaire at the entrance to the building does not require additional lighting	30-510

Table 6–1 *CEC* **requirements for lighting in dwelling units.**

needed to satisfy the minimum number of receptacles required by *Rule 26–712(a)*. This is achieved by removing the tab on the "hot" side of the receptacle (Figs. 6–3, 6–4, and 6–5).

This unit will identify the locations where the *CEC* requires 15-ampere, 125-volt (5-15R) and 20-ampere, 125-volt (5-20RA) receptacles. For information on special applications of receptacles or their branch circuits, see Units 11 to 20.

Table 6–2 and Figs. 6–6, 6–7, and 6–8 outline the minimum number of receptacles required in a dwelling unit and their locations.

BASICS OF WIRE SIZING AND LOADING

The *CEC* defines a *branch circuit* as "that portion of the wiring installation between the final overcurrent device protecting the circuit and the outlet(s)," Fig. 6–9.

The *CEC* defines a *feeder* as the circuit conductors between the service equipment and the final branch circuit overcurrent device. In the residence in this text, the wiring between Panel A and Panel B is a feeder.

The ampacity (current-carrying capacity) of a conductor must not be less than the rating of the overcurrent device protecting that conductor, *Rule 14–104*. An exception is a motor branch circuit, where it is common to have overcurrent devices (fuses or breakers) sized larger than the ampacity of the conductor. The *CEC* covers motors and motor circuits specifically in *Section 28*. The rating of the branch circuit is determined by the lesser of the rating of either the overcurrent device or the ampacity of the wire. For example, if a 20-ampere conductor is protected by a 15-ampere fuse, then the circuit is a 15-ampere branch circuit, *Rule 8–104(1)*.

The ampacity of branch circuit conductors must not be less than the maximum load to be served. Where the circuit supplies receptacle outlets, as is typical in houses, the conductor's ampacity must be rated to carry the load, *Rule 8–104(2)*. The load current must not exceed 80 percent of the rating of the overcurrent device for that circuit, *Rules 8–304* and *8–104(5)(a)*.

Where the ampacity of the conductor does not match up with a standard rating of a fuse or breaker,

Fig. 6–1 Lighting outlets required in a typical dwelling unit.

the next higher standard size overcurrent device may be used, provided the overcurrent device does not exceed 600 amperes, *Rule 14–104(a)*. This exception is not applicable to residential branch circuits because the standard overcurrent device sizes listed in *CEC Table 13* (Table 6–3, p. 104) correspond directly with the ampacities listed in *Table 2* for No. 14 and No. 12 gauge wire. *Table 13* shows that the maximum overcurrent protection is 15 amperes when No. 14 AWG copper wire is used.

Residential branch circuits are fed from panelboards having overcurrent devices that are rated to be run continuously at a maximum of 80 percent of their ampere rating. The conductors attached to these overcurrent devices may not be loaded to more than 80 percent of the ampere rating listed in *Table 2*.

For example, a 15-ampere circuit breaker has a maximum load of 12 amperes (15 amp × 0.8). Therefore, the wire must be rated for 15 amperes in *Table 2;* a No. 14 R90 XLPE is suitable.

The ampacities of conductors commonly used in residential occupancies are listed in *Table 2*. These are subject to *correction factors,* which must be applied where there are high ambient temperatures, for example, in attics. See *Note (1)* to *Table 5A*.

Conductor ampacities are also *derated* when four or more conductors are installed in a single raceway or cable. See *Table 5C* (Table 6–4).

A continuous load should not exceed 80 percent of the ampere rating of the branch circuit. The *CEC* defines a *continuous load* as one that persists under normal operation for "more than one hour in

ROOM OR AREA	PROVIDE	REFERENCE RULE
Living room, dining room, family room, bedroom	No point along the floorline of a wall may be more than 1.8 m horizontally from a receptacle, including isolated wall spaces 900 mm or longer. See rule for exclusions.	26-712-a,c
Hallway	No point in a hallway may be more than 4.5 m from a receptacle. Receptacles in rooms off the hallway may be used if the opening to the room from the hallway is not fitted with a door.	26-712-f
Laundry room	One receptacle for the washing machine. At least one additional receptacle.	26-710-e
Bathroom	At least one receptacle, within 1 m of the washbasin and GFCI protected.	26-710-f
Kitchens	One receptacle for a refrigerator. One receptacle for a gas range if gas piping has been provided. Counter receptacles such that no point measured along the back of the counter is more than 900 mm horizontally from a receptacle. Counter receptacles may be 15A, 3-conductor split circuit or 20A T slot. See Unit 14 for more detail. Isolated work surfaces 300 mm or more require a receptacle. At least one receptacle for a permanently installed island. At least one receptacle for counter peninsula. At least one receptacle in a dining area forming part of a kitchen. One receptacle for a microwave oven.	26-712-d
Utility room	At least one receptacle.	26-712-a
Unfinished basement areas	At least one receptacle.	26-710-e
Balcony, porch, veranda	At least one receptacle.	26-712-e
Outdoors	At least one receptacle for the use of outdoor appliances.	26-714-a
Garage, carport	One receptacle for each space.	26-714-b
Miscellaneous	One receptacle for a cord-connected central vacuum if ducting system is installed.	26-710-m

Table 6–2 *CEC* **requirements for receptacles in dwelling units.**

any two-hour period if the load does not exceed 225 amperes," *Rule 8–104 (3)(a)*. The *CEC* permits 100 percent loading on an overcurrent device *only* if that overcurrent device is *listed* for 100 percent loading. The CSA has *no* listing of circuit breakers of the moulded-case type used in residential installations that are suitable for 100 percent loading.

This calls for the judgment of the electrician and/or the electrical inspector. Will a branch circuit supplying a heating cable in a driveway likely be on for one hour or more? The answer is "probably." Good design practice and experience tell us to never load a circuit conductor or overcurrent protective device to more than 80 percent of its rating.

The branch-circuit rating shall *not* be less than any noncontinuous load plus 125 percent of the

continuous load, *Rule 8–104*. For residential applications, all loads are considered to be continuous except for the load on the service or feeder conductors, *Rule 8–200(3)*.

The *CEC* considers most receptacle outlets in a residence part of the basic load, and additional load calculations are *not* necessary. Appliance outlets with loads greater than 12 amperes are not considered part of the basic load; these larger loads are calculated individually. The power requirements for these receptacle outlets will be discussed in detail later in this text.

To determine the number of lighting circuits required, divide the total watts by the ampere rating of the circuit(s) to be used.

In this residence, all lighting circuits are rated at 15 amperes. The maximum loading of these cir-

THIS DUPLEX RECEPTACLE
MEETS THE REQUIREMENTS OF CSA
CONFIGURATION 5-20RA DIAGRAM 1.

THIS DUPLEX RECEPTACLE
MEETS THE REQUIREMENTS OF
CSA CONFIGURATION 5–15 R DIAGRAM 1.

Fig. 6–2   15 and 20 ampere, 125 volt (two pole, three wire) receptacle configurations.

cuits is 12 amperes (15 amperes × 0.8). Therefore, the minimum number of circuits for general lighting and power is

$$\frac{watts}{120 \ volts} = amperes$$

$$\frac{amperes}{12} = minimum \ number \ of \ circuits \ required$$

Note that the *CEC* provides the minimum number of branch circuits required. For various purposes, you may install more.

Load calculations for electric water heaters, ranges, motors, electric heat, and other specific loads are in addition to the general lighting load, and are discussed throughout this text as they appear.

VOLTAGE

All calculations throughout this text use voltages of 120 volts and 240 volts. *CEC Rule 8–100,* titled

Fig. 6–3   Duplex receptacle showing method for removing tab on hot side.

Fig. 6–4   Split-switched receptacle wiring diagram.

CEC TABLE 13
RATING OR SETTING OF OVERCURRENT DEVICES PROTECTING CONDUCTORS
(See *Rules 14–104* and *28–204*)

Ampacity of Conductor	Rating or Setting Permitted Amperes	Ampacity of Conductor	Rating or Setting Permitted Amperes
0–15	15	126–150	150
16–20	20	151–175	175
21–25	25	176–200	200
26–30	30	201–225	225
31–35	35	226–250	250
36–40	40	251–275	300
41–45	45	276–300	300
46–50	50	301–325	350
51–60	60	326–350	350
61–70	70	351–400	400
71–80	80	401–450	450
81–90	90	451–500	500
91–100	100	501–525	600
101–110	110	526–550	600
111–125	125	551–600	600

Table 6–3   *CEC Table 13*. Rating or setting of overcurrent devices protecting conductors. See *Rules 14–104* and *28–204*.

CEC TABLE 5C
AMPACITY CORRECTION FACTORS FOR TABLE 2

Number of Conductors	Ampacity Correction Factor
1–3	1.00
4–6	0.80
7–24	0.70
25–42	0.60
43 and up	0.50

Table 6–4   *CEC Table 5C*. See *Rules 4–004* and *12–2210* and *Tables 2* and *4*.

"Current calculation," states that the voltage divisors to use shall be 120, 208, and 240, as applicable. This voltage is used for calculation even though the actual voltage may be 220, 245, or 247. This provides us with a uniformity, to avoid misleading and confusing results in our computations.

LIGHTING BRANCH CIRCUITS

The *CEC* limits overcurrent protection for lighting branch circuits to a maximum of 15 amperes in

Fig. 6–5   Split-wired receptacle (split receptacle) on a two-pole 15-ampere breaker connected to a three-wire circuit for use in a kitchen.

Fig. 6–6   Locating receptacles on the first floor.

dwelling units, *Rule 30–104*. When a branch is loaded it will normally have a maximum of 12 outlets. *Rule 8–304*. If the **actual** load on the circuit is known, the number of lighting outlets may be increased as long as the load on the circuit does not exceed 80 percent of the rating of the circuit breaker or fuse protecting the branch circuit. It is normally impossible to tell what the **actual** load of each lighting fixture on a branch circuit is when doing the rough-in, unless a large number

Isolated work surfaces 300 mm or greater require a receptacle.

Countertop island requires a receptacle.

One receptacle in dining area of a kitchen.

Counter receptacles to be 15A split or 20A T slot.

397 mm

Receptacles within 1.5 m of sink must be GFCI protected.

Provide a receptacle for built-in microwave oven if space is provided.

One receptacle for refrigerator

Halls. No point in the hall to be more than 4.5 m horizontally from a receptacle.

No point along the back wall of the counter to be more than 900 mm from a receptacle.

Laundry room. Provide one receptacle for the washing machine and at least one additional receptacle.

Door openings and areas occupied by an open door are not counted as usable wall space.

Garage. Provide a minimum of one receptacle for each car space.

Provide at least one receptacle for the use of outdoor appliances.

One receptacle at the front of the house and one at the back controlled by an indoor wall switch is recommended.

Fig. 6–7   Locating receptacles in the kitchen, laundry room, and garage.

of recessed luminaires are installed; therefore the maximum number of outlets on a lighting branch circuit is 12.

When you install track lighting with movable fixtures, the authors recommend that you count every 600 mm as one outlet.

RECEPTACLE BRANCH CIRCUITS

Branch circuits in dwelling units can consist of lighting only, receptacles only, or a combination of lighting and receptacles. When a circuit consists of a combination of lighting and receptacles, the protection for the circuit is limited to 15 amperes, the same as it would be if the circuit had lighting only.

When a circuit contains receptacles only, the protection for the circuit cannot exceed the rating of the receptacle, *Rule 14–600*. Receptacles for general applications in dwelling units (convenience receptacles) may be CSA configuration 5-15R or 5-20RA (T slot) and are found in *CEC Diagram 1* (Fig. 6–2). If 15-ampere receptacles (5-15R) are used, the overcurrent protection is limited to 15 amperes. If 20-ampere receptacles are used (5-20RA), the overcurrent protection is permitted to be 20 amperes.

The general requirements for receptacle branch circuits are found in *CEC Sections 8* and *26*. Table 6–5 and Figs. 6–10, 6–11, and 6–12 outline the

ROOM OR AREA	PROVIDE	REFERENCE RULE
Living room, dining room, family room, hallway	Maximum of 12 outlets per circuit.	8-304
Bedroom	Maximum of 12 outlets per circuit. Receptacles in bedrooms must be AFCI protected. See Unit 9.	26-720
Laundry room, utility room	At least one circuit solely for receptacles in laundry rooms. At least one circuit solely for receptacles in utility room or areas.	26-720
Bathroom	No special requirements in *CEC*. Consult with your local electrical inspector for recommendations.	
Kitchen	One circuit for the receptacle for a refrigerator. A receptacle provided for a clock outlet is permitted on this circuit.	26-720
	A minimum of two circuits to be provided for counter receptacles, when more than one counter receptacle is provided. No more than two counter receptacles are permitted on a circuit except that receptacles provided for people with disabilities may be added.	26-722
	One circuit for the dining area of a kitchen. A receptacle for a gas range may be on the same circuit.	26-722
	One circuit when a receptacle has been supplied for a microwave oven.	26-720
Balcony porch, veranda	One circuit if for a single dwelling unit and accessible from grade level.	26-724
Outdoors	One circuit for receptacles located outdoors.	26-724
Garage, carport	One circuit for receptacles in garages or carports. The garage lighting and door opener may be on the same circuit.	26-724
Miscellaneous	One receptacle for a cord-connected central vacuum if ducting system is installed.	26-720

Table 6–5 *CEC* **requirements for branch circuits supplying receptacles in dwelling units.**

basic *CEC* requirements for branch circuits supplying receptacles in dwelling units. Units 11 to 20 give more information about receptacles and their branch circuits.

PREPARING A LIGHTING AND RECEPTACLE LAYOUT

Two blank full-size drawings of the floor plan are provided with this textbook to allow you to design and prepare your own electrical layout.

Begin by using Table 6–1 to identify where luminaires are required. Locate the luminaires on the floor plan. Add any extra luminaires (such as in closets) that you feel enhance the design. (Just remember that the lowest bid will normally get the job). Determine where you would like to locate the switches controlling the luminaires. Write S on the floor plan if you

want to control the luminaires in a room from one location. Write S_3 on the floor plan if you want to control the lights from two locations. If you want to control the luminaires from more than two locations, see Unit 8 (Switch Control of Lighting Circuits).

Use Table 6–2 to determine the location of receptacles required by the *CEC* and locate them on the floor plans. Use Figs. 6–6, 6–7, and 6–8 to assist you. Add any extra receptacles that you feel enhance the design.

When you have completed the layout of lighting and receptacle outlets, group the outlets into circuits. In general, 12 outlets are permitted on a circuit. There can be a mixture of lights and receptacles in most cases. Use Table 6–3 and Figs. 6–10, 6–11, and 6–12 to help you.

Further information about additional loads that will be added to the layout are found in later units.

Fig. 6–8   Locating receptacles in the basement.

Fig. 6–9   Branch circuit.

Fig. 6–10   Circuiting for receptacles.

Fig. 6–11   Circuiting for kitchen, laundry, and garage receptacles.

Receptacles within
1.5 m of a sink must
be GFCI protected.

FINISHED BASEMENT AREA

UNFINISHED BASEMENT

A maximum of 12 outlets
permitted per circuit.

Provide a separate circuit for
receptacles in a utility room.

Fig. 6–12   Circuiting for basement receptacles.

REVIEW

1. A living room wall is 6 m long. How many receptacles are required for the room?

   _____

2. A hallway is 3 m long and has a doorway fitted with a door at each end. Does this hallway require a receptacle?

   _____

3. A hallway is 6 m long and has an open doorway each end. The hall opens onto two rooms that each have one receptacle within 1 m of the doorway. Does this hallway require a receptacle?

   _____

4. How many receptacles are required in an unfinished basement with an area of 63 m²?

   _____

5. A kitchen counter is 3 m long. A sink that is 600 mm is located in the exact centre of the counter. How many receptacles are required for the counter?

   _____

6. How many of the receptacles in Question 5 require GFCI protection?

_____

7. A counter has an isolated work space of 600 mm between a sink and a built-in oven. Is a receptacle required for this space?

_____

8. How many luminaires or lampholders are required for an unfinished basement area that is 63 m²?

_____

9. How many receptacles are required in an attached two-car garage?

_____

10. How many luminaires are required in an attached two-car garage?

_____

11. When is a luminaire required for a stairway?

_____

12. When is switch control required at the top and bottom of a stairway?

_____

13. When a stairway leads to an unfinished basement that has an outside exit, is a switch required at the top and bottom of the stairs?

_____

14. When should a receptacle be provided for a central vacuum unit?

_____

15. A receptacle in a utility room is supplied from its own circuit. Can the wiring from the receptacle be extended to supply eight outlets in a finished family room?

_____

16. Why shouldn't a freezer be placed on a circuit protected by a GFCI?

_____

_____

17. What outlets in a dwelling unit must have AFCI protection?

_____

_____

18. A counter has three 5-20RA receptacles. What is the minimum number of branch circuits that may be used for these receptacles?

_____

19. What is the minimum number of circuits that may be run to a laundry room?

_____

20. Does the receptacle located next to a washbasin in a bathroom have to be on its own circuit?

_____

_____

UNIT 7

Conductor Sizes and Types, Wiring Methods, Wire Connections, Voltage Drop, Neutral Sizing for Services

OBJECTIVES

After studying this unit, the student will be able to

- define the terms used to size and rate conductors
- discuss aluminum conductors
- describe the types of cables used in most dwelling unit installations
- list the installation requirements for each type of cable
- describe the uses and installation requirements for electrical conduit systems
- describe the requirements for service grounding conductors
- describe the use and installation requirements for flexible metal conduit
- make voltage drop calculations
- make calculations using *CEC, Table D3.*
- understand the fundamentals of the markings found on the terminals of wiring devices and wire connectors
- understand why the ampacity of high-temperature insulated conductors may not be suitable at that same rating on a particular terminal on a wire device, breaker, or switch
- determine when a reduced neutral can be used on residential services

CONDUCTORS

Throughout this text, all references to conductors are for copper conductors, unless otherwise stated.

Wire Size

The copper wire used in electrical installations is graded for size according to the American Wire Gauge (AWG) standard. The wire diameter in the AWG standard is expressed as a whole number. AWG sizes vary from fine, hairlike wire used in coils and small transformers to very large-diameter wire required in industrial wiring to handle heavy loads.

The wire may be a single strand (solid conductor) or consist of many strands. Each strand of wire acts as a separate conducting unit. Conductors that are No. 8 AWG and larger generally are stranded. The wire size used for a circuit depends on the maximum current to be carried. The *CEC* states that the minimum conductor size permitted in house wiring is No. 14 AWG. (Exceptions to this rule are covered in *Section 16* and *Table 12* for the wires used in lighting fixtures, bell wiring, and remote-control low-energy circuits.) See the conductor application chart in Table 7–1.

Ampacity

Ampacity means "the current in amperes a conductor can carry continuously under the condition of use without exceeding its temperature rating." This value depends on the cross-sectional area of the wire. Since the area of a circle is proportional to the square of the diameter, the ampacity of a round wire varies with the diameter squared (d^2). The diameter of a wire is measured in mils, a *mil* defined as one-thousandth of an inch (0.001 in). A *circular mil* is the area of a circle one mil in diameter. The circular mil area (CMA) of a wire determines the current-carrying capacity of the wire. This can be expressed as a formula:

$$CMA = d^2$$

Therefore, the larger the CMA of a wire, the greater its current-carrying capacity.

AWG sizes are also expressed in mils and range from No. 40 (10 circular mils) to No. 4/0 (211 600 circular mils). Wire sizes larger than No. 4/0 are expressed in circular mils only. Since the letter "k" designates 1000, large conductors, such as 500 000 circular mils, can also be expressed as 500 kcmils. However, many texts and electricians

continue to use an older term, MCM, in which the first "M" is a roman numeral representing 1000. Thus, 500 MCM also means 500 000 circular mils.

Conductors must have an ampacity not less than the maximum load they are supplying, Fig. 7–1. All conductors of a specific branch circuit must have an ampacity of the branch circuit's rating, Figure 7–2. There are some exceptions to this rule, such as taps for electric ranges (Unit 15).

Fig. 7–1   Branch circuit conductors are required to have an ampacity not less than the maximum load to be served.

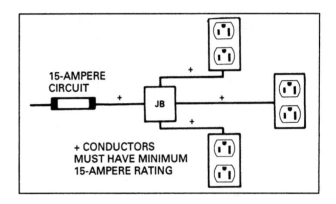

Fig. 7–2   All conductors in this circuit supplying readily accessible receptacles shall have an ampacity of not less than the rating of the branch circuit. In this example, 15-ampere conductors must be used, *Rule 8–104(1)*.

CEC Tables 1–4 list the allowable ampacities of conductors in free air or when not more than three are installed in a raceway or cable. *Table 12* lists the allowable ampacities of equipment wire and flexible cords. These ampacities are reduced if the conductors are installed in areas of high ambient (surrounding) temperature (*Table 5A*). Ampacities are also reduced when there are more than three conductors in a raceway or cable (*Table 5C*).

Circuit Rating

The rating of the overcurrent device (OCD) and the ampacity of the wire must be compared. The branch circuit ampacity rating is the lesser of these two values, Fig. 7–3.

CONDUCTOR SIZE	APPLICATIONS
No. 16, No. 18 AWG	cords, low-voltage control circuits, bell and chime wiring
No. 14, No. 12 AWG	normal lighting circuits, circuits supplying receptacle outlets
No. 10, No. 8, No. 6, No. 4 AWG	clothes dryers, ovens, ranges, cooktops, water heaters, heat pumps, central air conditioners, furnaces, feeders to subpanels
No. 3, No. 2, No. 1 (and larger) AWG	main service entrances, feeders to subpanels

Table 7–1   Conductor applications chart.

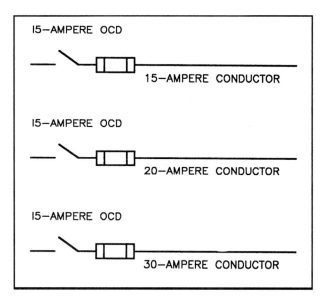

Fig. 7–3 These are all classified as 15-ampere circuits even though larger conductors were used for some other reason, such as solving a voltage drop problem. The lesser of the amperage rating of the overcurrent device (OCD) or the wire determines the rating of a circuit, *Rule 8–104(1)*.

Aluminum Conductors

The conductivity of aluminum wire is not as great as that of copper wire for a given size. For example, a No. 8 T90 Nylon copper wire has an ampacity of 45 amps according to *Table 2*. Aluminum wire has a significant weight advantage, but a No. 8 T90 Nylon aluminum wire has an ampacity of only 30 amperes, *Table 4*.

It is important to consider resistance when selecting a conductor for a long run. An aluminum conductor has a higher resistance than a copper conductor for a given wire size, which therefore causes a greater voltage drop.

voltage drop (E_d) = amperes(I) × resistance(R)

Although aluminium conductors are approved for branch circuit wiring, they are seldom used for this application. They are, however, widely used in such applications as apartment feeders and utility distribution systems. They are also found on such equipment as ballasts and luminaires.

Common Connection Problems

Problems often arise with aluminum conductors *that are not properly connected*.

- A corrosive action is set up when dissimilar wires come in contact with each other, especially if moisture is present.

- The surface of aluminum oxidizes as soon as it is exposed to air. If this oxidized surface is not broken through, a poor connection results. When installing aluminum conductors, particularly in large sizes, you must brush an inhibitor onto the aluminum conductor and scrape the conductor with a stiff brush where the connection is to be made. Scraping the conductor breaks through the oxidation, and the inhibitor keeps the air from coming into contact with the conductor, preventing further oxidation. Aluminum connectors of the compression type usually have an inhibitor paste factory-installed inside the connector.

- Aluminum wire expands and contracts more than copper wire for an equal load, another possible cause of a poor connection. Crimp connectors for aluminum conductors are usually longer than those for comparable copper conductors, resulting in greater contact surface of the conductor in the connector.

Proper Installation Procedures

Proper, trouble-free connections for aluminum conductors require terminals, lugs, and/or connectors that are suitable for the type of conductor being installed.

Terminals on receptacles and switches must be suitable for the conductors being attached. Table 7–2 shows how the electrician can identify these terminals. Listed connectors provide proper connection when properly installed. See *Rules 12–116* and *12–118*.

Wire Connections

When splicing wires or connecting to a switch, fixture, circuit breaker, panelboard, meter socket, or other electrical equipment, you must twist the wires together and then use a wire connector, *Rule 12–112*.

Wire connectors are known in the trade by such names as *screw terminal, pressure terminal connector, wire connector, Wing-Nut™, Wire-Nut™, Scotchlok™, split-bolt connector, pressure cable connector, solderless lug,* and *solder lug*. Solder-type lugs are not often used today. In fact, connections that depend on solder are *not* permitted for connecting service entrance conductors to service equipment. The labour costs make the cost of using solder-type connections prohibitive.

Solderless connectors, designed to establish connections by means of mechanical pressure, are quite common. Fig. 7–4 shows some types of wire connectors and their uses.

As with the terminals on wiring devices (switches and receptacles), only wire connectors marked "AL" may be used with aluminum conductors. This marking is found either on the connector itself or on (or in) the shipping carton.

Connectors marked "AL/CU" are suitable for use with aluminum, copper, or copper-clad aluminum conductors. This marking also appears on the connector itself or on (or in) the shipping carton.

Connectors not marked "AL" or "AL/CU" are for use with copper conductors only.

Conductors made of copper, aluminum, or copper-clad aluminum may not be used in combination in the same connector unless the conductor

CRIMP CONNECTORS USED TO SPLICE AND TERMINATE NO. 20 AWG TO 500 KCMILS ALUMINUM-TO-ALUMINUM, ALUMINUM-TO-COPPER, OR COPPER-TO-COPPER.	PROPERLY CRIMP, THEN TAPE
CONNECTORS USED TO CONNECT WIRES TOGETHER IN COMBINATIONS OF NO. 18 AWG THROUGH NO. 6 AWG. THEY ARE TWIST-ON, SOLDERLESS, AND TAPELESS.	WIRE CONNECTORS VARIOUSLY KNOWN AS WIRE NUT, WING NUT, AND SCOTCHLOK
CONNECTORS USED TO CONNECT WIRES TOGETHER IN COMBINATIONS OF NO. 16, NO. 14, AND NO. 12 AWG. THEY ARE CRIMPED ON WITH A SPECIAL TOOL, THEN COVERED WITH A SNAP-ON INSULATING CAP OR WRAPPED WITH INSULATING TAPE.	CRIMP-TYPE WIRE CONNECTOR AND INSULATING CAP
SOLDERLESS CONNECTORS ARE AVAILABLE IN NO. 14 AWG THROUGH 500 MCM. THEY ARE USED FOR ONE SOLID OR ONE STRANDED CONDUCTOR ONLY, UNLESS OTHERWISE NOTED ON THE CONNECTOR OR ON ITS SHIPPING CARTON. THE SCREW MAY HAVE A STANDARD SCREWDRIVER SLOT OR AN ALLEN WRENCH SOCKET.	SOLDERLESS CONNECTORS
COMPRESSION CONNECTORS ARE USED FOR NO. 8 AWG THROUGH 1000 MCM. THE WIRE IS INSERTED INTO THE END OF THE CONNECTOR, THEN CRIMPED ON WITH A SPECIAL COMPRESSION TOOL.	COMPRESSION CONNECTOR
SPLIT-BOLT CONNECTORS ARE USED FOR CONNECTING TWO CONDUCTORS TOGETHER, OR FOR TAPPING ONE CONDUCTOR TO ANOTHER. THEY ARE AVAILABLE IN NO. 10 AWG THROUGH 1000 MCM. THEY ARE USED FOR TWO SOLID AND/OR TWO STRANDED CONDUCTORS ONLY, UNLESS OTHERWISE NOTED ON THE CONNECTOR OR ON ITS SHIPPING CARTON.	SPLIT-BOLT CONNECTOR

Fig. 7–4   Types of wire connectors.

TYPE OF DEVICE	MARKING ON TERMINAL OR CONNECTOR	CONDUCTOR PERMITTED
15- or 20-ampere receptacle and switch	CO/ALR	aluminum, copper, copper-clad aluminum
15- and 20-ampere receptacle and switch	NONE	copper, copper-clad aluminum
30-ampere and greater receptacle and switch	AL/CU	aluminum, copper, copper-clad aluminum
30-ampere and greater receptacle and switch	NONE	copper only
Screwless pressure terminal connector of the push-in type	NONE	copper or copper-clad aluminum
Wire connector	AL/CU	aluminum, copper, copper-clad aluminum
Wire connector	NONE	copper only
Wire connector	AL	aluminum only
Any of the above devices	COPPER OR CU ONLY	copper only

Table 7–2   Approved equipment and conductor combinations.

is specially identified for the purpose either on the connector itself or on (or in) the shipping carton. The conditions of use will also be indicated. Such connectors are usually limited to dry locations.

Rule 12–118 covers termination and splicing of aluminum conductors.

Insulation of Wires

Rule 12-100 requires that insulation on conductors be suitable for the conditions in which they are installed. The insulation must completely surround the metal conductor, have a uniform thickness, and run the entire length of the wire or cable.

One type of insulation used on wires is a thermoplastic material (T90 Nylon or TEW), although thermoset (rubber) compounds are also widely used. A thermoplastic will deform and melt when excess heat is applied to it. Any insulation with a designation that starts with a "T," such as T90 Nylon or TEW, is a thermoplastic.

A thermoset material will not soften and deform when heat is applied to it. When heated above its rated temperature, it will char and crack. Any insulation that starts with "R," such as R90XLPE, is a thermoset material.

Table 19 lists the various conductor insulations and applications.

Table 2 gives the allowable ampacities of copper conductors in a cable for various types of insulation, Table 7–3. The ampacity of 60°C wire is found in Column 2; 75°C in Column 3, and 85–90°C in Column 4.

Table 4 gives the allowable ampacities of aluminum conductors in a cable. *Table 19* and *Table 11* list the conditions of use for equipment wires, portable power cables, and flexible cords.

The insulation covering wires and cables used in house wiring is usually rated at 300 volts or less. Exceptions to this are low-voltage bell wiring and fixture wiring.

Temperature Considerations

Conductors are also rated for the temperature they can withstand. For example:

Type TWN	60°C
Type TWN75	75°C
Type T90 Nylon	90°C
Type RW90-XLPE	90°C

Conductors with TWN75, T90 Nylon, or RW90-XLPE insulation are the types most commonly used in conduits because their relatively small size makes them easy to handle and allows more conductors in a given size of conduit. The last two are rated for 90°C so they can be used in high temperatures, such as those found in attics, near recessed light fixtures, and around cables buried in insulation.

The most commonly used cable is NMD90, a non-metallic-sheathed type. Easy to handle, it is rated for 90°C.

The permitted ampacity of all 90°C-rated conductors used in ceiling light outlet boxes is reduced to that of 60°C-rated conductor, *Rule 30–408*.

Conductor Temperature Ratings

Table 2 shows the types of insulations available on conductors for building wire used in most electrical installations. It is this table that electricians refer to most often when selecting wire sizes for specific loads. These are the types of conductors found in non-metallic-sheathed cable (Romex), armoured cable (BX), and conductors commonly installed in conduit.

TABLE 2
ALLOWABLE AMPACITIES FOR NOT MORE THAN 3 COPPER CONDUCTORS IN RACEWAY OR CABLE
(BASED ON AMBIENT TEMPERATURE OF 30°C*)
(See Rules 4–004, 8–104, 12–2210, 26–142, 42–008, and 42–016 and Tables 5A, 5C, 19, and D3.)

			Allowable Ampacity†‡‡			
	60°C‡	75°C‡	85–90°C‡	110°C‡	125°C‡	200°C‡
Size, AWG or kcmil	Type TW	Types RW75, TW75	Types R90, RW90, T90 NYLON Mineral-insulated cable**	See Note	See Note	See Note
14	15	15	15	30	30	30
12	20	20	20	35	40	40
10	30	30	30	45	50	55
8	40	45	45	60	65	70
6	55‡	65	65	80	85	95
4	70	85	85	105	115	120
3	80	100	105	120	130	145
2	100	115	120	135	145	165
1	110	130	140	160	170	190
0	125	150	155	190	200	225
00	145	175	185††	215	230	250
000	165	200	210	245	265	285
0000	195	230	235	275	310	340
250	215	255	265	315	335	—
300	240	285	295	345	380	—
350	260	310	325	390	420	—
400	280	335	345	420	450	—
500	320	380	395	470	500	—
Col. 1	Col. 2	Col. 3	Col. 4	Col. 5	Col. 6	Col. 7

*   See Table 5A for the correction factors to be applied to the values in Columns 2 to 7 for ambient temperatures over 30 °C.
†   The ampacity of aluminum-sheathed cable is based on the type of insulation used on the copper conductors.
‡   These are maximum allowable conductor temperatures for 1, 2, or 3 conductors run in a raceway, or 2 or 3 conductors run in a cable and may be used in determining the ampacity of other conductor types in Table 19, which are so run, as follows: From Table 19 determine the maximum allowable conductor temperature for that particular type; then from Table 2 determine the ampacity under the column of corresponding temperature rating.
** These ratings are based on the use of 90 °C insulation on the emerging conductors and for sealing. Where a deviation has been allowed in accordance with Rule 2–030, mineral-insulated cable may be used at higher temperatures without decrease in allowable ampacity, provided that insulation and sealing material approved for such higher temperature is used.
†† For 3-wire 120/240 V and 120/208 V service conductors for single dwellings, or for feeder conductors supplying single dwelling units of row housing of apartment and similar buildings, and sized in accordance with Rules 8–200(1), 8–200(2), and 8–202(1), the allowable ampacity for sizes No. 6 and No. 2/0 AWG shall be 60 A and 200 A respectively. In this case, the 5% adjustment of Rule 8–106(1) cannot be applied.
‡‡ See Table 5C for the correction factors to be applied to the values in Columns 2 to 7 where there are more than 3 conductors in a run of raceway or cable.

NOTE: These ampacities are applicable only under special circumstances where the use of insulated conductors having this temperature rating is acceptable.

Table 7–3   *CEC Table 2* extract.

Note that in the common building wire categories, we find 60°C, 75°C, and 90°C temperature ratings.

When selecting a conductor based on column 3 or 4 of *Table 2* (see also the table footnotes), make sure its ampacity is compatible with the equipment the conductors are being connected to. We can use the high temperature ratings when high temperatures are encountered, such as in attics, or when we must derate because of the number of conductors in a conduit, but we must always check the temperature ratings of terminals to make sure that we are not creating a "hot spot." Terminals and/or the label in the equipment might be marked "75°C only" or "60°/75°C."

Neutral Size

In Fig. 7–5, the grounded neutral conductor (N) is permitted to be a smaller AWG than the "hot" ungrounded phase conductor (D) (*Rule 4–022*) only if it can be proved that all three service entrance conductors are properly and adequately sized to carry the loads computed by *Rule 8–200*. Some utilities may not permit the reduction of the service neutral for services of less than 200 amperes. Therefore, check with the local utility and inspection authority before proceeding.

Rule 4–022(1) states that "the neutral conductor shall have sufficient ampacity to carry the unbalanced load." *Rule 4–022(2)* further states that "the maximum unbalanced load shall be the maximum connected load between the neutral and any one ungrounded conductor."

In this residence, a number of loads carry little or no neutral currents, including the electric water heater, electric clothes dryer, electric oven and range, electric furnace, and air conditioner (Fig. 7–5). These appliances often draw their biggest current through the "hot" conductors because they are fed with 240-volt Ⓒ or 120/240-volt circuits Ⓔ. Loads Ⓑ are connected line-to-neutral only.

Thus we find logic in the *CEC*, which permits reducing the neutral size on services, feeders, and branch circuits where the computations prove that the neutral will be carrying less current than the "hot" phase conductors.

VOLTAGE DROP

Low voltage can cause lights to dim, television pictures to shrink, motors to run hot, electric heaters to produce less than their rated heat output, and appliances to operate improperly.

Low voltage in a home can be caused by:

- wire that is too small for the load being served,

- a circuit that is too long,

- poor connections at the terminals

- conductors operating at high temperatures having higher resistance than when operating at lower temperatures.

A simple formula for calculating voltage drop on single-phase systems considers only the DC resistance of the conductors and the temperature of the conductor. The more accurate formulas consider AC resistance, reactance, three-phase installations, temperature, and spacing in metal conduit and in non-metallic conduit. Voltage drop is covered in great detail in *Electrical Wiring: Commercial*.

The simple voltage drop formula is more accurate with smaller conductors, but loses accuracy as conductor size increases. It is sufficiently accurate for the voltage drop calculations necessary for residential wiring.

To find voltage drop:

$$E_d = \frac{K \times I \times L \times 2}{CMA}$$

Fig. 7–5   Line-to-line and line-to-neutral loads.

where:

E_d = allowable voltage drop in volts

K = resistance of conductor at 75°C

- For copper conductors
  - about 12 ohms per circular mil foot
  - about 39.4 ohms per circular mil metre*
- For aluminum conductors
  - about 19 ohms per circular mil foot
  - about 63.3 ohms per circular mil metre*

I = the current in amperes flowing through the conductors

L = the length of the conductor from the beginning of the circuit to the load

- in feet if the circular mil foot is used for K
- in metres if the circular mil metre is used for K

2 is to multiply the length of the run $\times$ 2 = total length of wire

CMA = cross-sectional area of the conductor in circular mils (Table 7–4).

* derived from the circular mil foot

To find conductor size:

$$CMA = \frac{K \times I \times L \times 2}{E_d}$$

Code Reference to Voltage Drop

Rule 8–102 states the maximum allowed voltage drop percentages. The voltage drop shall not exceed 3 percent in any feeder or branch circuit (Fig. 7–6) and 5 percent overall.

Fig. 7–7 explains the requirements for a voltage drop that involves both the feeder and a branch circuit.

Nominal	
AWG	**CM Area**
14*	4 110
12*	6 530
10*	10 380
8	16 510
6	26 240
4	41 740
3	52 630
2	66 360
1	83 690
0	105 600
00	133 100
000	167 800
0000	211 600

*These sizes are customarily supplied with solid conductors.

Table 7–4   Cross-sectional area for building wire and cable.

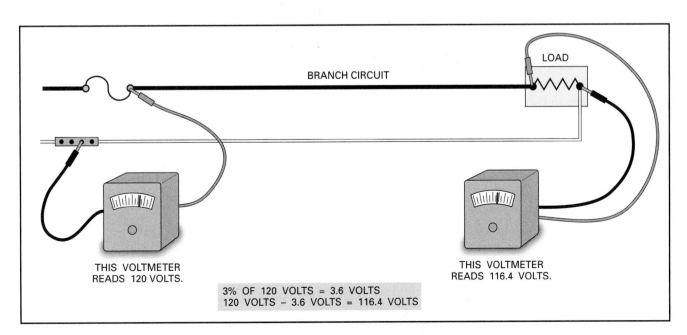

THIS VOLTMETER
READS 120 VOLTS.

THIS VOLTMETER
READS 116.4 VOLTS.

3% OF 120 VOLTS = 3.6 VOLTS
120 VOLTS − 3.6 VOLTS = 116.4 VOLTS

Fig. 7–6   Maximum allowable voltage drop on a branch is 3 percent, *Rule 8–102 1(c)*.

Fig. 7–7   *Rule 8–102 1(b)* states that the total voltage drop from the beginning of a feeder to the farthest outlet on a branch circuit should not exceed 5 percent. In this figure, if the voltage drop in the feeder is 3 percent, do not exceed 2 percent voltage drop in the branch circuit. If the voltage drop in the feeder is 2 percent, do not exceed 3 percent voltage drop in the branch circuit.

Caution Using High-Temperature Wire

When trying to use smaller conduits, be careful about selecting conductors on the ability of their insulation to withstand high temperatures. Reference to *Table 2* will confirm that the ampacity of high-temperature conductors is greater than the ampacity of lower-temperature conductors. For instance:

- No. 8 T90 nylon copper has an ampacity of 45 amperes.

- No. 8 TWU copper has an ampacity of 40 amperes.

Always calculate a voltage drop when your decision to use the higher-temperature-rated conductors was based on the fact that the use of smaller-size conduit would be possible. Select the proper size conductor to use for the job according to the load requirements, then complete a voltage-drop calculation to see that the maximum permitted voltage drop is not exceeded.

EXAMPLE 1: What is the approximate voltage drop on a 120-volt, single-phase circuit consisting of No. 14 AWG copper conductors where the load is 11 amperes and the distance of the circuit from the panel to actual load is 25.9 metres?

SOLUTION:

$$E_d = \frac{K \times I \times L \times 2}{CMA}$$

$$= \frac{39.4 \times 11 \times 25.9 \times 2}{4110}$$

$$= 5.46 \text{ volts drop}$$

Note: Refer to Table 7–4 for the CMA value.

This exceeds the voltage drop permitted by the *CEC*, which is:

3% of 120 volts = 3.6 volts

Using No. 12 AWG:

$$E_d = \frac{39.4 \times 11 \times 25.9 \times 2}{6530}$$

$$= 3.44 \text{ volts drop}$$

This meets code.

EXAMPLE 2: Find the wire size needed to keep the voltage drop to no more than 3 percent on a single-phase 240-volt circuit that feeds a 44-ampere air conditioner. The circuit originates at the main panel, which is located about 19.8 m from the air conditioner. No neutral is required for this equipment.

SOLUTION: First, size the conductors at 125 percent of the air-conditioner load.

$44 \times 1.25 = 55$-ampere conductors are required.

Checking *Table 2*, the conductors could be No. 6 TW75 or RW75 copper.
The permitted voltage drop is:

$$E_d = 240 \times 0.03$$

$$= 7.2 \text{ volts}$$

$$CMA = \frac{K \times I \times L \times 2}{E_d}$$

$$= \frac{39.4 \times 44 \times 19.8 \times 2}{7.2}$$

$$= 9535 \text{ circular mils minimum}$$

According to Table 7–4, this would be a No. 10 AWG conductor, but this size conductor would be too small for the load. So, you are back to No. 6 TW.

If you compare the calculation of voltage drop to the size of the wire or current, note that the larger wire gauge was from the current calculation. If the voltage drop calculation required a larger wire than the current table, the larger wire is always installed.

APPROXIMATE CONDUCTOR SIZE RELATIONSHIP

Rule One: For wire sizes up through 0000, every third size doubles or halves in circular mil area. A No. 1 AWG conductor is two times larger than a

No. 4 AWG conductor (83 690 versus 41 470), and a "0" wire is one-half the size of "0000" wire (105 600 versus 211 600).

Rule Two: For wire sizes up through 0000, every consecutive wire size is about 1.26 times larger or smaller than the preceding wire size. A No. 3 AWG conductor is about 1.26 times larger than a No. 4 AWG conductor (41 740 × 1.26 = 52 592). Thus, a No. 2 AWG wire is about 1.26 times smaller than a No. 1 AWG wire (66 360 ÷ 1.26 = 52 667).

A No. 10 AWG conductor has a cross-sectional area of 10 380 CM and a resistance of 1.2 ohms per 1000 ft. The resistance of aluminum wire is about 2 ohms per 1000 ft (300 m). By remembering these numbers, you will be able to perform voltage drop calculations without having the wire tables readily available.

EXAMPLE: What is the approximate CMA and resistance of a No. 6 AWG copper conductor?

SOLUTION:

WIRE SIZE	CMA	OHMS PER 1000 FEET (300 m)
10	10 380	1.2
9		
8		
7	20 760	0.6
6	26 158	0.0476

When the CMA of a wire is doubled, its resistance is cut in half. And inversely, when a given wire size is reduced to one-half, its resistance doubles.

CALCULATION OF MAXIMUM LENGTH OF CABLE RUN USING *TABLE D3*

CEC Table D3 (Table 7–5) gives a method of calculating the maximum length of a two-wire circuit. According to *Rule 8–102(1)*, this may be either 2 percent or 3 percent.

It considers:

- AWG of the wire,
- current on the wire (load),
- distance of the conductor run (e.g., 10 m from panel to outlet),
- rated conductor temperature,

CEC TABLE D3
DISTANCE TO CENTRE OF DISTRIBUTION FOR A 1% DROP IN VOLTAGE ON NOMINAL 120 V, 2-CONDUCTOR COPPER CIRCUITS
(See *Appendix B* Note to *Rule 4–004*.)

Current, A	Copper Conductor Size, AWG														
	18	16	14	12	10	8	6	4	3	2	1	1/0	2/0	3/0	4/0
	Distance to centre of distribution measured along the conductor run, m (calculated for conductor temperature of 60°C)														
1.00	24.2	38.5	61.4												
1.25	19.4	30.8	49.1												
1.6	15.1	24.1	38.4	61.0											
2.0	12.1	19.3	30.7	48.8											
2.5	9.7	15.4	24.6	39.0	62.0										
3.2	7.6	12.0	19.2	30.5	48.5										
4.0	6.1	9.6	15.3	24.4	38.8	61.7									
5.0	4.8	7.7	12.3	19.5	31.0	49.3									
6.3	3.8	6.1	9.7	15.5	24.6	39.1	62.2								
8.0	3.0	4.8	7.7	12.2	19.4	30.8	49.0								
10.0	2.4	3.9	6.1	9.8	15.5	24.7	39.2	62.4							
12.5		3.1	4.9	7.8	12.4	19.7	31.4	49.9	62.9						
16		2.4	3.8	6.1	9.7	15.4	24.5	39.0	49.1	62.0					
20			3.1	4.9	7.8	12.3	19.6	31.2	39.3	49.6	62.5				
25				3.9	6.2	9.9	15.7	24.9	31.4	39.7	50.0	63.1			
32					4.8	7.7	12.2	19.6	24.6	31.0	39.1	49.3	62.1		
40					3.9	6.2	9.8	15.6	19.7	24.8	31.3	39.4	49.7	62.7	
50						4.9	7.8	12.5	15.7	19.8	25.0	31.5	39.8	50.1	63.2
63						3.9	6.2	9.9	12.5	15.7	19.8	25.0	31.6	39.8	50.2
80						3.1	4.9	7.8	9.8	12.4	15.6	19.7	24.8	31.3	39.5
100							3.9	6.2	7.9	9.9	12.5	15.8	19.9	25.1	31.6
125							5.0	6.3	7.9	10.0	12.6	15.9	20.1	25.3	
160								4.9	6.2	7.8	9.9	12.4	15.7	19.8	
200									5.0	6.3	7.9	9.9	12.5	15.8	
250												6.3	8.0	10.0	12.6
320													6.2	7.8	9.9

NOTES: (1) *Table D3* is calculated for copper wire sizes No.18 AWG to No. 4/0 AWG and, for each size specified, gives the approximate distance in metres to the centre of distribution measured along the conductor run for a 1% drop in voltage at a given current, with the conductor at a temperature of 60°C. Inductive reactance has not been included since it is a function of conductor size and spacing.

(2) The distances for a 3% or 5% voltage drop are 3 or 5 times those for a 1% voltage drop.

(3) Because the distances in *Table D3* are based on conductor resistances at 60°C, these distances must be multiplied by the correction factors below according to the temperature rating of the conductor used and the percentage load with respect to the allowable ampacity. Where the calculation and the allowable ampacity fall between two columns, the factor in the higher percentage column shall be used.

Distance correction factor

Rated conductor temperature	Percent of allowable ampacity						
	100	90	80	70	60	50	40
60°C	1.00	1.02	1.04	1.06	1.07	1.09	1.10
75°C	0.96	1.00	1.00	1.03	1.06	1.07	1.09
85–90°C	0.91	0.95	1.00	1.00	1.04	1.06	1.08
110°C	0.85	0.90	0.95	1.00	1.02	1.05	1.07
125°C	0.82	0.87	0.92	0.97	1.00	1.04	1.07
200°C	0.68	0.76	0.83	0.90	0.96	1.00	1.04

(4) For other nominal voltages, multiply the distances in metres by the other nominal voltage (in volts) and divide by 120.

(5) Aluminum conductors have equivalent resistance per unit length to copper conductors that are smaller in area by two AWG sizes. *Table D3* may be used for aluminum conductors because of this relationship, i.e., for No. 6 AWG aluminum use the distances listed for No. 8 AWG copper in *Table D3*. Similarly, for No. 2/0 AWG aluminum use the distances for No. 1 AWG copper.

(6) The distances and currents listed in *Table D3* follow a pattern. When the current, for any conductor size, is increased by a factor of 10, the corresponding distance decreases by a factor of 10. This relationship can be used when no value is shown in the table. In that case, look at a current 10 times larger. The distance to the centre of distribution is then 10 times larger than the listed value.

(7) For multi-conductor cables, ensure wire size obtained from this table is suitable for ampacity from *Table 2* or *4*, and *Rule 4–004*.

(8) For currents intermediate to listed values use the next higher current value.

(9) Example on use of table:
Consider a two-conductor circuit of No. 12 AWG copper NMD90 carrying 16 A at nominal 240 V under maximum ambient of 30°C.
The maximum run distance from the centre of distribution to the load without exceeding a 3% voltage drop is:
Maximum run length for No. 12 AWG, 16 A, 1% voltage drop at nominal 120 V from Table is: 6.1m
Distance correction factor to be used is:
From *Table 2*, allowable ampacity for 2-conductor No. 12 AWG NMD90 (90°C rating per *Table19*) is 20 A. The given current is 16 A or 80% (16/20) of the allowable ampacity.
The distance correction factor to be used, from Note (3), 90°C row, 80% column, is 1.00.
The maximum run length is:

$$6.1 \text{ m} \times 3(\%) \times 1.00 \times \frac{240 \text{ V}}{120 \text{ V}} = 37 \text{ m}$$

If the distance is between 37 and 60.5 m, a larger size of conductor is required, e.g., No. 10 AWG (30 A allowable ampacity)
9.7 m × 3(%) × 1.04 × 240 V/120 V = 60.5 m.

Table 7–5 *CEC Table D3. See Rule 8–102 and Appendix B, Rule 4–004.*

- percentage of allowable ampacity,
- type of conductor, Cu or Al, *Table D3, Note 5,*
- percentage voltage drop on feeder or branch circuit (2% or 3%),
- voltage applied to circuit.

You can use this formula to find the maximum length of a run of wire using *Table D3*.

$$L = TBL\ D3\ \text{dist.} \times \%\ \text{V.D.} \times \text{D.C.F.} \times \text{volts}/120$$

Where:

L   = maximum length of two-wire Cu conductor run in metres

TBL D3 dist. = the distance shown in *Table D3* for the size of wire (AWG heading at top of table) and the actual load in amperes ("Current, A" column)

% V.D.   = the maximum percentage voltage drop allowed on the circuit (value should be shown as a whole number, not as a percentage)

D.C.F.   = distance correction factor, calculated by the distance correction factor table found in *Table D3, Note 3,* using "Percent of allowable ampacity"

$$\%\ \text{allowable ampacity} = \frac{\text{actual load amperes}}{\text{allowable conductor ampacity}} \times 100$$

and rated conductor temperature

Volts   = the voltage that the circuit operates at (e.g., 120, 208, 240)

EXAMPLE: What is the maximum distance that a No. 10/2 NMWU copper wire circuit may be run from the panel if it is carrying a load of 18 amperes at 240 volts with a 3 percent maximum voltage drop?

SOLUTION:
$$L = TBL\ D3\ \text{dist.} \times \%\ \text{V.D.} \times \text{D.C.F.} \times \text{volts}/120$$
$$= 7.8 \times 3 \times 1.07 \times 240/120$$
$$= 50.08\ \text{m}$$

NON-METALLIC-SHEATHED CABLE (*RULES 12–500* TO *12–526*)

Description

Nonmetallic-sheathed cable is a factory assembly of two or more insulated conductors having an outer sheath of moisture-resistant, flame-retardant non-metallic material. This cable is available with two, three, or four current-carrying conductors, ranging in size from No. 14 through No. 2 for copper or aluminum. Two-wire cable contains one black conductor, one white conductor, and one bare grounding conductor or green insulated conductor. Three-wire cable contains one black, one white, one red, and one bare grounding conductor or green insulated conductor.

The CSA lists three classifications of non-metallic-sheathed cable:

- Type NMD90 has a flame-retardant, moisture-resistant non-metallic covering over the conductors and is suitable for both dry and damp locations.

- Types NMW and NMWU have a flame-retardant, moisture-resistant, corrosion-resistant, fungus-resistant non-metallic covering over the conductors. NMWU is designed especially to use in wet and direct burial locations.

Electricians once referred to non-metallic-sheathed cable as *Romex* or *Loomex. Table 19* indicates that the conductors in type NMD90 cable are rated for 90°C (194°F), making it suitable for the high temperatures found in attics, near recessed lighting fixtures, and around cables buried in insulation.

Non-metallic-sheathed cable has an uninsulated copper conductor that is used for bonding purposes only (Fig. 7–8). This bonding conductor is not intended for use as a current-carrying circuit wire.

Equipment bonding requirements are specified in *Rules 10–400* to *408, 10–808* and *10–814*. These rules require that all boxes and fixtures in the residence be bonded to ground.

Fig. 7–8   Non-metallic-sheathed cable with an uninsulated copper conductor. ① is a white grounded conductor; ② is a bare equipment bonding conductor; ③ is a black, ungrounded ("hot") conductor.

TABLE 16
MINIMUM SIZE CONDUCTORS FOR BONDING RACEWAYS AND EQUIPMENT
(See *Rules 10–204, 10–626, 10–814, 10–816, 10–906, 12–1814, 24–104, 24–202, 30–1030, 68–058,* and *68–406.*)

Ampacity of largest ungrounded conductor in the circuit or equivalent for multiple parallel conductors, A	Size of bonding conductor	
	Copper wire, AWG	Aluminum wire, AWG
20	14	12
30	12	10
40	10	8
60	10	8
100	8	6
200	6	4
300	4	2
400	3	1
500	2	0
600	1	00
800	0	000
1000	00	0000
1200	000	250 kcmil
1600	0000	350 kcmil
2000	250 kcmil	400 kcmil
2500	350 kcmil	500 kcmil
3000	400 kcmil	600 kcmil
4000	500 kcmil	800 kcmil
5000	700 kcmil	1000 kcmil
6000	800 kcmil	1250 kcmil

Table 7–6   *CEC Table 16.*

Table 16 (Table 7–6) lists the sizes of the bonding conductors used in cable assemblies. Note that the copper bonding conductor shown is a smaller size than the circuit conductors for 15-, 20-, and 30-ampere ratings.

Table 7–7 shows the uses permitted for type NMD90 and type NMW or NMWU cable. Refer also to conditions of use in *Table 19.*

Installation

Non-metallic-sheathed cable is the least expensive of the wiring methods. It is relatively light in weight and easy to install. It is widely used for dwelling unit installations on circuits of 300 volts or less. The installation of all types of non-metallic-sheathed cable must conform to the requirements of *Rules 12–500* to *12–526* (Fig. 7–9).

- The cable must be strapped or stapled not more than 300 mm from a box or fitting. See Fig. 7–10 for permitted fasteners.

- The intervals between straps or staples must not exceed 1.5 m.

- The cable must be protected against physical damage where necessary.

- The cable must not be bent or stapled so that the outer covering or the wires are damaged.

FOR TYPICAL RESIDENTIAL WIRING TYPE NMD90, NMW, AND NMWU CABLE	TYPE NMD90	TYPE NMW	TYPE NMWU
• May be used on circuits of 300 volts or less	Yes	Yes	Yes
• Has flame-retardant and moisture-resistant outer covering	Yes	Yes	Yes
• Has fungus-resistant and corrosion-resistant outer covering	No	Yes	Yes
• May be used to wire one- and two-family dwellings or multifamily dwellings	Yes	Yes	Yes
• May be installed exposed or concealed in dry or damp locations	Yes	Yes	Yes
• May be embedded in masonry, concrete, plaster, fill	No	No	No
• May be used for concealed wiring in Category 1 and Category 2 locations, *Rule 22–200, 202*	No	Yes	Yes
• May be installed in wet locations	No	No	Yes
• May be used for direct burial locations	No	No	Yes
• May be used as service entrance cable	No	No	No
• Must be protected against damage	Yes	Yes	Yes

Table 7–7   Uses permitted in typical residential wiring for type NMD90, NMW, and NMWU non-metallic-sheathed cable.

Fig. 7–9   Installation of non-metallic-sheathed cable.

- The cable must not be used in circuits of more than 300 volts between conductors.

- Wherever non-metallic-sheathed cable is less than 1.5 metres above a floor, *Rule 12–518* requires that the cable be protected by rigid conduit, EMT, or other means (Fig. 7–11). A fitting (bushing or connector) must be used at both ends of the conduit to protect the cable from abrasion, *Rule 12–906*. A metal conduit must be bonded to ground.

- When cables are "bundled" together for distances longer than 600 mm (Fig. 7–12), the heat generated by the conductors cannot easily dissipate. These conductors *must be derated* according to *CEC Table 5C*.

When non-metallic-sheathed cable is run close to a heating source, an air space must be maintained to minimize the transfer of heat to the cable.

The spacing requirements are:

- 25 mm to heating ducts

Fig. 7–10   Devices used for attaching NMSC cable to wood surfaces, *Rule 12–510 (1)*.

Fig. 7–11   Installation of exposed non-metallic-sheathed cable where passing through a floor.

Fig. 7–12   Bundled cables. When cables are bundled for distances over 600 mm, the correction factor in *CEC Table 5C* applies.

- 50 mm to masonry chimneys
- 150 mm to chimney and flue cleanouts, *Rule 12-506(4)(c)*.

The spacing may be reduced if an approved thermal barrier is placed between the conductor and the heat source that will maintain an ambient conductor temperature of 30°C.

Fig. 7–13 shows the *CEC* requirements for securing non-metallic-sheathed cable.

Rule 10–906(3) requires that all metal boxes for use with non-metallic cable must have provisions for attaching the bonding conductor (Fig. 7–14). Fig. 7–15 shows a gangable switch (device) box that is tapped for a screw by which the bonding conductor may be connected underneath.

ARMOURED CABLE *(RULES 12–600 TO 12–618)*

Description

The *CEC* describes the construction of armoured cable in *Table D1*. It lists three types: AC90 (commonly called "BX"), ACWU90, and TECK90. AC90 is approved for dry locations only. At one time BX was a trademark owned by the General Electric Company: it is now a generic term.

Armoured cable is an assembly of insulated conductors in a flexible metallic enclosure, Fig. 7–16. Armoured cable may be used in a variety of locations depending on the type of jacket on top of the armoured cable and the location for the cable.

All three types may be used for a service entrance above grade in dry locations, but ACWU90 and TECK90 are approved for service entrance below grade. In all instances, verify with the local supply authority which cables you can use for service entrances.

Armoured cable is generally available with two, three, or four conductors in sizes from No. 14 AWG to 2000 kcmil inclusive:

- two wire = one black, one white + one bare
- three wire = one black, one red, one white + one bare
- four wire = one black, one red, one blue, one white + one bare

Armoured cable must have an internal bonding wire of copper or aluminum to meet the bonding requirements of *Rule 10–502*.

Fig. 7–13 *Rule 12–514* states how non-metallic-sheathed cable must be run in unfinished basements.

Use and Installation

Armoured cable can be used in more applications than non-metallic-sheathed cable. *Rules 12–600* to *12–618* govern the use of armoured cable in dwelling unit installations. AC90 and TECK90 armoured cable:

- may be used on circuits and feeders for applications of 600 volts or less;

- may be used for open and concealed work in dry locations;

- may be run through walls and partitions;

- may be laid on the face of masonry walls and buried in plaster if the conductors are not larger than #10 AWG and used for the extension of existing outlets;

- must be secured within 300 mm of every outlet box or fitting and at intervals not exceeding 1.5 m;

- must not be bent so that the radius of the curve of the inner edge is less than six times the external diameter of the cable;

- must have an approved fibre or plastic insulating bushing (antishort) at the cable ends to protect the conductor insulation, Fig. 7–17.

Armoured cable (types ACWU90 and TECK90) is approved for use in these situations:

- underground installations;

- burying in masonry or concrete during construction with minimum 50 mm of covering;

Fig. 7–14 *Rule 12–524* requires boxes to be of a type for use with NMSC.

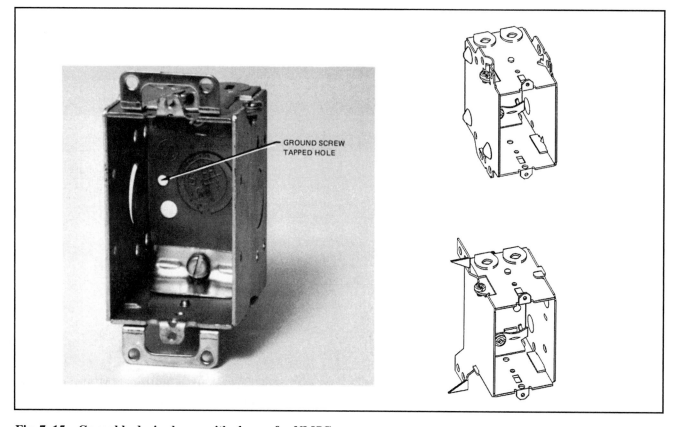

Fig. 7–15 Gangable device boxes with clamps for NMSC.

- installation in any location exposed to weather;
- installation in any location exposed to oil, gasoline, or other materials that have a destructive effect on the insulation.

To remove the outer metal cable armour, you should use a tool like the one shown in Fig. 7–18A or a hacksaw (Fig. 7–18B) to cut through one of the raised convolutions of the cable armour. Be

Fig. 7–16 Flexible armoured cable.

Fig. 7–17   Antishort bushing prevents the sharp metal armour from cutting the conductor insulation.

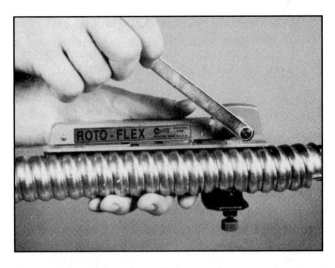

Fig. 7–18A   This tool precisely cuts the outer armour of armoured cable, making it easy to remove with a few turns of the handle. Courtesy of Seatek Co., Inc.

Fig. 7–18B   A hacksaw can be used to cut through a raised convolution of the cable armour.

careful to not cut too deep or you will cut into the conductor insulation. Then bend the cable armour at the cut. It will snap off easily, exposing the desired amount of conductor.

To prevent cutting of the conductor insulation by the sharp metal armour, insert an antishort bushing at the cable ends (Fig. 7–17).

Both non-metallic-sheathed cable and armoured cable have advantages that make them suitable for particular types of installations. However, the type of cable to be used in a specific situation depends largely on the wiring method permitted or required by the local building code.

INSTALLING CABLES THROUGH WOOD AND METAL FRAMING MEMBERS (*RULE 12–516*)

To complete the wiring of a residence, you must run cables through studs, joists, and rafters. One method is to run the cables through holes drilled at the approximate centres of wood building members, or at least 32 mm from the nearest edge. Holes bored through the centre of standard 2 × 4s meet the requirements of *Rule 12–516* (Fig. 7–19).

Where cables and certain types of raceways are run concealed or exposed parallel to a stud, joist, rafter, or other building framing member, the cables and raceways must be supported and installed so that there is at least 32 mm from the edge of the stud, joist, or other framing member to the cable

or raceway. If the 32 mm clearance cannot be maintained, a steel plate or equivalent at least 1.59 mm thick must be used to protect the cable or raceway from damage that can occur when nails or screws are driven into the wall or ceiling, *Rule 12–516*, Figs. 7–19 and 7–20.

This additional protection is not required when the raceway is rigid metal conduit, rigid PVC conduit, or EMT.

You cannot install additional protection inside the walls or ceilings of an existing building where the walls and ceilings are already closed in. *Rule 12–520* permits armoured cables to be "fished" between boxes or other access points without additional protection. The logic is similar to the exception that permits non-metallic-sheathed cable to be fished between outlets without requiring supports as stated in *Rule 12–520*.

(1) CABLES   RUN THROUGH HOLES DRILLED IN CENTRES OF STUDS.
EDGE OF BORED HOLE SHALL BE NOT LESS THAN 32 mm
FROM THE NEAREST EDGE OF THE STUD, OTHERWISE THE CABLE
SHALL BE PROTECTED BY A STEEL PLATE AT LEAST
1.59 mm THICK.

LESS THAN 32 mm

32 mm

CABLES

NOTCHED STUDS

2 x 2

(2) METAL PLATE, 1.59 mm THICK AND
MEASURING ABOUT 76 mm × 41 mm, USED
TO PROTECT CABLES LAID IN NOTCH IN STUD.

2 x 4

Fig. 7–19   Methods of protecting cables, *Rule 12–516*. This rule does not apply to rigid metal conduit and EMT.

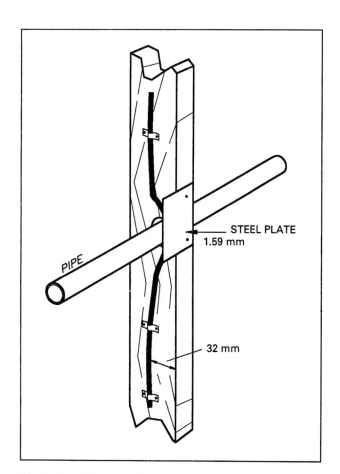

STEEL PLATE
1.59 mm

PIPE

32 mm

Fig. 7–20   Where cables or certain types of raceways are run parallel to a joist or stud (any framing member), keep the cable 32 mm from the edge of the framing member. Where this clearance is impossible to maintain, install a steel plate at least 1.59 mm thick, *Rule 12–516*.

The recreation room of this residence requires adequate protection against physical damage for the wiring. The walls in the recreation room are to be panelled. Therefore, if the carpenter uses 1 × 2 or 2 × 2 furring strips (Fig. 7–21A), non-metallic-sheathed cable would require the additional 1.59 mm steel plate protection up the walls from the receptacle outlet and switch. This could be quite costly and time-consuming, and a good case could be made for doing the installation in EMT.

Watch out for any wall partitions where 2 × 4 studs are installed "flat," Fig. 7–21B. In this case, you can either provide mechanical protection as required by *Rule 12–516(2)* or install the wiring in EMT.

Steel framing is used in a wide variety of residential applications, including single and multi-dwelling buildings. The framing is galvanized sheet steel that is pre-punched to accommodate the installation of electrical and mechanical piping and cabling. Steel framing is used for non-structural (light gauge) and structural (heavy gauge) applications.

When installing metal boxes on steel framing, you can use self-piercing screws on light-gauge framing, but self-drilling screws will be required on heavy-gauge framing. A minimum of three threads should penetrate the framing. Steel studs do not provide as much support as wooden studs; therefore, extra support must be provided to prevent

Fig. 7–21A   When 1 × 2 or 2 × 2 furring strips are installed on the surface of a basement wall, additional protection must be provided.

Fig. 7–21B   When 2 × 4 studs are installed "flat," such as might be found in non-load-bearing partitions, a cable running parallel to or through the studs must be protected against nails or screws being driven through the cable. This means protecting the cable with 1.59 mm steel plates or equivalent for its entire length, or installing the wiring in a metallic raceway system.

movement of the box after the drywall has been installed, *Rule 12–3010.*

When running NMSC through metal framing, you must protect the cable from mechanical injury by an approved insert where it passes through a framing member, *Rule 12–516* (Fig. 7–22).

INSTALLATION OF CABLE IN ATTICS

Wiring in the attic must be done in cable and meet the requirements of *Rules 12–500* to *12–526* for *non-metallic-sheathed* cable.

In accessible attics, Fig. 7–23(A), cables must not be installed:

- across the top of floor or ceiling joists ①,
- run across the face of studs ② or rafters ③ that have a vertical distance between the floor or floor joists exceeding one metre.

Guard strips are not required if the cable is run along the sides of rafters, studs, or floor joists ④, provided *Rule 12–516(1, 2)* is met.

Fig. 7–23 (C) illustrates a cable installation that most electrical inspectors consider to be safe. Because the cables are installed close to the point where the ceiling joists and the roof rafters meet, they are protected from physical damage. It would be very difficult for a person to crawl into this space,

METAL
STUDS

SHARP
EDGES

Fig. 7–22 **Steel stud installation showing approved bushings for running NMSC through steel studs.**

or to store cartons in an area with a clearance of less than one metre, Fig. 7–23(B). Although the plans for this residence show a 600-mm-wide cat-walk in the attic, the owner may decide to install additional flooring in the attic to obtain more storage space. Because of the large number of cables required to complete the circuits, it would interfere with flooring to install guard strips wherever the cables run across the tops of the joists. Fig. 7–24. However, the cables can be run through holes bored in the joists and along the sides of the joists and rafters. In this way, the cables do not interfere with the flooring. Note that most building codes will not allow manufactured roof systems to be drilled. This includes roof trusses.

When running cables parallel to framing members, maintain at least 32 mm between the cable and the edge of the framing member, *Rule 12–516 (1, 2)*. This minimizes the possibility of driving nails into the cable.

INSTALLATION OF CABLES THROUGH DUCTS

Rule 2–124 is strict about what types of wiring methods are permitted for installing cables through ducts or plenum chambers. These stringent rules are for fire safety.

The *CEC* permits type NMD90 cable to be installed in joist and stud spaces (i.e., cold air

Fig. 7–23   Protection of cable in attic.

returns) in dwellings. It recommends that type NMD90 pass through such spaces only if they are run perpendicular to their long dimensions (Fig. 7–25).

CONNECTORS FOR INSTALLING NON-METALLIC-SHEATHED AND ARMOURED CABLE

The connectors shown in Figs. 7–26A and 7–26B (pp. 135 and 136) are used to fasten non-metallic-sheathed cable and armoured cable to the boxes and panels in which they terminate. These connectors clamp the cable securely to each outlet box. Many boxes have built-in clamps (Fig. 7–26C) and do not require the separate connectors. Knockouts and pryouts on the sides of boxes allow installation of cables and connectors, Fig. 7–26D.

Unless the CSA listing of a specific cable connector indicates that the connector has been tested for use with more than one cable, the rule is *one cable, one connector*.

NOT PERMITTED

GUARD STRIPS

RUNNING BOARD

CABLES MAY BE RUN THROUGH HOLES BORED IN THE JOISTS, *RULE 12–516(1)*. HOWEVER, CHECK LOCAL BUILDING CODES SINCE HOLES MAY NOT BE DRILLED IN ANY PART OF A PREFABRICATED ROOF SYSTEM. CABLES SHOULD BE RUN WHERE ROOF CLEARANCE IS LESS THAN 1 m.

Fig. 7–24   Methods of protecting cable installations in attics.

ELECTRICAL METALLIC TUBING (*RULE 12–1400*), RIGID METAL CONDUIT (*RULE 12–1000*), AND RIGID PVC CONDUIT (*RULES 12–1100* TO *12–1122*)

In a building constructed of cement block, cinder block, or poured concrete, it is necessary to make the electrical installation in conduit.

According to the *CEC* installation requirements, EMT and rigid metal conduit:

- may be used for open or concealed work,
- may be buried in concrete or masonry (with certain provisions),
- may be used in combustible or noncombustible construction.

In general, EMT must be supported within 1 m of each box or fitting and at 1.5 m intervals along runs for 16-mm and 21-mm conduit, Fig. 7–27, *Rule 12–1010.*

The number of conductors permitted in EMT and rigid conduit is given in *Table 6.* Heavy-wall rigid conduit provides greater protection against mechanical injury to conductors than does EMT, which has a thinner wall but is more expensive.

The residence specifications in this text indicate that a meter base is to be supplied and installed by the electrical contractor. Main Panel A, a combination panel, is located in the workshop. The meter base is located on the outside of the house as indicated on the electrical plan. The electrical contractor must furnish and install a 53-mm rigid PVC conduit between the meter base and Main Panel A.

Rule 12–906 requires that insulating bushings or equivalent be used where No. 8 or larger conductors enter a raceway. Also, if rigid metallic conduit is used between the meter base and the combination panel, bonding-type bushings must be used on the service entrance conduit. (See *Rule 10–606* for methods of bonding at service equipment.)

The plans indicate that EMT is to be run from the workshop to the attic.

Properly installed EMT provides good continuity for equipment bonding, a means of withdrawing or pulling in additional conductors, and excellent protection against physical damage to the conductors.

RIGID PVC CONDUIT (*RULES 12–1100* TO *12–1122*)

PVC conduit is easily handled, stored, cut, and joined. It is commonly used for service entrance conduit in residential services in both overhead and underground applications. Easy to work with, it provides a rigid, watertight conduit for the consumer's service conductors.

When joining PVC to a fitting, you must cut, ream, clean, and glue the conduit. To cut PVC, use a hacksaw; for larger sizes of conduit, use a mitre box so that the cut is square. *Measure twice, cut once.* After cutting PVC conduit, ream out the burrs with a knife and clean the end of the pipe and inside of the fitting with an approved PVC pipe cleaner. Then apply a liberal coat of solvent cement to both surfaces, slide together, and give a quarter turn to make sure the solvent is spread evenly on the material. Hold together for a few seconds, and the joint is made. See Figs. 7–28(A) and (B).

Fig. 7–25 Non-metallic-sheathed cable may pass through cold air return or joist, or stud spaces in a dwelling.

Fig. 7–26A Various cable connectors for non-metallic-sheathed cable.

Fig. 7–26B   Various cable connectors for armoured cable and flexible conduit.

Fig. 7–26C   Typical cable clamps for boxes.

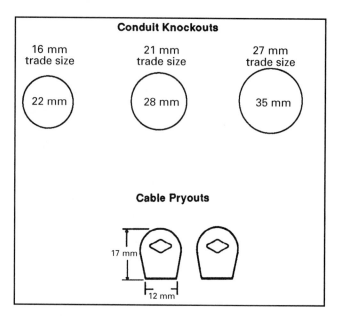

Fig. 7–26D   Knockouts and pryouts. Note the difference between the nominal trade sizes of the knockouts and the actual diameters.

Fig. 7–27   *CEC* requirements for support of rigid metal conduit *(Rule 12–1010)* and EMT *(Rule 12–1404)*. For supporting rigid PVC conduit, refer to *Rule 12–1114*.

A.

B.

C.

D.

E.

Fig. 7–28   (A) Clean conduit and apply solvent cement. (B) Insert conduit into fitting and rotate a quarter turn. (C) Heat conduit until soft using approved heater or heat gun (D) for bending conduit. (E) Use template for 90° bends. *Rule 12–1108* and *Appendix B.*

Do not use an open flame to bend PVC, *Rule 12–1108, Appendix B*. Instead, use an approved heat gun, heat blanket, or other flameless heat source to heat the pipe to about 125°C, Figs. 7–28(C) and (D). Always heat an area of at least ten times the diameter of the conduit before attempting to bend it. If not heated enough, the pipe will kink or collapse. When you make a 90° bend, the centre radius of the bend must meet the minimum radius requirements of *Table 7, Rule 12–922*. Thus, for 16-mm conduit, the minimum radius is 102 mm. If the radius is ten times the internal diameter, it is a lot easier to fish and pull in wires after the conduit is installed. Draw a template of the bend first, and then heat the conduit and shape it to the template, Fig. 7–28(E).

To provide adequate support for the conduit system, you must use proper straps. The two-hole straps in Fig. 7–29 allow the conduit to move as it expands or contracts. Never clamp the conduit tightly, *Rule 12–1114(3)*. The spacing for supports is about half the distance for other types of conduit; therefore, twice the number of straps are needed. See *Rule 12–1114* for maximum support spacing.

Rule 12–1118 and *Appendix B* give a formula for calculating the linear expansion of various materials. Since PVC has one of the highest coefficients of linear expansion, check that the maximum expansion of the conduit due to temperature change will not exceed 45 mm, *Rule 12–1118*. The formula is:

$$\Delta l = L \times \Delta T \times C$$

where:

Δl = the change in length, in millimetres, of a run of rigid PVC conduit, due to the maximum expected variation in temperature

L = the length of the run of conduit in metres

ΔT = the maximum expected temperature change, in degrees Celsius

C = the coefficient of linear expansion

EXAMPLE: A 25-m length (L = 25) of rigid PVC conduit is run outside where the minimum expected temperature is –40°C and the maximum expected temperature is 40°C (ΔT = 80). The coefficient of linear expansion for PVC conduit is 0.0520 mm per m per °C (C = 0.052), *Rule 12–1118, Appendix B*.

SOLUTION:

$$\begin{aligned} \Delta l &= L \times \Delta T \times C \\ &= 25 \times 80 \times 0.052 \\ &= 104 \text{ mm} \end{aligned}$$

The expansion and contraction of the conduit exceeds the 45-mm maximum permitted in *Rule 12–1118*; therefore, an "O" ring expansion joint with 104 mm of expansion will be required, Fig. 7–30. Generally, where the expected temperature change is more than 14°C, you should use expansion joints.

FLEXIBLE CONNECTIONS (*RULES 12–1000* AND *12–1300*)

Some equipment installations require flexible connections, both to simplify the installation and to stop the transfer of vibrations. In residential wiring, flexible connections are used to wire attic fans, food waste disposers, dishwashers, air conditioners, heat pumps, recessed fixtures, and similar equipment.

The three types of flexible conduit used for these connections are:

- flexible metal conduit, Fig. 7–31(A);
- liquidtight flexible metal conduit, Fig. 7–31(B); and
- liquidtight flexible non-metallic conduit, Fig. 7–31(C).

Fig. 7–32 shows types of connectors used with flexible conduit.

FLEXIBLE METAL CONDUIT (*RULES 12–1002* TO *12–1014*)

Rule 12–1002 covers the use and installation of flexible metal conduit. This wiring method is similar to armoured cable, except that the conductors are installed by the electrician. For armoured cable, the cable armour is wrapped around the conductors at the factory to form a complete cable assembly.

Fig. 7–33 shows some installations that commonly use flexible metal conduit. The flexibility required to make the installations is provided by the flexible metal conduit, and a separate bonding conductor is required.

A. SERVICE/ENTRANCE/CAP

B. METER HUB

C. OFFSET CONNECTOR

D. RIGID METAL TO PVC CONDUIT ADAPTOR

E. F. EXPANSION JOINT

G. TYPE 'LB'

H. TYPE 'LL'

I. TYPE 'LR'

J. TYPE 'T'

K. TYPE 'C'

L. TYPE 'E'

M. TYPE 'TA'

N. PVC COUPLING

O. 2-HOLE STRAP

Fig. 7–29 Various PVC fittings: access fittings, a TA fitting for connection to a panel or a box, a coupling, and a two-hole PVC strap for support.

Fig. 7–30 PVC "O" ring expansion joint.

Fig. 7–31 (A) Flexible metal conduit. (B) Liquidtight flexible metal conduit. (C) Liquidtight flexible non-metallic conduit.

Fig. 7–32 Fittings for (A) flexible metal conduit, (B) liquidtight flexible metal conduit, and (C) liquidtight flexible non-metallic conduit.

Summarizing *Rules 12–1002* to *12–1014* concerning flexible metal conduit:

- Do not install in wet locations unless the conductors are suitable for that use (RW90, TWN75). Ensure that water will not enter the enclosure or other raceways to which the flex is connected.

- Do not bury in concrete.

- Do not bury underground.

- Do not use in locations subject to corrosive conditions.

- Do not use the metal armour of flexible conduit as a bonding means; a separate bonding conductor must be installed inside flexible conduit, *Rule 12–1306*.

Fig. 7–33 Some common places where flexible metal conduit may be used.

- • Support conduit within 300 mm of box, cabinet, or fitting.

- • If flexibility is needed, support conduit within 900 mm of termination.

- • Do not conceal angle-type fittings because access will be needed when pulling wires in.

- • You may use conduit for exposed or concealed installations.

- • Same conductor fill rules as regular conduit, *Table 6*. The cross-sectional area of 12-mm trade size conduit is 118 mm², *Rule 12–1304(2)*.

- • Use fittings approved for use with flexible metal conduit.

LIQUIDTIGHT FLEXIBLE METAL CONDUIT (*RULES 12–1300* TO *12–1306*)

The use and installation of liquidtight flexible metal conduit are described in *Rules 12–1300* to *12–1306*. Liquidtight flexible metal conduit has a "tighter" fit to its spiral turns than standard flexible metal conduit, as well as a thermoplastic outer jacket that is liquidtight. It is commonly used as the flexible connection to a central air-conditioning unit located outdoors.

Fig. 7–34 shows the *CEC* rules for the use of liquidtight flexible metal conduit.

Liquidtight flexible metal conduit:

- • May be used for exposed and concealed installations.

- • Runs of 12-mm trade size flexible metal conduit are not permitted to be longer than 1.5 m, *Rule 12–1004*. Fixture "drops" are a good example. (See Recreation Room, Unit 19, for a discussion of fixture drops.)

- • Support conduit every 1.5 m.

A. RUNS OF 12 mm NOT LONGER THAN 1.5 m ARE PERMITTED FOR CONNECTIONS TO EQUIPMENT, *RULE 12–1302(2)*.

B. LIQUIDTIGHT FLEXIBLE METAL CONDUIT MAY NOT BE USED AS A BONDING MEANS IN ANY SIZE, *RULE 12–1306*.

Fig. 7–34 Liquidtight flexible metal conduit rules, *12–1300* to *12–1306*.

- May be buried directly in the ground if so listed and marked.

- May not be used where subject to physical abuse.

- May not be used if ambient temperature and heat from conductors will exceed 60°C unless marked for a higher temperature.

- If 12-mm trade size, is not permitted longer than 1.5 m.

- Must have the same conductor fill as regular conduit, *Rule 12–1014.*

- Must have conductor fill for 12-mm size based on cross-sectional area of 118 mm², *Rule 12–1304(2).*

- Must be used only with approved fittings.

- Must not be used when flexing at low temperatures will cause injury.

- Must not be used as general purpose raceway.

- Is not acceptable as a grounding means, *Rule 10–510(3).*

- Must not have concealed angle-type fittings.

LIQUIDTIGHT FLEXIBLE NON-METALLIC CONDUIT (*RULES 12–1300* TO *12–1306*)

Summarizing *CEC* rules pertaining to liquidtight flexible non-metallic conduit:

- Do not use in direct sunlight unless specifically marked for use in direct sunlight.

- You may use it in exposed or concealed installations.

- It can become brittle in extreme cold applications.

- You may use it outdoors when listed and marked as suitable for this application.

- You may bury it directly in the earth when listed and marked as suitable for this application.

- You may not be use it where subject to mechanical damage.

- You may not use it if ambient temperature and heat from the conductors will exceed the temperature limitation of the non-metallic material.

- 12-mm trade size is allowed for enclosing the leads to a motor, *Rule 12–1302(2).*

- Conductor fill is same as for regular conduit, *Table 6, Rule 12–1014.*

- It must be used with approved fittings.

- A separate bonding conductor is required and should be sized with reference to *Table 16* and installed according to *Rule 10–808.*

REVIEW

Note: Refer to the *CEC* or the plans where necessary.

1. What is the largest size of solid wire that is commonly used for branch circuits and feeders?

    _____

2. What is the minimum size branch-circuit conductor that may be installed in a dwelling unit?

    _____

3. What exceptions, if any, are there to the answer for Question 2? _____

    _____

4. Define the term *ampacity.* _____

    _____

5. What is the maximum voltage rating of all NMSC? _____

6. Indicate the ampacity of these type T90 Nylon (copper) conductors. Refer to *Table 2*.

   a. 14 AWG _____ amperes   d. 8 AWG _____ amperes

   b. 12 AWG _____ amperes   e. 6 AWG _____ amperes

   c. 10 AWG _____ amperes   f. 4 AWG _____ amperes

7. What is the maximum operating temperature of these conductors? Give the answer in Celsius. (Use *Tables 11* and *19*.)

   a. Type TEW _____   c. Type LVT _____

   b. Type R90 _____   d. Type DRT _____

8. What are the colours of the conductors in non-metallic-sheathed cable for

   a. two-wire cable? _____

   b. three-wire cable? _____

9. For non-metallic-sheathed (type NMD90) cable, can the uninsulated conductor be used for purposes other than bonding? _____

10. Under what condition may non-metallic-sheathed cable (type NMD90) be fished in the hollow voids of stud walls? _____

11. a. What is the maximum distance permitted between straps on a cable installation?

    _____

    b. What is the maximum distance permitted between a box and the first strap in a cable installation? _____

12. What is the difference between type AC90 and type TECK90 cable?

    _____

13. Type AC90 cable may be bent to a radius of not less than _____ times the diameter of the cable, measured to the (inside, outside) edge of the cable. Circle the correct answer.

14. When armoured cable is used, what protection is provided at the cable ends? _____

    _____

    _____

    _____

15. What protection must be provided when installing a cable in a notched stud or joist?

    _____

    _____

    _____

    _____

    _____

16. Cables passing through a stud where the edge of the bored hole is less than _____ mm from the edge of the stud require additional protection.

17. a. Is non-metallic-sheathed cable permitted in your area for residential occupancies?

_____

b. From what source is this information obtained? _____

18. Is it permitted to use flexible metal conduit over 2 m long as a bonding means? (Yes) (No) Circle one.

19. Liquidtight flexible metal conduit (may) (may not) serve as a bonding means. Circle one.

20. The allowable current-carrying capacity (ampacity) of aluminum wire is less than that of an equivalent copper wire. Use *Rule 4-002* and *Tables 2, 4,* and *13* to complete the following table.

WIRE	COPPER		ALUMINUM	
	Ampacity	Overcurrent Protection	Ampacity	Overcurrent Protection
No. 12 R90				
No. 10 R90				
No. 3 TW75				
0000 TW				
500 kcmil RW75				

*ENTER BOTH AMPACITY AND MAXIMUM OVERCURRENT PROTECTION VALUES.

21. All solderless wire connectors are approved to connect aluminum and copper conductors together in the same connector. (True) (False) Circle one.

22. Terminals of switches and receptacles marked CO/ALR are suitable for use with _____ and _____ conductors.

23. When non-metallic-sheathed cables are bunched or bundled together for distances longer than 600 mm, what happens to their current-carrying ability?

_____

_____

24. A 120-volt branch circuit supplies a resistive heating load of 10 amperes. The distance from the panel to the heater is about 43 metres. Calculate the voltage drop using (a) No. 14, (b) No. 12, (c) No. 10, and (d) No. 8 AWG copper conductors.

25. In Question 24, it is desired to keep the voltage drop to 3 percent maximum. What is the minimum size wire that would be installed to accomplish this 3 percent maximum voltage drop?

_____

_____

_____

Switch Control of Lighting Circuits, Receptacle Bonding, and Induction Heating Resulting from Unusual Switch Connections

OBJECTIVES

After studying this unit, the student will be able to

- identify the grounded and ungrounded conductors in cable or conduit (colour coding)
- identify the types of toggle switches for lighting circuit control
- select a switch with the proper rating for the specific installation conditions
- describe the operation that each type of toggle switch performs in typical lighting circuit installations
- demonstrate the correct wiring connections the *CEC* requires for each type of switch
- understand the various ways to bond wiring devices to the outlet box
- understand how to design circuits to avoid heating by induction

The electrician installs and connects various types of lighting switches. To do this, the electrician must know both the operation and method of connection of each type of switch and understand the meanings of the current and voltage ratings marked on lighting switches and the *CEC* requirements for installing them.

CONDUCTOR IDENTIFICATION (*RULE 4–036*)

Before making any wiring connections to devices, the electrician must be familiar with the ways in which conductors are identified. For AC circuits,

the *CEC* requires that the grounded (identified) circuit conductor have an outer covering that is either white or grey. The *CEC* also recognizes that installations may require that identified conductors from different systems be installed in the same raceway, box, or enclosure. In this case, the identified conductors of the other system shall be permitted to be white with a coloured stripe, other than green, along the insulation as indicated in *Rule 4–028(2)*. In multiwire circuits, the grounded circuit conductor is also called a *neutral* conductor, *Rule 4–028* and *Section O*.

The ungrounded (unidentified) conductor of a circuit must be marked in a colour other than green, white, or grey. This conductor generally is called the *hot* conductor. You feel a shock if you hold this conductor and the grounded conductor at the same time, or touch this conductor and a grounded surface such as a water pipe at the same time.

Neutral Conductor

In all residential wiring, the grounded neutral conductor's insulation is white or natural grey. It may also be white with a coloured stripe as permitted by *Rule 4–028(2)*. The grounded circuit conductor is also called a neutral conductor when it is part of a multiwire branch circuit.

When a two-wire circuit is used, the white grounded circuit conductor is not truly a neutral conductor unless:

- it is the conductor that carries only the unbalanced current from the other conductors, as in the case of multiwire circuits of three or more conductors, *Rule 4–004(3)*,

- it is the conductor where the voltage from every other conductor to it is equal under normal operating conditions.

Therefore, the white grounded circuit conductor of a two-wire circuit is not really a "neutral," even though many electricians refer to it as such. See Figs. 8–1 and 8–1A.

Colour Coding (Cable Wiring)

Non-metallic-sheathed cable (Romex) and armoured cable (BX) are colour-coded as follows:

Two-wire: one black ("hot" phase conductor)
one white (grounded "identified" conductor)
one bare (equipment bonding conductor)
Or
one black ("hot" phase conductor)
one red ("hot" phase conductor)

Fig. 8–1   Definition of a true neutral conductor.

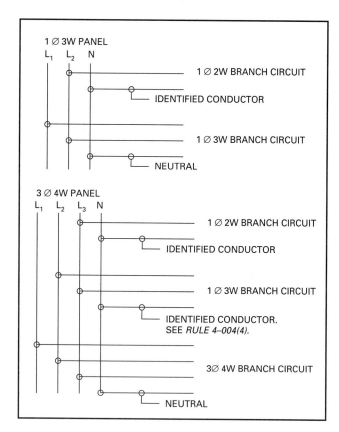

Fig. 8–1A   The difference between an identified conductor and a neutral conductor in single-phase three-wire and three-phase four-wire panels.

	one bare (equipment bonding conductor)
Three-wire:	one black ("hot" phase conductor) one white (grounded "identified" conductor) one red ("hot" phase conductor) one bare (equipment bonding conductor)
Four-wire:	one black ("hot" phase conductor) one white (grounded "identified" conductor) one red ("hot" phase conductor) one blue ("hot" phase conductor) one bare (equipment bonding conductor)

Colour Coding (Conduit Wiring)

When the installation is conduit, *Rule 4–036(2)* permits the electrician to use any of the above colours for the hot phase conductor except:

green	reserved for use as a bonding conductor only
white or grey	reserved for use as the grounded identified circuit conductor

Changing Colours

Should it become necessary to change the actual colour of the conductor to meet *CEC* requirements, the electrician may change the colours as indicated in Table 8–1.

For cable installations only, *Rule 4–034(2)* permits you to use the white conductor as a switch "loop" for switch drops, but states that the white conductor must feed the switch. Therefore, you must make up the connection so that the return conductor from the switch is a coloured or unidentified conductor. When the connections are done in this way, the *CEC* does not require re-identification, but instead accepts the splices and attachment of the conductors to the terminals of the switch as adequate identification. Therefore, if a white wire is used in way that makes it a "hot" conductor, such as in a switch circuit, it should be wired so that it is *always* "hot." It should be the feed to the switch, never the return.

Fig. 8–6 illustrates re-identification where the white wire running from the octagon box to the switch box is a "hot" phase conductor.

Although *Rule 4–034(2)* states that re-identification is *not* required, some electrical inspectors may require that these white wires be rendered permanently unidentified to avoid any

FROM	TO	DO
red, black, blue	bonding conductor	For conductors larger than No. 2 AWG, strip off insulation to make it bare—or paint it green where exposed in the box—or mark exposed insulation with green tape, *Rule 4–036(1)b.*
red, black, blue	grounded identified conductor	Re-mark with coloured tape or paint white or grey. For conductors larger than No. 2 AWG, *Rule 4–030.*
white, grey, or green	red, black, blue	Re-identify with coloured tape or paint, Fig. 8–2, *Rule 4–036 (3).*

Table 8–1 Changing colour of wire.

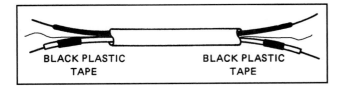

Fig. 8–2   You may wrap a white wire with a piece of black plastic tape so that everyone will know that this wire is *not* a grounded conductor. Use red tape if you desire to make the conductor red, *Rule 4–036(3)*.

confusion, Fig. 8–2. *Acceptable* is defined as "acceptable to the authority enforcing this code."

When using two-conductor non-metallic-sheathed cable for applications that do not require an identified conductor other than switch lines, the authority enforcing the code may require the use of a cable that has a red conductor in place of the white. In residential applications, this would be for loads that operate at 240 volts, such as electric heaters and hot water tanks. This cable usually has a red jacket so that the inspection authority can easily identify it.

Many electricians choose to use non-metallic-sheathed cable having different coloured jackets. For example, the 12/2 cable used for the 20-ampere T slot receptacles may have a yellow jacket, or the 14/2 cable used for the AFCI-protected circuit may have a blue jacket. Different coloured cables for specific applications aid the electrician when the conductors are terminated in the panel, clearly identifying the ampacity and type of circuit breaker required.

TOGGLE SWITCHES (*RULES 14–500* TO *14–514*)

The most frequently used switch in lighting circuits is the toggle flush switch, Fig. 8–3. When mounted in a flush switch box, the switch is concealed in the wall with only the insulated handle or toggle protruding.

Fig. 8–4 shows how switches and receptacles are weatherproofed.

Toggle Switch Ratings

The CSA classifies toggle switches for lighting circuits as *general-use switches*. These switches are

Fig. 8–3   Toggle flush switches.

divided into three categories.

Category 1 contains general-use switches used to control:

- alternating-current (AC) or direct-current (DC) circuits,
- resistive loads not to exceed the ampere rating of the switch at rated voltage,
- inductive loads not to exceed one-half the ampere rating of the switch at rated voltage unless the switch is marked with the letter "F,"
- tungsten filament lamp loads not to exceed the ampere rating of the switch at 125 volts when the switch is marked with the letter "T."

A tungsten filament lamp draws a very high momentary inrush current at the instant the circuit is energized because the *cold resistance* of tungsten is very low. For instance, the cold resistance

Fig. 8–4   Receptacle outlet and toggle switch are protected by a weatherproof cover. (*Right:* © Brian Cameron.)

of a typical 100-watt lamp is about 9.5 ohms. This same lamp has a *hot resistance* of 144 ohms when operating at 100 percent of its rated voltage.

Normal operating current would be

$$I = \frac{E}{R} = \frac{120}{144} = 0.83 \text{ amperes}$$

But, *maximum* instantaneous inrush current could be as high as

$$I = \frac{E}{R} = \frac{170 \text{ (peak voltage)}}{9.5} = 17.9 \text{ amperes}$$

This inrush current drops off to normal operating current in about six cycles (0.10 seconds). The contacts of T-rated switches are designed to handle these momentary high inrush currents.

The AC/DC general-use switch normally is not marked AC/DC. However, it is always marked with the current and voltage rating, such as 10 A–125 V or 5 A–250 V–T, *Rule 14–508(b)(iii)*.

Category 2 contains those AC general-use switches used to control:

- alternating currents only,

- resistive, inductive, and tungsten filament lamp loads not to exceed the ampere rating of the switch at 120 volts,

- motor loads not to exceed 80 percent of the ampere rating of the switch at rated voltage, but not exceeding two horsepower.

AC general-use snap switches are marked "AC only" in addition to identifying their current and voltage ratings. A typical switch marking is 15 A, 120–277 V AC. The 277-volt rating is required on 277/480-volt systems.

Category 3 contains those 347 V AC switches used to control:

- AC loads only,

- resistive and inductive loads not to exceed the ampere rating of the switch at 347 V.

In addition, the current rating shall not be less than 15 A, 347 V. The switches are designed so they cannot be mounted in boxes for Category 1 and Category 2 switches: Their mounting holes are 6.4 mm farther apart than those in standard switches.

If they are grouped or ganged in the same box, AC general-use switches operating in circuits exceeding 300 V to ground must have permanent barriers installed between them. A typical switch is 15 A 347 V. The 347 V rating is required on 347/600 V systems. Switches and devices rated at 347 volts will not be installed in dwelling units, as the maximum allowable voltage is 150 volts to ground, *Rule 2–106*.

Terminals of switches rated at 15 or 20 amperes, when marked *CO/ALR*, are suitable for use with aluminum, copper, and copper-clad aluminum conductors. Switches not marked *CO/ALR* are suitable for use with copper and copper-clad conductors only.

Screwless pressure terminals of the conductor push-in type may be used with copper and copper-clad aluminum conductors only. These push-in type terminals are not permitted for use with ordinary aluminum conductors, *Rule 12–118(6)*.

Further information on switch ratings is given in *Rule 14–508* and in *CSA Standard C22.2 No. 111, General-use Switches*.

Table 7–2 on page 116 in Unit 7 lists the markings on wiring devices indicating the types of conductors the devices are used for.

Toggle Switch Types

Toggle switches are available in four types: single-pole, three-way, four-way, and double-pole.

Single-Pole Switch: A single-pole switch is used when a light or group of lights or other load is to be controlled from one switching point. The switch is identified by its two terminals and the toggle marked ON/OFF. The single-pole switch is connected in series with the ungrounded or hot wire feeding the load.

Fig. 8–5 shows a single-pole switch controlling a light from one switching point. The 120-volt source feeds directly through the switch location. The identified white wire continues directly to the load, and the unidentified black wire is broken at the single-pole switch.

In Fig. 8–6, the 120-volt source feeds the light outlet directly, and a two-wire cable with black and white wires is used as a switch loop between the light outlet and the single-pole switch.

Fig. 8–5   Single-pole switch in circuit with feed at switch.

Fig. 8–6   Single-pole switch in circuit with feed at light.

Rule 4–034(2) permits the use of a white wire in a single-pole switch loop. The unidentified or black conductor must connect between the switch and the load.

Fig. 8–7 shows another application of a single-pole switch. The feed is at the switch that controls the light outlet. The receptacle outlet is independent of the switch.

Three-Way Switch: The term *three-way* is somewhat misleading, since a three-way switch is used to control a light from two locations. A three-way

switch has one terminal to which the internal switch mechanism is always connected. This terminal is called the *common terminal.* The two other terminals are called the *traveller wire terminals.* The switching mechanism alternately switches between the common terminal and either one of the traveller terminals. Fig. 8–8 shows the two positions of the three-way switch, which is actually a single-pole, double-throw switch.

A three-way switch differs from a single-pole switch in that the three-way switch does not have an ON or OFF position. Thus, the switch handle does not have ON/OFF markings, Fig. 8–9, and cannot be used as a disconnecting means, *Rule 2–304(2).* The three-way switch can be identified further by its three terminals. The common terminal is darker in colour than the two traveller wire terminals, which are usually natural brass in colour.

Three-Way Switch Control with Feed at Switch:
Three-way switches are used when a load is to be controlled from two different switching points. Fig. 8–10 shows how two three-way switches are used. Note that the feed is at the first switch control point.

Three-Way Switch Control with Feed at Light:
The circuit in Fig. 8–11 uses three-way switch control with the feed at the light. For this circuit, the white wire in the cable must be used as part of the three-way switch loop. The unidentified or black wire is used as the return wire to the light outlet, *Rule 4–034(2).* This exception makes it unnecessary to paint, tape, or otherwise re-identify the white wire at the switch location when wiring single-pole, three-way, or four-way switch loops.

Alternative Configuration for Three-Way Switch Control with Feed at Light: Fig. 8–12 shows another arrangement of components in a three-way switch circuit. The feed is at the light with cable runs from the ceiling outlet to each of the three-way switch control points located on either side of the light outlet.

Three-way switches are needed in rooms that have entry from more than one location. In this residence, the garage lighting, entry hall lighting, living room receptacle outlets, rear outdoor lighting located between the living room sliding doors and the kitchen sliding doors, and receptacle outlets in the study require three-way switches.

Fig. 8–7   Ceiling outlet controlled by single-pole switch with live receptacle outlet and feed at switch.

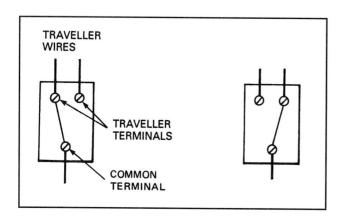

Fig. 8–8   Two positions of a three-way switch.

Conductor Colour Coding for Switch Connections

The colour coding for the travellers (messenger) for three-way and four-way switch connections when using cable can vary. *Rule 30–602* requires that the white (identified) conductor always be connected to the white (silver) terminal of a lampholder or receptacle. Because of the many ways that three-way or four-way switches can be connected, the choice of colours to use for the travellers is left to the electrician. Most electrical contractors and electricians establish some sort of colour coding that works well for them.

Fig. 8–9   Toggle switch: three-way flush switch.
© Brian Cameron.

One option is to establish white and red for the travellers and make up the electrical splices accordingly, Fig. 8–11. This is not a *CEC* requirement.

Fig. 8–10(A) shows another option: The white conductor is spliced straight through to the lampholder; the black and red are the travellers.

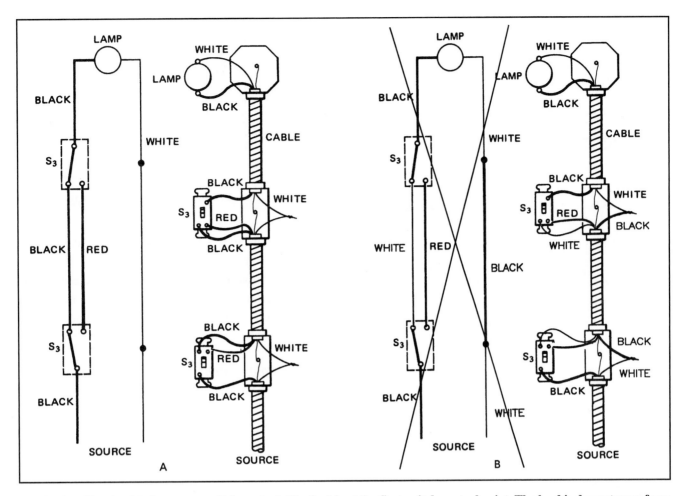

Fig. 8–10   Circuit with three-way switch control. The feed is at the first switch control point. The load is downstream from the second switch control point. Diagram A uses black and red for the travellers. Diagram B uses the red and white for the travellers. The method used in diagram B is not acceptable as it requires the use of the black as the identified conductor. Be sure that the white conductor at the lampholder is connected to the lampholder's white (silver) terminal. See *Rules 14-604* and *4-034(2)*.

Fig. 8–11   Circuit with three-way switch control and feed at light.

Fig. 8–12 Alternative circuit with three-way switch control and feed at light.

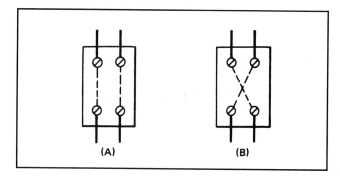

Fig. 8–13 Two positions of a typical four-way switch.

This issue does not arise when using conduit, because a variety of conductor insulation colours exist for the travellers and switch returns.

Four-Way Switch: The four-way switch is constructed so that the switching contacts can alternate their positions, Fig. 8–13. The four-way switch has two positions, but neither is ON or OFF. This switch can be readily identified by its four terminals and the absence of ON or OFF markings.

Four-way switches are used when a load must be controlled from more than two switching points. To do this, three-way switches are connected to the source and to the load. The switches at all other control points, however, must be four-way switches.

Fig. 8–14 shows how a lamp can be controlled from any one of three switching points. Take care in connecting the traveller wires to the proper terminals of the four-way switch. The two traveller wires from one three-way switch are connected to the two terminals on the top of the four-way switch, and the other two traveller wires from the second three-way switch are connected to the two terminals on the bottom of the four-way switch.

Some four-way switches have the connections from side to side on the switch. The electrician must verify this using a continuity tester before connecting the switch.

Take care when installing the flat-type four-way switches generally known as "Decora" or "Designer." Many of these switches use a different terminal connection layout than a toggle-type four-way switch. Check the packaging or use a continuity tester to verify connections before installing.

Double-Pole Switch: A double-pole, or two-pole, switch may be used when two separate circuits must be controlled with one switch, Fig. 8–15. A double-pole switch may also be used to provide a double-pole disconnecting means for a 24-volt load, Fig. 8–16.

Double-pole toggle switches are not commonly used in residential work. Double-pole disconnect switches, however, are used quite often in residences for the furnace, water pump motors, and other 240-volt feeders.

Fig. 8–14 Circuit with switch control at three locations.

Fig. 8–15   Application of a double-pole switch.

Switches with Pilot Lights: Sometimes a pilot light is desired at the switch location, as is the case for the attic lighting in this residence. Pilot light switches are available in several styles, Figs. 8–17, 8–18, and 8–19.

BONDING AT RECEPTACLES

Ground-fault protection is covered in detail in Unit 9. However, providing bonding of the equipment, bonding of the conductor to a metal box and to the receptacle's bonding terminal, is important.

To wire bonding-type receptacles with armoured cable, the cable must contain a separate bonding connector to bond the metal outlet box and the bonding terminal of the receptacle. If non-metallic-sheathed cable is used, it too must contain a separate bonding conductor, *Rules 10–500* and *10–502*. Fig. 8–20 shows the bonding conductor of non-metallic-sheathed cable attached to the receptacle, *Rule 10–808(2).*

Almost all bonding terminals of the type shown in Fig. 8–20 are bonded to the metal yoke of the receptacle. However, this connection might not provide reliable continuity between a bonded outlet box and the bonding circuit of the receptacle.

An outlet box is said to be bonded when it is connected to a bonding conductor that is part of a cable or when it is connected to a bonded metal raceway, *Rule 10–502.* When the receptacle is fastened to a bonded device box, the bonding terminal may well be bonded through the threads of the No. 6-32 mounting screws. For a 102-mm square outlet box, bonding to the raised cover could take place through the threads of the No. 8-32 cover mounting screws. However, it is doubtful that the continuity of the bonding circuit is reliable under these conditions. Therefore, *Rule 10–906(6)* requires the installation of a separate bonding jumper between all receptacles and the device box bonding connection, whatever wiring method is used: metal conduit, armoured cable, or non-metallic-sheathed cable with a bonding conductor. The metal outlet box must be bonded in addition to using the bonding terminal of the receptacle, Fig. 8–20.

To ensure the continuity of the equipment bonding conductor path, *Rule 10–808(2)* requires that when more than one bonding conductor enters a box, they shall be secured under bonding screws or connected using solderless connectors. Splices shall not depend on solder, *Rule 10–906(2).* The splicing must be done so that if a receptacle or other wiring device is removed, the continuity of the equipment bonding path shall not be interrupted, as clearly shown in Figs. 8–21 and 8–22. When wiring with non-metallic-sheathed cable, the equipment bonding conductor is a noninsulated,

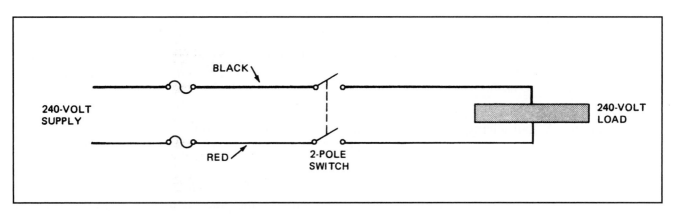

Fig. 8–16   Double-pole (two-pole) disconnect switch.

Fig. 8–17   In diagram A, the switch is in the OFF position: The neon lamp in the handle of the switch glows, and the load is off. This is a series circuit. The neon lamp has extremely high resistance. In a series circuit, voltage divides proportionately to the resistance. Therefore, the neon lamp "sees" line voltage (120 volts) for all practical purposes, and the load "sees" zero voltage. The neon lamp glows.

In diagram B, the switch is in the ON position: The neon lamp is shunted out and therefore has zero voltage across it. Thus, the neon lamp does not glow. Full voltage is supplied to the load. This type of switch might be referred to as a *locator* because it glows when the load is *off* and does not glow when the load is *on*.

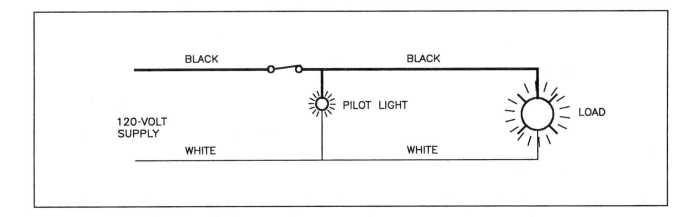

Fig. 8–18   A true pilot light. The pilot lamp is an integral part of the switch. When the load is turned on, the pilot light is also on. When the load is turned off, the pilot light is also off. This switch has three terminals because it requires a grounded neutral circuit conductor.

bare conductor. When wiring with conduit, an equipment-bonding conductor, when required, must be a green insulated conductor, sized per *Table 16.*

Splices must be made in accordance with *Rule 10–808(1)* and *(2).*

A bonding jumper, Fig. 8–20, is required between all receptacles and the device box bonding connection, *Rule 10–906(6).* The No. 6-32 mounting screws of most receptacles and switches are held captive in the yoke by a small fibre or cardboard washer. Removing this washer gives additional

Fig. 8–19 This pilot light switch is a true pilot light. The pilot light is on when the load is on, and is off when the load is off. This is possible through its internal electronic circuit. This type of switch can be used on any 120-volt, AC grounded electrical circuit. It uses the system's equipment ground as its reference to permit the pilot light to glow. It does *not* require a grounded neutral conductor. The pilot light "pulsates" about 60 times per minute. Photo courtesy of Hubbell Corporation.

Fig. 8–20 Connections for bonding-type receptacles. See *Rules 10–808(2), 10–906(6).*

Fig. 8–21 An approved method of connecting the bonding conductor using a special bonding-type wire connector. See *Rule 10–808(2).*

Fig. 8–22 Method of attaching bonding wires to device box or octagon box. See *Rule 10–808(2)*.

Fig. 8–23 Bonding-type receptacle detailing parts of the receptacle. Courtesy of Pass & Seymour, Inc.

metal-to-metal contact; however, this is not adequate as a bonding means.

Some outlet and switch-box manufacturers use washer head screws so that a bonding conductor can be placed under them. This method is useful for installing armoured cable or non-metallic-sheathed cable, as it provides a means to terminate the bonding conductor in the metal outlet or switch box.

Rule 10–906(3) states that the bonding jumper must be secured to every metal box even if the receptacle has a special self-bonding strap, Figs. 8–23 and 8–24. When you install such a receptacle, you must wire a bonding jumper from the receptacle to the bonded outlet box.

NONBONDING-TYPE RECEPTACLES

Modern appliances and portable hand-held tools have a three-wire cord and a three-wire bonding-type attachment plugcap. Frequently, these tools and appliances are plugged into nonbonding-type receptacles with only two prongs. In these circumstances, the appliance is not bonded and the user has probably cut off or bent the grounding prong to use the appliance. **This is an incorrect and**

unsafe practice. As a result, it will later be impossible to bond the appliance even when it is plugged into a properly connected three-wire bonding-type receptacle.

In these circumstances *Rule 10–408(3)* and *(4)* permits two methods of connection:

- the use of "double insulated" tools and appliances, *Rule 10–408(3)* and *Appendix B*.
- the use of a portable double-insulated GFCI receptacle to supply the equipment.

In neither case is the equipment bonded to ground.

In many cases, it may be necessary to replace existing nonbonding-type receptacles due to damage or because a three-wire bonding-type attachment plug is to be used. A bonding-type receptacle may be used if the receptacle is bonded to ground by one of these methods:

- If the device box is bonded, attach a bonding conductor from the device to the device box.
- Bond the receptacle to the system grounding conductor with a separate bonding conductor, *26–700(7)*.

Fig. 8–24 Exploded view of a bonding-type receptacle, showing all internal parts. Courtesy of Pass & Seymour, Inc.

- Connect to a metal cable sheath or raceway that is bonded to ground, *26–700(7)(a)*.

- Bond the device to a metal cold-water line, *26–700(7)(c)*.

If it is not possible to use one of these methods, the new receptacle can be protected by a GFCI breaker, GFCI receptacle, or connection to the load side of a feed through type GFCI receptacle, *26–700(8)(c)*.

INDUCTION HEATING

When AC circuits are run in metal raceways or trenches, and through openings in metal boxes, the circuiting must be arranged to prevent *induction heating* of the metal, *Rule 12–904(1)*. This means that the conductors of a given circuit must be arranged so that the magnetic flux surrounding each conductor will cancel that of the other(s), Fig. 8–25. This also means that when individual conductors of the same circuit are run in trenches, they should be kept together, not spaced far apart, Figs. 8–26, 8–27, and 8–28, and *Rule 12–012(4)*.

This situation is rarely encountered in house wiring, *but* some electricians have learned that they can cut corners and save substantial lengths of three-wire cable in hooking up three-way and four-way switches in unusual ways. They use two-wire cable and make connections as shown in Fig. 8–29. This unusual hook-up is not permitted when metallic boxes are used, violating *Rules 12–904(1)(b)* and *2–108*. Before attempting to use this type of circuitry, check with the electrical inspection authority since it is in violation of *Rule 2–108*.

Fig. 8–25   Arranging circuitry to avoid induction heating, according to *Rule 12–904(1)*, when metal raceway or armoured cables (BX) are used. Always ask yourself, "Is the same amount of current flowing in both directions in the metal raceway?" If the answer is "no," induction heating can damage the insulation on the conductors.

Fig. 8–26   Arrangement of conductors in trenches.

Fig. 8–27   Induction heating can occur when conductors of the same circuit are run through different openings of a metal box.

Fig. 8–28   To reduce the inductive effects of Fig. 8–27, a plate of aluminum or other nonmagnetic metal should be used. *Rule 4–008* and *Appendix B.*

Fig. 8–29   Violation of *Rule 12–904(1)(b)*, when metal switch boxes are used.

REVIEW

Note: Refer to the *CEC* or the plans where necessary.

1. The identified grounded circuit conductor may be which colour? Circle one.

   a. green   b. white   c. yellow   d. grey   e. green with yellow stripes

   Rule _____

   _____

   _____

2. A T-rated switch may be used to its _____ current capacity when controlling an incandescent lighting load.

3. What switch type and rating are required to control five 300-watt tungsten filament lamps on a 120-volt circuit? Show calculations.

   _____

   _____

   _____

   _____

   _____

4. To control a lighting load from one control point, what type of switch would you use?

   _____

5. Single-pole switches are always connected to the _____ wire.

6. Complete the connections in the following arrangement so that both ceiling light outlets are controlled from the single-pole switch. Assume the installation is in cable.

7. a. Complete the connections for the diagram. Installation is cable.

   b. Which conductor of the cable feeds the switch? _____

   c. Which conductor is used as the return wire? _____

   d. From which wire does the switch feed tap? _____

   e. What are the colours of the conductors connected to the fixture? _____

   _____

8. What type of switch is installed to control a lighting fixture from two control points? How many switches are needed? _____

   _____

   _____

9. Complete the connections in the following diagram so that the lamp may be controlled from either three-way switch.

10. Show the connections for a ceiling outlet to be controlled from any one of three switch locations. The 120-volt feed is at the light, *Rule 4–034(2)*. Label the conductor colours. Assume the installation is in cable.

11. Match the following switch types with the correct number of terminals for each.

   Three-way switch   Two terminals

   Single-pole switch   Four terminals

   Four-way switch   Three terminals

12. When connecting single-pole, three-way, and four-way switches, they must be wired so that all switching is done in the _____ circuit conductor.

13. If you had to install an underground three-wire feeder to a remote building using three individual conductors, which of the following installations "meets code"? Circle correct installation.

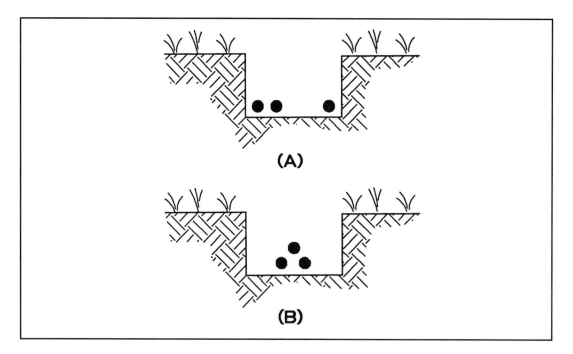

(A)

(B)

14. Is it always necessary to attach the bare equipment bonding conductor of a non-metallic-sheathed cable to the green hexagonal bonding screw on a receptacle? Explain.

_____

_____

_____

_____

_____

15. When two non-metallic-sheathed cables enter a box, is it permitted to bring both bare bonding conductors directly to the bonding terminal of a receptacle, using the terminal as a splice point? _____

Ground-Fault Circuit Interrupters, Arc Fault Circuit Interrupters, Transient Voltage Surge Suppressors, and Isolated Ground Receptacles

OBJECTIVES

After studying this unit, the student will be able to

- understand the theory of ground-fault circuit-interrupting (GFCI) devices
- explain the operation and connection of GFCIs
- explain why GFCIs are required
- discuss locations where GFCIs must be installed in homes
- discuss the *CEC* rules relating to the replacement of existing receptacles with GFCI receptacles
- understand the *CEC* requirements for GFCI protection on construction sites
- understand the logic of the exemptions to GFCI mandatory requirements for receptacles in certain locations in kitchens, garages, and basements
- discuss the *CEC* requirements for arc fault circuit interupters (AFCIs)
- understand *CEC* requirements for the installation of GFCIs and AFCIs
- understand and the discuss the basics of transient voltage surge suppressors and isolated ground receptacles

ELECTRICAL HAZARDS

The most important consideration in the installation of an electrical system is that it does not kill anyone. Many lives have been lost because of electrical shock from an appliance or a piece of equipment that is "hot." This means that the "hot" circuit conductor in the appliance is contacting the metal frame of the appliance. This condition may be due to the breakdown of the insulation because of wear and tear, defective construction, or accidental misuse of the equipment.

The shock hazard exists whenever the user can touch both the defective equipment and grounded surfaces, such as water pipes, metal sink rims, grounded metal lighting fixtures, earth, concrete in contact with the earth, water, or any other grounded surface.

Fig. 9–1 shows a time/current curve that indicates the amount of current that a normal healthy adult can stand for a given time. Just as in overcurrent protection for branch circuits, motors, appliances, and so on, it is always a matter of *how much* for *how long*.

CODE REQUIREMENTS FOR GROUND-FAULT CIRCUIT INTERRUPTERS (GFCIs)

To provide supplementary protection against shock hazard the *CEC* requires that GFCIs be provided for the receptacle outlets of dwellings in these situations:

• For *all* 125-volt single-phase 15- or 20-ampere receptacles in bathrooms and washrooms a Class A-type GFCI-protected receptacle is required. This receptacle, where

practical, shall be located at least 1 m, but in no case less than 500 mm, from the bathtub or shower stall. A bathroom/washroom is defined in *Section 0, Rule 26–700(11)* and *26–710(g)*.

• For *all* 125-volt 15- or 20-ampere receptacles installed within 1.5 m of a sink, GFCI protection is required. See *Rule 26–700, Appendices B and I,* for exceptions.

• For *all* 125-volt single-phase 15-ampere receptacles located outdoors and within 2.5 m of ground level, a Class A-type GFCI-protected receptacle is required. Accordingly, a GFCI-protected receptacle is required for a receptacle that may be used for an electric lawn mower or trimmer, *Rule 26–710(n)* and *26–714(a)(b)*.

• In boathouses, all 125-volt, single-phase, 15-ampere receptacles *must* have GFCI protection.

The *CEC* does *not* require GFCI receptacles for:

• the receptacle for laundry equipment;

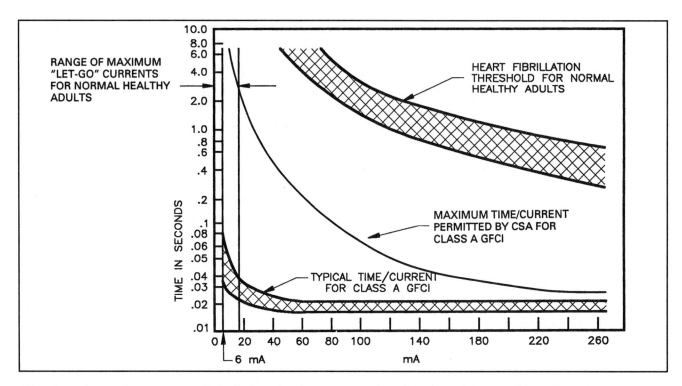

Fig. 9–1   The time/current curve shows the tripping characteristics of a typical Class A GFCI. If you follow the 6-mA line vertically to the crosshatched typical time/current curve, you find that the GFCI will open in from about 0.035 seconds to 0.08 seconds. One electrical cycle is 1/60 of a second (0.0167 seconds).

- a single-outlet receptacle that supplies a permanently installed sump pump;

- a single-outlet receptacle that is supplied by a dedicated branch circuit located and identified for specific use by a cord-and-plug-connected appliance, such as a refrigerator or freezer;

- habitable, finished rooms in basements;

- the receptacle installed in the ceiling of a garage and designed for the overhead garage door opener;

- the receptacle installed for a cord-connected appliance occupying a dedicated space, such as a freezer;

- a receptacle installed below the sink for plug-in connection of a food waste disposal;

- a receptacle installed solely for a clock;

- on a second floor balcony, such as in a condominium or apartment.

The *CEC* does not permit GFCIs to be used to protect circuits containing smoke detectors.

Section 68 gives the electrical requirements for swimming pools and the use of GFCI devices. (See Unit 23.)

The *CEC* requirements for ground-fault circuit protection can be met in several ways. The two most common types of GFCI protection are in the form of receptacles and circuit breakers, Fig. 9–2. The circuit breakers are available in either single-pole or double-pole configurations for 120-volt, 240-volt, or 120/240-volt applications.

Fig. 9–3 illustrates a Class A GFCI circuit breaker installed on a circuit. A fault or current of six milliamperes (6 mA) or more will shut off the entire circuit. For example, a ground fault at any point on circuit B10 will shut off the washroom and

GROUND-FAULT CIRCUIT INTERRUPTER AS AN INTEGRAL PART OF A DUPLEX GROUNDING-TYPE CONVENIENCE RECEPTACLE.

A SINGLE-POLE GROUND-FAULT CIRCUIT-INTERRUPTER CIRCUIT BREAKER. DOUBLE-POLE GFCI CIRCUIT BREAKERS ARE ALSO AVAILABLE.

Fig. 9–2   (A) A ground-fault circuit-interrupter receptacle, courtesy of Pass & Seymour Inc., and (B) a ground-fault circuit-interrupter circuit breaker. The switching mechanism of a GFCI receptacle opens both the ungrounded "hot" conductor and the grounded conductor. The switching mechanism of a GFCI circuit breaker opens the ungrounded "hot" conductor only. Courtesy of Schneider Electric.

bathroom receptacles together with any other devices that are connected to the outlets on this circuit.

When a GFCI receptacle is installed, only that receptacle and downstream receptacles are shut

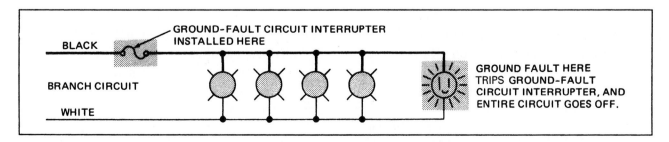

GROUND-FAULT CIRCUIT INTERRUPTER INSTALLED HERE

BLACK

BRANCH CIRCUIT

WHITE

GROUND FAULT HERE TRIPS GROUND-FAULT CIRCUIT INTERRUPTER, AND ENTIRE CIRCUIT GOES OFF.

Fig. 9–3   A GFCI as a part of the branch-circuit overcurrent device.

off when a ground fault of 6 mA or more occurs, Fig. 9–4. Upstream receptacles are not affected.

Regardless of the location of the GFCI in the circuit, it must open the circuit when the current to ground exceeds 6 mA (0.006 ampere).

A GFCI does not limit the magnitude of ground-fault current. It limits the time that a ground-fault current will flow. A GFCI does not provide protection against shock hazard should a person make contact with two of the normal circuit conductors on the load side of the GFCI (e.g., line and neutral).

Fig. 9–5 shows how a ground-fault circuit interrupter operates. Fig. 9–6 shows the internal wiring of a GFCI.

The GFCI monitors the current balance between the hot conductor and the identified conductor. As soon as the current in the identified conductor is less than the current in the hot conductor, the GFCI senses this imbalance and opens the circuit. The imbalance indicates that part of the current in the circuit is being diverted to some path other than the normal return path along the identified conductor. Thus, if the GFCI trips off, it is an indication of a possible shock hazard from a line-to-ground fault.

Nuisance tripping of GFCIs is known to occur. Sometimes this can be attributed to extremely long runs of cable for the protected circuit. Consult the manufacturer's instructions to determine if maximum lengths for a protected circuit are recommended. Some electricians advise using non-metallic staples or non-metallic straps instead of metallic to prevent nuisance tripping.

PRECAUTIONS FOR GROUND-FAULT CIRCUIT INTERRUPTERS

Unit 4 covers *interrupting ratings* for overcurrent devices. For life safety reasons, electricians and electrical inspectors should always concern themselves with the interrupting ratings of fuses and circuit breakers, particularly at the main service entrance equipment of a residence, condo, or apartment building. A GFCI receptacle *also* has the ability to interrupt current when a fault occurs. According to the CSA standard for GFCI receptacles, the current-withstand rating is 5000 amperes, RMS symmetrical.

So, unless the overcurrent protective device protecting the circuit that feeds the GFCI receptacle can limit the line-to-ground fault current to a max-

imum of 5000 amperes RMS symmetrical, installation of any GFCI receptacles close to the main panel should be done only after a short-circuit study has been made to ensure that the available line-to-ground fault current is 5000 amperes or less.

Without this precaution, the GFCI sensing and tripping mechanisms could be rendered inoperable should a line-to-ground fault occur. When the GFCI is called upon to prevent injury or electrocution of a person, it might not operate safely. This is a very important reason for periodically testing all GFCI devices. Instructions furnished with GFCIs emphasize "operate upon installation and at least monthly and record the date and results of the test on the form provided by the manufacturer." *Never* test a GFCI receptacle by shorting out line-to-neutral. The mechanism will be damaged and become inoperable.

On single-phase systems, the line-to-ground fault current can exceed the line-to-line fault current. See Unit 4.

Check fuse and circuit-breaker manufacturers' current-limiting charts to determine their current-limiting ability.

A GFCI does *not* protect against shock when a person touches both circuit conductors at the same time (two hot wires, or one hot wire and the identified wire).

Do not reverse the line and load connections on a feedthrough GFCI. This would result in the receptacle still being live even though the GFCI mechanism has tripped.

A GFCI receptacle *does not* provide overload protection for the circuit conductor. It provides *ground-fault protection only*.

GROUND-FAULT CIRCUIT INTERRUPTER IN RESIDENCE CIRCUITS

In this residence, the receptacle outlets installed outdoors, in the bathrooms, on the kitchen counter near the sink, and specific ones in the basement are protected by GFCIs, Fig. 9–7. These receptacle outlets are connected to the circuits listed in the panel schedule in Unit 4 and identified on the floor plans.

Swimming pools also have special requirements for GFCI protection. These requirements are covered in Unit 23.

Never ground a system neutral conductor except at the service equipment; otherwise, the GFCI will be inoperative, *Rule 10–204(1)*.

Fig. 9–4 A GFCI as an integral part of a receptacle outlet.

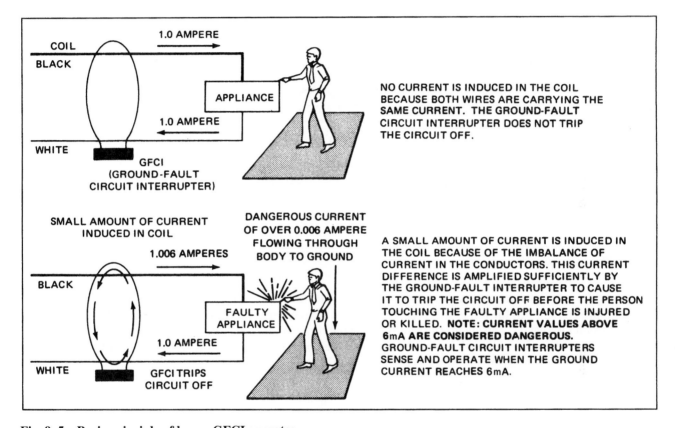

Fig. 9–5 Basic principle of how a GFCI operates.

Never connect the neutral of one circuit to the neutral of another circuit.

When a GFCI feeds an isolation transformer (separate primary and secondary windings), as might be used for swimming pool underwater fixtures, the GFCI *will not* detect any ground faults on the secondary of the transformer.

Long branch-circuit runs can cause nuisance tripping of the GFCI due to leakage currents in the circuit wiring. GFCI receptacles at the point of use tend to minimize this problem. A circuit supplied by a GFCI circuit breaker in the main panel could cause nuisance tripping if the branch circuit is long: Some electricians say 15 m or more can cause nuisance tripping of the GFCI breaker.

In older houses that were wired with knob and tube wiring or early forms of non-metallic-sheathed cable (which did not include a bonding conductor), non-bonding receptacles were used. *Rule 26–700(7)* allows a bonding receptacle to be used to replace older non-bonding receptacles, provided the receptacle bonding terminal is connected to ground by one of these means:

- bonding to the nearest metallic cold water pipe (Fig. 9–8)

- running a separate bonding conductor to the system ground

- connecting to a metallic raceway

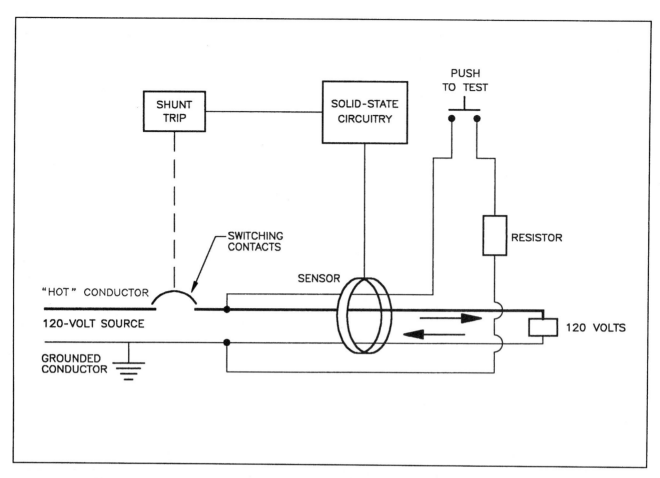

Fig. 9–6 **GFCI internal components and connections. Receptacle-type GFCIs switch both the phase ("hot") and grounded conductors. When the test button is pushed, the test current passes through the sensor and the test button, then back around (bypasses, outside of) the sensor to the opposite circuit conductor. This is how the imbalance is created and then monitored by the solid-state circuitry to signal the GFCI's contacts to open. Since both "load" currents pass through the sensor, no imbalance exists under normal receptacle use.**

Fig. 9–7 **Ground-fault circuit interrupters (GFCIs). Courtesy of Pass & Seymour/Legrand.**

Rule 26–700(8) also permits a bonding receptacle to replace an existing non-bonding receptacle if each replacement receptacle is protected by a GFCI and no bonding conductors are extended from the receptacle to any other outlet, *Rule 26–700(9).*

FEEDTHROUGH GROUND-FAULT CIRCUIT INTERRUPTER

The decision to use more GFCIs rather than trying to protect many receptacles through one GFCI becomes one of economy and practicality. GFCI receptacles are more expensive than regular receptacles. You must make the decision for each installation, keeping in mind that GFCI protection is a safety issue recognized and clearly stated in the *CEC*. However, the actual circuit layout is left up to the electrician.

Fig. 9–9 illustrates how a feedthrough GFCI receptacle supplies many other receptacles. Should a ground fault occur anywhere on this circuit, all 11 receptacles lose power—not a good circuit layout. Attempting to locate the ground-fault problem, unless obvious, can be very time-consuming. Using more GFCI receptacles is generally the more practical approach.

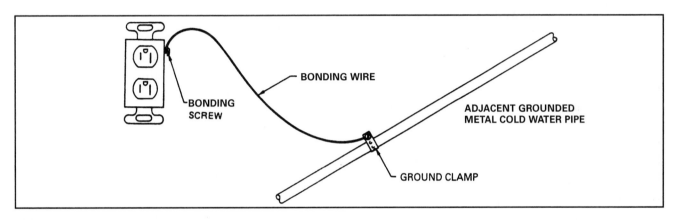

Fig. 9–8 In existing installations that do not have bonding wire as part of the branch-circuit wiring, the *CEC* permits replacing an old-style non-bonding-type receptacle with a bonding-type receptacle, *Rule 26–700(7).* The bonding terminal on the receptacle must be properly bonded. One acceptable way to bond the bonding terminal is to run a conductor to an effectively grounded water pipe. See *Rule 26–700(7)* for other acceptable means of properly bonding the receptacle's bonding terminal.

Fig. 9–9 One GFCI feedthrough receptacle protecting ten other receptacles. Although this might be an economical way to protect many receptacle outlets with only one GFCI feedthrough receptacle, the GFCI will probably nuisance-trip because of leakage currents. Use common sense and the manufacturer's recommendations as to how many devices should be protected by one GFCI.

Fig. 9–10 shows how the feedthrough GFCI receptacle is connected into the circuit. When connected in this manner, the feedthrough GFCI receptacle and all downstream outlets on the same circuit will have ground-fault protection, Fig. 9–11.

The most important factors to consider are the continuity of electrical power and the economy of the installation.

CAUTION: Do not connect the freezer receptacle in the workshop or the refrigerator receptacle in the kitchen to a GFCI-protected circuit. These devices are not required to be connected to a GFCI-protected circuit, since any nuisance-tripping of the circuit could result in spoiled food. The leakage current tolerances allowed by the CSA approach the minimum trip settings of Class A GFCIs (4 to 6 mA). Nuisance-tripping can shut off the power to the refrigerator or freezer. Coming home to spoiled food is not a pleasant experience.

A receptacle that is installed specifically for the washing machine in a combined laundry/bathroom is not required to be GFCI-protected if this receptacle is located behind the machine and is not more than 600 mm above the finished floor, *Rule 26–700, Appendices B and I.*

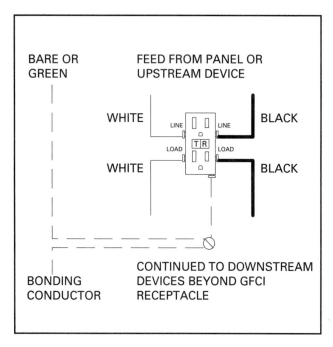

Fig. 9–10   Connecting a feedthrough GFCI into circuit.

IDENTIFICATION, TESTING, AND RECORDING OF GFCI RECEPTACLES

Rule 68–068(5) requires warning signs to be placed beside switches that control circuits protected by GFCIs. The signs will alert the user to the fact the circuits controlled by the switches are GFCI-protected and will specify how often the GFCIs should be tested, Fig. 9–12.

Fig. 9–2 shows the GFCI receptacle "T" (Test) and "R" (Reset) buttons. Pushing the test button places a small ground fault on the circuit. If operating properly, the GFCI receptacle should trip to the OFF position. The operation is the same for the GFCI circuit breakers. Pushing the reset button will restore power.

Detailed information and testing instructions are furnished with GFCI receptacles and GFCI circuit breakers. Monthly testing is recommended to ensure that the GFCI mechanism will operate properly if a human being is subjected to an electrical shock.

- Will the homeowner remember to do the monthly testing?
- Will the homeowner recognize and understand the purpose and function of the GFCI receptacle?
- Should all of the bathroom GFCI receptacles be identified?
- Will the homeowner recognize a receptacle if protected by a GFCI circuit breaker located in an electrical panel far from the actual receptacle?

These are important questions and a possible reason why the *CEC* requires identification, and why the CSA requires detailed installation and testing instructions to be included in the packaging for the GFCI receptacle or circuit breaker. Instructions are not only for the electrician but also for the homeowner, and must be left in a conspicuous place so that homeowners can familiarize themselves with the receptacle, its operation, and its need for testing. Fig. 9–13 shows a chart the homeowner can use to record monthly GFCI testing.

Fig. 9–11   GFCI receptacle protecting downstream receptacles.

THIS RECEPTACLE PROVIDES
PROTECTION AGAINST
ELECTRICAL SHOCK

THIS RECEPTACLE IS
GROUND-FAULT PROTECTED

GROUND-FAULT PROTECTED
Instructions at power panel

Fig. 9–12   GFCI warning signs.

REPLACING EXISTING RECEPTACLES

The *CEC* is very specific on the type of receptacle permitted as a replacement for an existing receptacle, *Rule 26–700*.

Replacing Existing Receptacles Where Bonding Means Exist (Refer to Fig. 9–14)

When replacing an existing receptacle (A) where the wall box is properly bonded (E) or where the branch-circuit wiring contains an equipment bonding conductor (D), at least two easy choices are possible for the replacement receptacle:

1. The replacement receptacle must be of the bonding type (B) unless . . .

2. The replacement receptacle is of the GFCI type (C).

In either case, be sure that only the bonding conductor of the circuit is connected to the receptacle's green hexagon bonding terminal.

Do not connect the white grounded circuit conductor to the green hexagon bonding terminal of the receptacle.

Do not connect the white grounded conductor to the wall box.

OCCUPANT'S TEST RECORD

TO TEST, depress the "TEST" button; the "RESET" button should extend. Should the "RESET" button not extend, the GFCI will not protect against electrical shock. Call qualified electrician.
TO RESET, depress the "RESET" button firmly into GFCI unit until an audible click is heard. If reset properly, the RESET button will be flush with the surface of the test button.
This label should be retained and placed in a conspicuous location to remind the occupants that for maximum protection against electrical shock, each GFCI should be tested monthly.

YEAR	JAN	FEB	MAR	APR	MAY	JUN	JUL	AUG	SEP	OCT	NOV	DEC

Fig. 9–13 Homeowner's testing chart for recording GFCI testing dates, as recommended in *Rule 26–440, Appendix B.*

Fig. 9–14 Where box (E) is properly bonded, an existing receptacle of the type shown in (A) *must* be replaced with a bonding-type receptacle (B). It is also acceptable to replace (A) with a GFCI receptacle (C). See text for further discussion.

The white conductor is the identified conductor and may be connected only to the identified terminal on the receptacle.

Replacing Existing Receptacles Where Bonding Means Do Not Exist (Refer to Figs. 9–15 and 9–16)

When replacing an existing non-bonding-type receptacle (A), where the box is not bonded (E), and an equipment bonding conductor has not been run with the circuit conductors (D), four choices are possible for selecting the replacement receptacle:

1. The replacement receptacle may be a non-bonding-type (A).

2. The replacement receptacle may be a GFCI type (C).

• The green hexagon terminal of the GFCI replacement receptacle does not have to be bonded to ground. It must be left "unconnected." The GFCI's trip mechanism will operate properly when ground faults occur anywhere on the load side of the GFCI replacement receptacle. Ground-fault protection is still there, Fig. 9–6.

Fig. 9–15 Where box (E) is *not* bonded and a bonding conductor has *not* been run with the circuit conductors (D), an existing non-bonding-type receptacle (A) may be replaced with a non-bonding-type receptacle (A), a GFCI-type receptacle (C), a bonding-type receptacle (B) if supplied through a GFCI receptacle (C), or a grounding-type receptacle (B) if a separate bonding conductor is run from the receptacle to a cold-water pipe or other effective grounding means, *Rule 26–700(7, 8)*.

Fig. 9–16 *Do not* connect a bonding conductor between the above receptacles. The text explains why this is not permitted, *Rule 26–700(9)*.

- Do not connect a bonding conductor from the green hexagon bonding terminal of a replacement GFCI receptacle (C) to any other downstream receptacles that are fed through the replacement GFCI receptacle, *Rule 26–700(9).*

   The reason this is not permitted is that if someone later saw the conductor connected to the green hexagon bonding terminal of the downstream receptacle, they would immediately assume that the other end of that conductor had been properly connected to an acceptable grounding point of the electrical system. The fact is that the "so-called" bonding conductor had been connected to the replacement GFCI receptacle's green hexagon bonding terminal that was not bonded in the first place. We now have a false sense of security—a real shock hazard.

3. The replacement receptacle may be a bonding type (B) if it is protected by a class A GFCI (C), *Rule 26–700(8).*

   - The green hexagon equipment bonding terminal of the replacement bonding-type receptacle (B) need not be connected to any grounding means. It may be left unconnected. The upstream feedthrough GFCI receptacle (C) trip mechanism will work properly when ground faults occur anywhere on its load side. Ground-fault protection is still there, Fig. 9–6.

4. The replacement receptacle may be a bonding type (B) if:

   - A bonding conductor sized per *Table 16* is run between the replacement receptacle's green hexagon terminal and a grounded water pipe or other effective bonding means, *Rule 26–700(8)* and *10–808.* The acceptable means of connecting the equipment bonding conductor to the water pipe is described in *Rule 10–906.*

GROUND-FAULT PROTECTION FOR CONSTRUCTION SITES

Safety on construction sites is of paramount importance to workers and employers. Safety regulations such as *The Occupational Health and Safety Act* governs how certain aspects of a building's construction will be completed. Temporary lighting and power are both referenced in the Act. Along with the *Canadian Electrical Code,* this determines the minimum requirements for the temporary wiring on a construction site.

Because of the nature of construction sites, the continual presence of shock hazard can lead to serious personal injury or death through electrocution. Workers are standing in water, standing on damp or wet ground, or in contact with steel framing members. Electric cords and cables are lying on the ground, subject to severe mechanical abuse. All of these conditions spell "danger."

GFCI receptacles provide protection for workers in the event of a ground fault. It is important that workers not attempt to bypass this protection. If the GFCI trips, it is an indication that there is a ground fault. The GFCI has just protected you from a ground fault. If you bypass the GFCI you no longer have the protection offered by the GFCI. You may not get the chance to make the same mistake twice. Fig. 9–17 illustrates some portable GFCI devices which provide ground-fault protection when plugged into non-GFCI receptacles.

The permanent site office and any temporary lighting and/or power panels must fully comply with the *CEC.* They must be inspected by the electrical inspection authority before connection can take place, *Rule 76–002.*

ARC-FAULT CIRCUIT INTERRUPTERS (AFCIs)

About 10 percent of fires are electrical fires. One-third of electrical fires are caused by arcing at poor connections or between conductors due to a breakdown in electrical insulation. For example, if the strands in a conductor of the cord for an electric iron break, a few strands may still be touching, and there may be enough of a path to permit current to flow. When the iron draws current, arcing will occur at the point of the conductor break. The current drawn by the iron will not trip the circuit breaker or fuse protecting the circuit but can develop intense heat at the break in the wire. Arcing faults can result from such things as poor terminations during

Fig. 9–17   Plug- and cord-connected portable GFCI devices are easy to carry around and use anywhere when working with electrical tools and extension cords. These devices operate independently. These inexpensive portable GFCI in-line cords should always be used with portable electric tools. Courtesy of Hubbell Incorporated, Wiring Device Division.

installation, equipment failure, and the failure of conductor insulation due to circuit overloading or improperly sized overcurrent protection.

Types of Arcs

A series arc is one that occurs when a current-carrying conductor breaks or there is a poor connection between the conductor and a terminal screw. The current from a series arc follows the normal circuit path.

Parallel arcing occurs between two conductors. The arcing may occur from line to line, line to neutral, or line to ground. Parallel arcing current does not follow the normal circuit path.

How AFCIs Work

An AFCI is designed to sense the fluctuations in voltage and current that occur during an arcing sequence and de-energize the circuit if dangerous arcing occurs. AFCIs are designed to sense both series and parallel arcing and can distinguish between the arcing that occurs in the normal operation of an electrical system and the unwanted arcing that can lead to a fire. The arc that occurs when the filament of an incandescent lamps breaks, when a switch is turned on and off, or when appli-

ances that are turned on are plugged into a receptacle are examples of things that produce arcs that are normally considered harmless.

Installing and Using AFCIs

AFCIs are required to protect receptacles located in bedrooms or sleeping facilities, *Rule 26–722(f)*. Installing an AFCI is similar to installing a GFCI. With a single-pole AFCI, the white pigtail on the breaker is connected to the neutral bus in the panel, and two circuit conductors black (hot), and white (identified), are connected to the respective brass and silver-coloured terminals of the breaker. See Fig. 9–19.

Some manufacturers incorporate arc-fault protection and ground-fault protection on a single breaker.

If an AFCI detects a fault, unplug and/or disconnect all appliances connected to the circuit. Reset the AFCI. If the AFCI trips with the appliances removed from the circuit, this indicates a faulty AFCI breaker or a fault in the wiring system. Disconnect the wiring from the AFCI circuit breaker and retest the breaker. If the breaker still detects a fault, the breaker is faulty or has been damaged by the arc fault. Before replacing the breaker, perform an insulation test on the wiring system.

Fig. 9–18   An arc-fault circuit interrupter (AFCI) circuit breaker. Courtesy of Square D/Schneider Electric.

Fig. 9–19   Connection diagram for single-pole AFCI circuit breaker.

If the AFCI is reset and it doesn't trip, reconnect the appliances that were connected to the circuit, one at a time. If the AFCI still does not trip, it is possible that you have a loose connection or that the arc cleared the fault.

Performing an Insulation Test

An insulation test is normally performed with an insulation tester since it applies a voltage of around 500 volts to the system. If an insulation tester is not available, use your multimeter on its highest resistance scale. Follow all safety procedures. Turn off the power to the circuit. Isolate the wiring system from its source by disconnecting the wiring from

the AFCI and disconnect all loads from the portion of the wiring system undergoing the test. (It is best to remove all the loads on the circuit, *particularly electronic loads*.) Isolate the ends of the wires at both ends of the run. Connect the insulation tester (megger) to the circuit. The resistance between the conductors should be one megohm or greater.

TRANSIENT VOLTAGE SURGE SUPPRESSION (TVSS)

In today's homes we find many electronic appliances (televisions, stereos, personal computers, facsimile equipment, word processors, video cassette recorders, digital stereo equipment, microwave ovens), all of which contain many sensitive, delicate electronic components (e.g., printed circuit boards, chips, microprocessors, transistors).

Voltage transients, called *surges* or *spikes*, can stress, degrade, and/or destroy these components; they can cause loss of memory in the equipment or "lock up" the microprocessor.

The increased complexity of electronic integrated circuits makes this equipment an easy target for "dirty" power that can and will affect the performance of the equipment.

Voltage transients cause abnormal current to flow through the sensitive electronic components. This energy is measured in joules. A *joule* is the energy of one ampere passing through a one-ohm resistance for one second. In other words, a joule is the amount of energy used by a one-watt load in one second. Hence, one watt-hour equals 3600 joules.

Line surges can be line-to-neutral, line-to-ground, and line-to-line.

Transients

Transients are generally grouped into two categories:

- *Ring wave* transients originate within the building and are caused by welders, elevators, copiers, motors, air conditioners, or other inductive loads.

- *Impulse* transients originate outside the building and are caused by utility company switching, lightning, and so on.

To minimize the damaging results of these transient line surges, service equipment, panels, load centres, feeders, branch circuits, and individual receptacle outlets can be protected with a *transient voltage surge suppressor* (TVSS), Fig. 9–20.

A TVSS contains one or more metal-oxide varistors (MOVs) that clamp the transients by absorbing the major portion of the energy (joules) created by the surge, allowing only a small, safe amount of energy to enter the connected load.

The MOV clamps the transient in less than one nanosecond (one billionth of a second) and limits the voltage spike passed through to the connected load to a maximum range of 400 to 500 peak volts.

Typical TVSS devices for homes are available as an integral part of receptacle outlets that mount in the same wall boxes as do regular receptacles. They look the same as a normal receptacle, and may have an audible alarm that sounds when an MOV has failed. The TVSS device may also have a visual indication, such as an LED (light-emitting diode), which glows continuously until an MOV fails. Then the LED starts to flash on and off. See Fig. 9–21.

TVSS devices are also available in plug-in strips, Fig. 9–22.

One TVSS on a branch circuit will provide surge suppression for nearby receptacles on the same circuit.

Noise

Low-level non-damaging transients can also be present. These can be caused by fluorescent lamps and ballasts, electronic equipment in the area, X-ray equipment, motors running or being switched on and off, improper grounding, and so forth. Although not physically damaging to the electronic equipment, they can cause computers to lose memory or perform wrong calculations. Their intended programming can malfunction.

"Noise" comes from *electromagnetic interference* (EMI) and *radio frequency interference* (RFI). EMI is usually caused by ground currents of very low values from motors, utility switching loads, or lightning, and is transmitted through metal conduits. RFI causes the buzzing heard on a car radio when you are driving under a high-voltage transmission line. This interference radiates through the air from the source and is picked up by the grounding system of the building.

You can reduce this undesirable noise by reducing the number of ground reference points on a system, by installing an isolated bond receptacle.

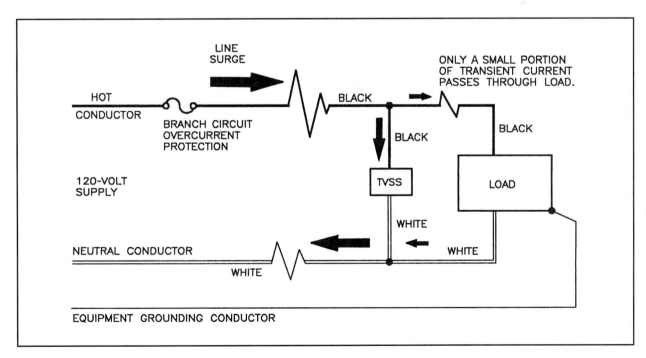

Fig. 9–20   A transient voltage surge suppressor absorbs and bypasses (shunts) transient currents around the load. The MOV dissipates the surge in the form of heat.

Fig. 9–21   Surge suppressor. Courtesy of Leviton.

Fig. 9–22   Surge suppressors. Courtesy of Leviton.

ISOLATED GROUND RECEPTACLE

In a standard receptacle outlet, Fig. 9–23, the green bonding hexagon screw, the bonding contacts, the yoke (strap), and the metal wall box are all connected together and bonded to the building's equipment grounding system. If you picture all the bonding conductors in the building, you can see that many outlets could be connected to ground by multiple paths.

In an isolated ground receptacle, the green bonding hexagon screw and the bonding contacts of the receptacle are isolated from the metal (strap) of the receptacle and also from the building's bonding system. A separate green insulated bonding conductor is then installed from the green hexagon screw on the receptacle all the way back to the nearest distribution panel where it will be connected to the bond bus in the panel. See Fig. 9–23. *Rule 10–906(8).*

Fig. 9–23   Standard conventional receptacle and isolated ground receptacle. Courtesy of Hubbell Corporation.

REVIEW

Note: Refer to the *CEC* or the plans where necessary.

1. Explain the operation of a GFCI. Why are GFCI devices used? Where are GFCI receptacles required? _____

   _____

   _____

   _____

   _____

   _____

   _____

   _____

   _____

   _____

   _____

   _____

   _____

   _____

   _____

   _____

   _____

   _____

   _____

2. Residential GFCI devices are set to trip a ground-fault current of _____ .

3. What are the residential applications of GFCI receptacles? _____

   _____

   _____

   _____

   _____

   _____

   _____

4. The *CEC* requires GFCI protection for certain receptacles in the bathroom. Explain where these are required. _____

   _____

   _____

   _____

5. Is it a *CEC* requirement to install GFCI receptacles in a fully carpeted, finished recreation room in the basement? (Yes) (No) Circle one.

6. A homeowner calls in an electrical contractor to install a separate circuit in an unfinished basement for a freezer. Is a GFCI receptacle required? _____

    _____

7. GFCI protection is available as (a) a branch circuit breaker GFCI, (b) a feeder circuit breaker GFCI, (c) an individual GFCI receptacle, (d) a feedthrough GFCI receptacle. What type would you install for residential use? Explain your choice.

    _____

8. Extremely long circuit runs connected to a GFCI branch circuit breaker might result in (circle the letter of the correct answer)

    a. nuisance-tripping of the GFCI

    b. loss of protection

    c. the need to reduce the load on the circuit

9. If a person comes in contact with the hot and grounded circuit conductors of a two-wire branch circuit that is protected by a GFCI, will the GFCI trip? Why?

    _____

    _____

    _____

10. What might happen if the line and load connections of a GFCI receptacle are reversed?

    _____

    _____

    _____

    _____

11. May a GFCI receptacle be installed on a replacement in an old installation where the two-wire circuit has no bonding conductor? (Yes) (No) Circle one.

12. What two types of receptacles may be used to replace a defective receptacle in an older home with knob and tube wiring where a bonding means does not exist in the box?

    _____

    _____

    _____

13. You are asked to replace a receptacle. Upon checking the wiring, you find that the wiring method is conduit and that the wall box is properly bonded. The receptacle is of the older style two-wire type that does not have a bonding terminal. You remove the old receptacle and replace it with (circle the letter of the correct answer)

    a. the same type of receptacle as the type being removed

    b. a receptacle of the bonding type

    c. a GFCI receptacle

14. What colour are the terminals of a standard bonding-type receptacle? _____

_____

_____

_____

15. What special shape are the bonding terminals of receptacle outlets and other devices?

_____

_____

_____

16. Why are GFCI receptacles installed on construction sites? _____

_____

_____

_____

_____

17. In your own words, explain why the *CEC* does not require receptacle outlets in garages
    and basements to be GFCI-protected. _____

_____

_____

_____

_____

_____

_____

_____

_____

_____

18. What do the letters TVSS stand for? _____

_____

_____

19. Transients (surges) on a line can cause spikes or surges of energy that can damage del-
    icate electronic components. A TVSS device contains one or more _____
    _____ that bypass and absorb the energy of the transient.

20. Undesirable noise on a circuit can cause computers to lock up or lose their memory,
    and/or can cause erratic performance of the computer. This noise does not damage the
    equipment. The two types of this noise are EMI and RFI. What do these letters mean?

_____

_____

_____

21. Can TVSS receptacles be installed in standard device boxes? _____

_____

22. When an isolated ground receptacle is installed, the *CEC* permits the separate equipment bonding conductor to be carried back through the raceway or cable to the distribution panel. What section of the *CEC* references this topic? _____

23. Briefly explain the difference between a GFCI breaker and an AFCI breaker.

_____

_____

_____

UNIT 10

Lighting Fixtures and Ballasts

OBJECTIVES

After studying this unit, the student will be able to

- understand fixture terminology, such as Type IC, Type NON IC, suspended ceiling fixtures, recessed fixtures, and surface-mounted fixtures
- connect recessed fixtures, both prewired and non-prewired types, according to *CEC* requirements
- specify insulation clearance requirements
- discuss the importance of temperature effects while planning recessed fixture installations
- describe thermal protection for recessed fixtures
- understand Class P ballasts
- discuss the *CEC* requirements for track lighting

TYPES OF LIGHTING FIXTURES

This unit begins an in-depth coverage of recessed fixtures commonly installed in homes. It also considers other types of available lighting fixtures.

CEC Section 30 sets forth the requirements for the installation of lighting fixtures. Although not involved in the actual manufacture of lighting fixtures, the electrician must "meet code" when installing fixtures, including mounting, supporting, grounding, live-parts exposure, insulation clearances, supply conductor types, and maximum lamp wattages.

The homeowner, interior designer, architect, or electrician has thousands of fixtures from which to choose. Specific needs, space requirements, and price considerations are among the factors to consider when selecting fixtures.

It is essential for the electrician to know early in the roughing-in stage of wiring a house what

types of fixtures are to be installed, particularly for recessed-type fixtures. The electrician must work closely with the general building contractor, carpenter, plumber, heating contractor, and the other building tradespeople to ensure that requirements for location, proper and adequate support, clearances from piping, ducts, combustible material, and insulation are complied with during construction.

The CSA provides the safety standards to which a fixture manufacturer must conform. Common fixture types are:

FLUORESCENT	INCANDESCENT
• surface	• surface
• recessed	• recessed
• suspended ceiling	• suspended ceiling

IMPORTANT: Always carefully read the label on the fixture. The information will help you to conform to *Section 30* of the *CEC* and complete a safe installation. The label provides such information as:

- for wall mount only
- ceiling mount only
- maximum lamp wattage
- type of lamp
- access above ceiling required
- suitable for air handling use
- for chain or hook suspension only
- suitable for operation in ambient temperatures not exceeding _____ °C
- suitable for installation in poured concrete
- for installation in poured concrete
- for line volt-amperes, multiply lamp wattage by 1.25
- suitable for use in suspended ceilings
- suitable for use in noninsulated ceilings
- suitable for use in insulated ceilings
- suitable for damp locations (such as bathrooms and under eaves)
- suitable for wet locations
- suitable for use as a raceway
- suitable for mounting on low-density cellulose fibreboard
- for supply connections, use wire rated for at least _____ °C
- not for use in dwellings
- thermally protected fixture
- type IC
- type NON IC
- inherently protected

The *Canadian Electrical Code, Part II* and the fixture manufacturers' catalogues and literature are excellent sources of information. The CSA tests, lists, and labels fixtures for conformance to its standards and to the *Canadian Electrical Code, Parts I and II.*

Recessed fixtures (Figs. 10–1A and 10–1B) in particular have an inherent heat problem. Therefore,

they must be suitable for the intended application and must be properly installed.

To protect against overheating, the CSA requires that some types of recessed fixtures be equipped with an integral thermal protector, Fig. 10–2. This device shuts the fixture off if it overheats, then resets once the fixture cools down.

Because of the potential fire hazard when mounting lighting fixtures near combustible materials, *Rule 30–200* states that fixtures must be equipped with shades or guards to limit the temperature to which such materials may be exposed to a maximum of 90°C. If lights are installed over combustible material, each fixture must be controlled by an individual wall switch if it is located less than 2.5 m above the floor or if the fixture is not guarded so that the lamps cannot be readily removed or damaged, *Rule 30–200(3).*

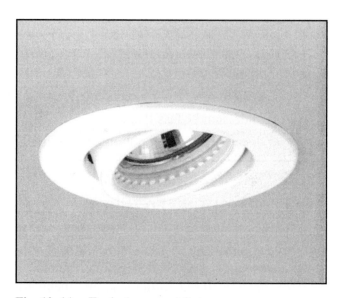

Fig. 10–1A   Typical recessed fixture.

Fig. 10–1B   Recessed fixture mounted in ceiling.

Fig. 10–2   Recessed fixture thermal cutout.

LOCATION	TYPE
Closets	Incandescent
Recreation room	Fluorescent
Wet bar	Incandescent
Above kitchen sink	Incandescent
Bathrooms (heat/vent/light)	Incandescent
Hall leading to kitchen	Incandescent

Fig. 10–3   When mounted on low-density ceiling fibre-board, surface-mounted fluorescent fixtures must be installed so as to limit the temperature of the fibreboard to less than 90°C. See *Rule 30–200(1)*.

Fig. 10–3 shows a fluorescent fixture mounted on low-density cellulose fibreboard—sheets, panels, and tiles made of bonded plant fibre material that have a density of 20 pounds per cubic foot or less. This excludes solid or laminated wood, fibreboard with a density of over 20 pounds per cubic foot and material that has been integrally treated with fire-retarding chemicals to meet specific standards.

All branch-circuit conductors used for wiring a ceiling outlet box must have insulation approved to 90°C (e.g., NMD90), but their ampacity is derated to that of 60°C wire, *Table 2, Column 2, Rule 30–408(2)*. This applies to most residential applications.

CEC REQUIREMENTS FOR INSTALLING RECESSED FIXTURES

CEC Rules 30–308, 30–408, and *30–900 to 30–910* list requirements for the installation and construction of recessed fixtures. Of particular importance are the restrictions on conductor temperature ratings, fixture clearances from combustible materials, and maximum lamp wattages. **Recessed fixtures generate a great deal of heat within the enclosure, making them a fire hazard if they are not wired and installed properly.** Fig. 10–4 shows the roughing-in box of a recessed fixture with mounting brackets and junction box.

The branch-circuit conductors are run to the junction box for the recessed fixture. Here they are connected to conductors whose insulation can handle the temperature at the fixture lampholder. The junction box is placed at least 300 mm from the fixture, so heat radiated from the fixture cannot overheat the wires in the junction box. Conductors rated for higher temperatures must run in a metal raceway between the fixture and the junction box. This raceway must be 450 mm to 2 m long. Any heat conducted from the fixture along the metal raceway will be reduced before reaching the junction box. Many recessed fixtures are factory-equipped with a flexible metal raceway containing high-temperature (150°C) wires that meet the requirements of *Rule 30–910(1)*. Fig. 10–5 illustrates clearance requirements for installing recessed lighting fixtures.

Suspended Ceiling Fixtures

The recreation room of this residence has a dropped 13-mm gypsum board ceiling. The fixtures installed in this ceiling are a "lay in" fixture, suitable for recess installation. In most residential applications, these fixtures would still be classified as recessed and the *CEC* requirement of 13 mm clearance from structural members must be maintained, *Rule 30–902*.

Fig. 10–4   Roughing-in box of a recessed fixture with mounting brackets and junction box.

ALL INCANDESCENT RECESSED FIXTURES MUST BE MARKED IC AND
APPROVED WHERE THEY ARE BLANKETED BY THERMAL INSULATION.
RULE 30–906.

BRANCH CIRCUIT CONDUCTORS
WITH INSULATION SUITABLE FOR
TEMPERATURE ENCOUNTERED
MAY BE RUN DIRECTLY INTO
APPROVED JUNCTION BOX ON
PREWIRED FIXTURES,
RULE 30–910(2).

JUNCTION BOX
AT LEAST 30 cm
FROM FIXTURE,
RULE 30–910(4).

AT LEAST 450 mm, BUT NOT MORE THAN
2 m, OF SUITABLE METAL RACEWAY
WITH CONDUCTORS HAVING INSULATION
SUITABLE FOR TEMPERATURE ENCOUNTERED,
RULE 30–910(3).

IF 12-mm FLEX, MAY NOT
BE MORE THAN 2 m.
RULE 30–910(3).

PREWIRED
FIXTURE
WITH
APPROVED
JUNCTION
BOX

THIS BOX MUST
BE ACCESSIBLE,
RULE 30–910(4)(a).

FIELD-WIRED
FIXTURE
WITHOUT
APPROVED
JUNCTION
BOX

AT LEAST 13 mm
CLEARANCE FROM
COMBUSTIBLE MATERIAL,
EXCEPT AT POINT
OF SUPPORT,
RULE 30–902.

KEEP INSULATION AWAY
FROM FIXTURE UNLESS
FIXTURE IS SUITABLE
FOR DIRECT CONTACT
WITH INSULATION,
RULE 30–902.

Fig. 10–5   Clearance requirements for installing recessed lighting fixtures.

Fig. 10–6   Suspended ceiling fixtures.

Support of Suspended Ceiling Fixtures

When framing members of a suspended ceiling grid
are used to support lighting fixtures, the grid installer
must securely fasten all members together and to
the building structure. All suspended ceiling fixtures must be securely fastened to the ceiling grid
members by bolts, screws, or rivets. Most fixture
manufacturers supply special clips suitable for fastening the fixture to the grid. The fixture must
NEVER be supported by the ceiling tile.

Some inspection authorities interpret the intent of *Rule 30–302* to mean that the recessed fixture must be supported independent of a suspended ceiling structure. This is usually accomplished by using a chain attached to the fixture and the building structure above the ceiling. Large fixtures such as fluorescents require at least two chain supports.

The logic behind all of the "securely fastening" requirements is that in the event of a major problem (e.g., earthquake or fire), the lighting fixtures will not fall down and injure someone.

Connecting Suspended Ceiling Fixtures

The most common way to connect suspended ceiling lay-in lighting fixtures is to complete all of the wiring above the suspended ceiling using normal wiring methods, EMT, non-metallic-sheathed cable, armoured cable, or whatever type of wiring method is acceptable by the governing electrical code, with outlet boxes strategically placed above the ceiling near the intended location of the fixtures. Then make a flexible connection between the outlet box and the fixture.

The flexible connection can be up to 2 m long and is usually made by installing either of these:

- Flexible metal conduit with conductors suitable for the temperature requirement as stated on the fixture label, usually 90°C, *Rule 30–910(1)*. *Rule 12–1010(3)* specifies that when flexibility is required, flexible metal conduit does not require a clip within 300 mm of the outlet box.
- Armoured cable containing conductors suitable for the temperature requirements as stated on the fixture label, usually 90°C. Be sure to staple the armoured cable within 300 mm of the outlet box. See *Rules 12–618* and *12–510(1)* and Figs. 10–7A, 10–7B and 10–7C.

Fig. 10–7A   Clips for supporting 12-mm flex or AC-90 cable.

Fig. 10–7B   A suspended ceiling fixture supplied by not over 2 m of flexible metal conduit. *Rule 12–1010(3)* indicates that if the conduit is less than 900 mm long, it does not require clipping. Remember that 12-mm flexible conduit cannot be more than 2 metres long, *Rule 30-910(3)*.

Fig. 10–7C A suspended ceiling fixture supplied by armoured cable not over 2 m long. Cable must be secured (stapled) within 300 mm of the outlet box, *Rule 12–510*.

In the electrical trade, these flexible connections between an outlet box and a fixture are called *fixture whips* or *fixture tails*.

It is necessary for the electrician installing recessed fixtures to work with the installer of the insulation to be sure that the clearances around the fixture are maintained. If the ceiling has insulation blanketing the fixture, it must be a type IC fixture, *Rule 30–906*, Fig. 10–8.

Recessed fixtures of the type installed in ceilings usually have a factory-mounted junction box on the side of the fixture. See Fig. 10–5. The branch-circuit conductors can be run directly into this box, where they are then connected to the conductors entering the fixture, *Rule 30–910*.

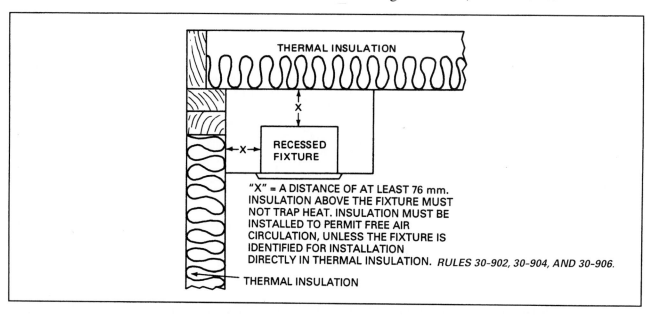

Fig. 10–8 Clearances for recessed lighting fixtures installed near thermal insulation. Fixtures must be CSA approved for use in thermal insulation, *Rule 30–906*.

When the only access to the junction box is through the fixture opening, the opening must be at least 180 cm² with no dimension less than 150 mm unless the entire fixture and junction box can be removed from the ceiling, *Rule 30–910(5)*. Many small, low-voltage spotlights installed in today's homes do not meet this requirement. Therefore, an alternative access method would be required, such as through an attic, or a removable fixture must be installed.

Prewired fixtures do not require additional wiring, Fig. 10–9. *Rule 30–910(8)* states that branch-circuit wiring shall not be passed through an outlet box that is an integral part of an incandescent fixture unless the fixture is approved and marked for through wiring.

For a recessed fixture that is not prewired, the electrician must check the fixture for a label indicating what cable insulation temperature rating is required.

The cables to be installed in the residence have 90°C insulation. If the temperature will exceed this value, conductors with other types of insulation must be installed, *Table 19* and *Table D1, Note 10*.

BALLAST PROTECTION

All approved fluorescent ballasts installed indoors (except simple reactance-type ballasts) for both new and replacement installations must have built-in thermal protection from the manufacturer, Fig. 10–10. Ballasts provided with built-in thermal protection are listed by the CSA as Class P ballasts.

Under normal conditions, the Class P ballast has a case temperature not exceeding 90°C. The thermal protector must open within two hours when the case temperature reaches 110°C.

Some Class P ballasts also have a nonresetting fuse integral with the capacitor to protect against capacitor leakage and violent rupture. The Class P ballast's internal thermal protector will disconnect the ballast from the circuit in the event of a high temperature. Excessive temperatures can be caused by abnormal voltage and improper installation, such as being covered with insulation.

The reason for thermal protection is to reduce the fire hazard from an overheated ballast. Ballasts can overheat if they are shorted, grounded, covered with insulation, or lack air circulation.

LIGHTING FIXTURE VOLTAGE LIMITATIONS

The maximum voltage allowed for residential lighting fixtures is 150 volts between a conductor and ground, *Rules 30–102* and *2–106*.

Rule 30–706 makes a further restriction in stating that for dwelling occupancies, no electrical discharge lighting equipment shall be used if it operates with an open-circuit voltage over 300 volts, Fig. 10–11. Fluorescent and neon lights are examples of discharge lighting. *Rule 30–706* addresses the increasing use of neon lighting for decorative purposes. Some important CSA and *CEC* requirements for installing recessed fixtures are summarized in Fig. 10–12 on page 193.

Fig. 10–9 Installation permissible only with prewired recessed fixtures with approved junction box, *Rule 30–910(8)*.

Fig. 10–10   Fluorescent ballasts installed on new or replacement installations are required to have built-in thermal protection. These ballasts are called Class P ballasts, *CEC Part II.*

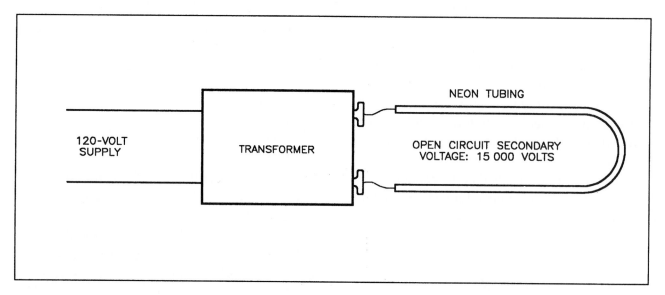

Fig. 10–11   It is NOT permitted to use neon lighting fixtures in residences where the open circuit voltage is over 300 volts, unless there is no access to live parts when changing the lamps, *Rule 30–706.*

Fig. 10–12 Some of the most important CSA and *CEC* requirements for recessed fixtures. Always refer to the CSA standards, the *CEC,* and the manufacturer's label and/or instructions furnished with the fixture.

REVIEW

Note: Refer to the *CEC* or the plans where necessary.

1. Is it permissible to install a recessed fixture directly against wood ceiling joists? Explain why or why not. _____
   _____

2. If a recessed fixture without an approved junction box is installed, what extra wiring must be provided? _____
   _____

3. Recessed fixtures are available for installation in direct contact with thermal insulation. These fixtures bear the CSA mark "Type _____."

4. Unless specially designed, all recessed incandescent fixtures must be provided with factory-installed _____.

5. Plans require the installation of a surface-mounted fluorescent fixture on the ceiling of a recreation room that is finished with low-density ceiling fibreboard. What sort of mark would you look for on the label of the fixture? _____
   _____

6. If a recessed fixture bears no marking that it is listed for branch-circuit feedthrough wiring, is it permitted to run the circuit conductors from fixture to fixture? What section of the *CEC* covers this? _____
   _____

7. Fluorescent ballasts for all indoor applications must be _____ type. These ballasts contain internal _____ protection to protect against overheating.

8. You are called upon to install a number of lighting fixtures in a suspended ceiling. The ceiling will be dropped about 200 mm from the ceiling joists. Briefly explain how you might go about wiring these fixtures. _____
   _____
   _____
   _____

9. The *CEC* places a maximum open circuit voltage on lighting equipment in homes. The maximum voltage is (300) (600) (750) (1000). (Circle the correct answer.) Where in the *CEC* is this voltage maximum referenced? _____
   _____
   _____

UNIT 11

Branch Circuits for the Bedrooms, Study, and Hall

OBJECTIVES

After studying this unit, the student will be able to

- explain the factors that influence the grouping of outlets into circuits
- estimate loads for the outlets of a circuit
- draw a cable layout and a wiring diagram based on information given in the residence plans, specifications, and *CEC* requirements
- select the proper wall box for a particular installation
- explain how wall boxes are bonded to ground
- list the requirements for the installation of fixtures in clothes closets
- make connections for three-way switches

GROUPING OUTLETS

A branch circuit is all of the wiring from the final overcurrent device in the system to the point of utilization. The grouping of outlets into branch circuits must conform to *CEC* standards and good wiring practices. There are many possible combinations of groupings of outlets. In most residential installations, the electrician usually does the circuit planning. In larger, more costly residences, the architect may complete the circuit layout and include it on the plans. In either case, the electrician must ensure that the branch circuits conform with the *CEC*.

An electrician plans circuits that are economical without sacrificing the quality of the installation. See Fig. 11–1. For example, some electricians prefer to have more than one circuit feed a room. Should

one circuit have a problem, the second circuit would still supply power to the other outlets in that room, Fig. 11–2.

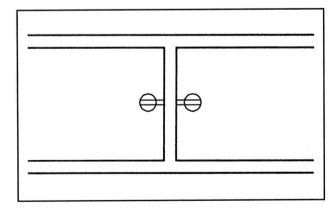

Fig. 11–1   Receptacle outlets connected back to back can reduce the cost of the installation because of the short distance between the outlets.

Some electricians consider it poor practice to include outlets on different floors on the same circuit. Here, too, the decision can be a matter of personal choice. Some local building codes limit this type of installation to lights at the head and foot of a stairway.

Residential Lighting

Residential lighting is a personal thing. The homeowner, builder, and electrical contractor must meet to decide on what types of lighting fixtures are to be installed in the residence. Many variables, including cost, personal preference, and construction obstacles, must be considered.

Residential lighting can be divided into four groups.

General Lighting: General lighting provides overall illumination for an area, such as the front entry hall, bedroom hall, and garage.

Accent Lighting: Accent lighting provides a "focus" or accent on an object or area in the home. This might be recessed spot lights over the fireplace or track lighting on the ceiling of the living room, accenting a picture, photo, painting, or sculpture.

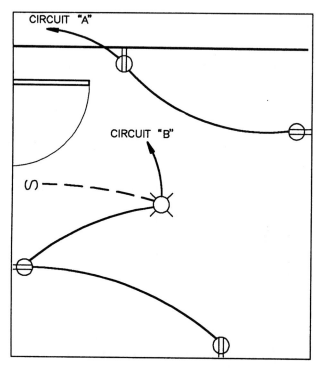

Fig. 11–2 Wiring layout showing one room that is fed by two different circuits. If one of the circuits goes out, the other circuit still provides electricity to the room.

Task Lighting: Task lighting provides proper lighting where tasks are performed. In this residence, the fluorescent fixtures over the workbench, the lighting inside the range hood in the kitchen, and the recessed fixture over the kitchen sink are all examples.

Security Lighting: Security lighting generally includes outdoor lighting (such as post lamps, wall fixtures, walkway lighting) and all other lighting used for security and safety. Security lighting is often provided by the normal types of lighting fixtures found in the typical residence. In this residence, the outdoor bracket fixtures in front of the garage and next to the entry doors, and the post light might be considered to be security lighting, even though they add to the beauty of the residence.

A visit to the showroom of one of the many manufacturers or distributors of lighting fixtures offers the buyer an opportunity to select from hundreds—in some cases thousands—of lighting fixtures. Most of the finer lighting showrooms are staffed with individuals highly qualified to make recommendations to homeowners so they get the most value for their money.

The lighting provided throughout this residence conforms to accepted trade practice. Certainly many other variations are possible. For the purposes of studying an entire wiring installation for the residence, all load calculations, wiring diagrams, and so on use the fixture selection indicated on the plans and in the specifications.

CABLE RUNS

Studying the great number of wiring diagrams throughout this text, you will find many ways to run the cables and make up the circuit connections. Sizing of boxes that will conform to the *CEC* for the number of wires and devices is covered in great detail. When a circuit is run through a recessed fixture like the type shown in Fig. 10–1B, that recessed fixture must be approved for "through wiring," a subject discussed in Unit 10. When you are unsure whether a fixture can accomodate the extra wiring for a run to another fixture or some other part of the circuit, it is best to design the circuit so that the wiring ends at the recessed fixture with just the conductors needed to connect it.

The grouping of outlets into circuits must satisfy the requirements of the *CEC,* local building codes, good wiring practices, and common sense. A good wiring practice is to:

- divide all loads as evenly as possible among the circuits
- ensure that no more than 12 outlets are placed on a two-wire circuit
- keep cable runs as short as possible
- use boxes that have adequate space for conductors and devices.

See *Rules 8–104* and *8–304.*

ESTIMATING LOADS FOR OUTLETS

When calculating loads, remember that:

$$\text{volts} \times \text{amperes} = \text{volt-amperes}$$

Yet, many times we say that:

$$\text{volts} \times \text{amperes} = \text{watts}$$

What we really mean to say is that:

$$\text{volts} \times \text{amperes} \times \text{power factor}$$
$$= \text{watts (true power)}$$

In a pure resistive load, such as a simple light bulb, a toaster, an iron, or a resistance electric heating element, the power factor is 100 percent. Then:

$$\text{volts} \times \text{amperes} \times 1$$
$$= \text{watts (true power)}$$

With transformers, motors, ballasts, and other "inductive" loads, wattage is not necessarily the same as volt-amperes.

EXAMPLE: Calculate the wattage and volt-amperes of a 120-volt, 10-ampere resistive load.

SOLUTIONS:

   a. $120 \times 10 \times 1 = 1200$ watts true power

   b. $120 \times 10 \qquad = 1200$ volt-amperes
                             (apparent power)

EXAMPLE: Calculate the wattage and volt-amperes of a 120-volt, 10-ampere motor load at 50 percent power factor.

SOLUTIONS:

   a. $120 \times 10 \times 0.5 = 600$ watts (true power)

   b. $120 \times 10 \qquad = 1200$ volt-amperes
                             (apparent power)

Therefore, to be sure that branch circuit wiring has adequate ampacity and to provide for feeder sizing and service caculations, the *volt-amperes* calculation would reflect the true current draw of the AC load.

In most code calculations, *watts* are used in calculating the demand on service or feeder conductors because the CSA requires that all electrical equipment be marked with a *wattage* or *kilowatt* rating, *Rule 2–100.*

Most residential loads are resistive; therefore, the power factor of 1 gives the same value to the demand *watts* as *volt-amperes,* and kW ratings are equivalent to kVA values. Therefore, this text uses the terms *wattage* and *volt-amperes* when calculating and/or estimating loads.

Building plans typically do not specify the ratings in watts for the outlets shown. When planning circuits, the electrician must consider the types of fixtures that may be used at the various outlets, based on the general uses of the receptacle outlets in the typical dwelling.

The *CEC* specifies that the maximum number of receptacle and lighting outlets that may be connected to one circuit in a residence is 12, *Rule 8–304(1).*

However, *Rule 8–304(3)* could permit more outlets if there is a maximum loading of 80 percent on the overcurrent device and the load on the outlets is known.

The layout of the receptacles on the walls must satisfy *Rule 26–712.* This should virtually eliminate the use of extension cords, one of the highest reported causes of electrical fires. Rarely, if ever, would all of the outlets be fully loaded at the same time.

Circuit Loading Guidelines (*Rules 8–104(5)* and *8–304*)

A good rule to follow in residential wiring is *never load the circuit to more than 80 percent of its capacity.* A 15-ampere, 120-volt branch circuit would be calculated as:

$$15 \times 0.80 = 12 \text{ amperes}$$

or

$$12 \text{ amperes} \times 120 \text{ volts} = 1440 \text{ volt-amperes}$$

Certain fixtures, such as recessed lights and fluorescent lights, are marked with their maximum lamp wattage; fluorescent fixtures also list their ballast current. Other fixtures, however, are not marked, and the electrician does not know the size of the lamps that will be installed in them. Furthermore, the electrician does not know the exact load that will be connected to the receptacle outlets as it is difficult to anticipate what the homeowner may do after the installation is complete. The room in which the outlets are located gives some indication of their possible uses; therefore, the electrician should plan the circuits accordingly.

Load Estimation

The electrician can estimate the lamp loads in the lighting fixtures. By determining the lamp wattages that will probably be needed to provide adequate lighting for an area, many lighting manufacturers publish useful information and recommendations in catalogues and on the Internet.

Estimating the Number of Outlets by Assigning an Amperage Value to Each

One method of determining the number of lighting and receptacle outlets to be included on one circuit is to assign a value of one ampere to each outlet, to a total of 12 amperes. Thus, a total of 12 outlets can be included in a 15-ampere circuit.

Although the *CEC* limits the number of outlets on one circuit to 12, local building codes may specify fewer. Therefore, before planning any circuits, the electrician must check the local building code requirements.

All outlets will not be required to deliver one ampere. For example, closet lights, night lights, and clocks use only a small portion of the allowable current. A 60-watt closet light draws less than one ampere:

$$I = \frac{W}{E} = \frac{60}{120} = 0.5 \text{ amperes}$$

If low-wattage fixtures are connected to a circuit, it is quite possible that 15 or more lighting outlets could be connected to the circuit without a problem. On the other hand, if the load consists of high-wattage lamps, the number of outlets would be less. In this residence, estimated loads were calculated at 120 volt-amperes (one ampere) for those receptacle outlets intended for general use.

To estimate the number of 15-ampere lighting branch circuits desired for a new residence where the total count of lighting and receptacle outlets is, for example, 80:

$$\frac{80}{12} = 6.7 \text{ (seven 15-ampere branch circuits)}$$

We divide by 12 because each outlet counts as one ampere. Twelve amperes is 80 percent, or the maximum for a 15-ampere circuit, as $15 \times 80\% = 12$ *(Rule 8–304)*.

In residential occupancies there is great diversity in the loading of lighting branch circuits.

CAUTION: This procedure for estimating the number of lighting and receptacle outlets is for branch circuits only. Where the branch circuits in the kitchen, laundry area, and other areas supply heavy concentrations of plug-in appliances, the one ampere per receptacle outlet assumption is not applicable. These circuits are discussed in more detail in Unit 14.

Review of How to Determine the Minimum Number of Lighting Circuits

1. Connect 12 outlets per circuit. The required number of outlets should be divided by 12; this will show the minimum number of branch circuits, *Rule 8–304*.

2. Estimate the probable loading for each lighting and receptacle outlet (do not include the small-appliance receptacle outlets covered by *Rule 26–720 to 724*). Try to have a total load not over 1440 volt-amperes for each 15-ampere lighting circuit.

3. Alternatively, add up the actual loading of outlets in a circuit; verify that the overcurrent device is not loaded to more than 80 percent of its capacity, and the number may, in this instance, exceed 12, *Rule 8–304(3)*. Including outlets for smoke detectors is a good example

of a situation in which the number could exceed 12. The wattage of a 120-volt smoke detector is approximately $\frac{1}{2}$.

Methods 1 and 3 are laid out in the *CEC*. Method 2 is only a guideline because the general lighting loads in homes are so varied. As the electrician lays out the circuits of a residence, these guidelines will generally provide adequate circuitry for both lighting outlets and receptacle outlets that are intended for general lighting, *not* appliance circuit receptacle outlets. Be practical. Consider the diversity of uses.

Divide Loads Evenly

It is mandatory to divide loads evenly among the circuits to avoid overloading some while leaving others lightly loaded. At a main service panel, do not connect the branch circuits and feeders to result in, for instance, 120 amperes on phase A and 40 amperes on phase B. Look at the probable and/or calculated loads to attain a balance of loads on the A phase and the B phase.

SYMBOLS

The symbols used on the cable layouts in this text are the same as those found in the legend of symbols on the electrical plans for the residence.

Pictorial illustrations are used on all wiring diagrams in this text to make it easy for the reader to understand the diagrams, Fig. 11–3.

DRAWING THE WIRING DIAGRAM OF A LIGHTING CIRCUIT

The electrician must convert information from the building plans to the form most useful for planning an installation that is economical, conforms to all code regulations, and follows good wiring practice. To do this, the electrician makes a cable layout and a wiring diagram to clearly show each wire and all connections in the circuit.

After many years of experience, skilled electricians will not always prepare wiring diagrams for most residential circuitry because they have the ability to "think" the connections through. An unskilled individual should make detailed wiring diagrams.

The following steps will guide you in preparing wiring diagrams. (You will be required to draw

wiring diagrams in later units of this text.) As you work on the wiring diagrams of the many circuits provided in this text, note that the receptacles are positioned exactly as though you were standing in the centre of the room. Then, as you turn to face each wall, you will find the receptacles positioned in that wall with the grounded conductor slot on the top, as shown in Fig. 11–4.

1. Refer to the plans and make a cable layout of all lighting and receptacle outlets, Fig. 11–5.

2. Draw a wiring diagram showing the traveller conductors for all three-way switches, if any, Fig. 11–6.

3. Draw a line between each switch and the outlet it controls. For three-way switches, do this for one switch only. This is the "switch leg."

4. Draw a line from the neutral bus bar on the lighting panel to each current-consuming outlet. This line may pass through switch boxes, but must not be connected to any switches. This is the white grounded circuit conductor, often called the "neutral" or identified conductor. It is not truly a neutral conductor unless it is part of a multiwire (three-wire) circuit.

NOTE: An exception to Step 4 may be made for double-pole switches. For these switches, all conductors of the circuit are opened simultaneously. They are rarely used in residential wiring.

5. Draw a line from the ungrounded ("hot") terminal on the panel to connect to each single-pole switch and unswitched outlet. Connect to common of one three-way switch only.

6. Show splices as small dots where the various wires are to be connected. In the wiring diagram, the terminal of a switch or outlet may be used for the junction point of wires. However, the *CEC* does not permit more than one wire to be connected to a terminal unless the terminal is a type approved for more than one conductor. The standard screw-type terminal is *not* approved for more than one wire.

7. Mark the colour of the conductors, Fig. 11–6. Note that the colours selected—black (B), white (W), and red (R)—are the colours of two- and three-conductor cables. You should

DUPLEX RECEPTACLE OUTLET GROUNDING TYPE

DUPLEX SPLIT-CIRCUIT RECEPTACLE OUTLET GROUNDING TYPE

CEILING OR WALL FIXTURE OUTLET

PILOT LIGHT

SINGLE-POLE SWITCH

THREE-WAY SWITCH

FOUR-WAY SWITCH

Fig. 11–3   Pictorial illustrations used on wiring diagrams in this text.

Fig. 11–4   This top view of a room shows the relative positioning of receptacle outlets so that the ground terminal slot is at the top. This provides for added safety if a metal face plate loosens or if a metal clip or other object falls on top of the blades of an attachment plug cap that is not plugged in all the way. All of the wiring diagrams throughout this text will illustrate the receptacles this way.

use coloured pens or markers for different conductors when drawing the wiring diagram. Use a pencil to indicate white conductors. Note this on the diagram, *Rule 4–028.*

Fig. 11–5   Typical cable layout.

Estimating Cable Lengths

The length of cable needed to complete an installation can be estimated roughly. It may be possible to run the cable in a straight line directly from one wall outlet to another. However, obstacles such as steel columns, sheet-metal ductwork, and plumbing may require the cable routing to follow a longer path.

To ensure that the estimate is not short, make all measurements "square." For example, measure from the ceiling outlet straight to the wall, and finally measure straight over to the wall outlet. Add 300 mm of cable at each cable termination. This allowance is for the free conductor needed to make connections in the junction and outlet boxes. Check the plans carefully to determine where the cables can be routed.

LIGHTING BRANCH CIRCUIT A16 FOR FRONT BEDROOM AND STUDY

Receptacles in bedrooms are required to be protected by arc fault circuit interrupters (AFCIs). This protection is normally provided by an AFCI breaker in

Fig. 11–6   Wiring diagram of circuit shown in Fig. 11–5. For simplicity, no bonding is shown.

the panel. Because of the AFCI breaker's high cost, bedroom receptacles are often placed on their own circuit to reduce the number of AFCI breakers required. The receptacles in the front bedroom and study/bedroom are connected to a single circuit along with the two lighting outlets in the front bedroom for a total of 12 outlets.

The receptacles in the bedrooms and living room are split-switched receptacles. Half the receptacle (normally the bottom half) is controlled by a wall switch in the room, and the other half is energized continuously. Receptacles in bedrooms, living rooms, and most other areas in a dwelling must be tamper-resistant, as per *Rule 26–712(g)*. These receptacles must be clearly marked "Tamper-Resistant" or "TR."

Radio, television, and other electrical items not intended to be controlled by the wall switch will be plugged into the unswitched ("hot" continuously) half of the duplex receptacle.

The recessed closet light is controlled by a single-pole switch outside and to the right of the closet door.

Note that the outside weatherproof receptacle outside the study/bedroom is connected to circuit A24, a separate circuit from the front bedroom circuit, A18, Figure 11–7.

Checking the actual electrical plans, we find one television outlet and one telephone outlet are to be installed in each bedroom. Television and telephones are discussed in Unit 24. Table 11–1 summarizes the outlets in the front bedroom and the estimated load.

BRANCH CIRCUIT A24 OUTSIDE RECEPTACLE

This is a 15-ampere circuit providing power to the outside receptacle outlet at the front of the house. The receptacle is GFCI protected with a weatherproof cover on the receptacle outlet box.

Rule 26–710(n) states that all outdoor receptacles 2.5 m or less above ground or grade level shall be GFCI protected. Ordinary receptacles may be installed above 2.5 m, but must incorporate a weatherproof cover, *Rule 26–702(1)*.

Fig. 11–7   Cable layout for front bedroom, circuit A18.

DESCRIPTION	QUANTITY	WATTS	VOLT-AMPERES
Receptacles @ 120 watts each	10	1200	1200
Closed recessed fixture	1	60	60
Ceiling outlet	1	120	120
TOTALS	12	1380	1380

Table 11–1   Front bedroom and study/bedroom receptacles: outlet count and estimated load, circuit A18.

THE "PIGTAILS" CONNECTED TO THE RECEPTACLE NEED NOT BE COUNTED WHEN DETERMINING THE CORRECT BOX SIZE. *RULE 12–3034 (I)(c)*

PIGTAIL

WHITE

BLACK

RED

Fig. 11–8   Tab must be broken between red and black conductor terminals. The box size is determined according to the number of conductors. This box would be required to be 102 × 38 mm complete with a plaster ring for the receptacle.

DETERMINING THE WALL BOX SIZE

The electrician must consider several factors when determining the size of the wall box:

- the number of conductors entering the box
- the types of boxes available
- the space allowed for the installation of the box.

Sizing Boxes According to the Number of Conductors in a Box *(Rule 12–3036)*

To find the proper box size for any location, you must determine the total number of conductors entering the box (Fig. 11–8). For example:

1. Count the No. 14 AWG circuit conductors:
   $2 + 3 + 3 =$ 8
2. Add one conductor for each pair of wire connectors   2
3. Add two conductors for the receptacle   2

Total   12 conductors

The pigtails connected to the receptacle need *not* be counted when determining the correct box size (Fig. 11–8). *Rule 12–3034(1)(c)* states, "a conductor, of which no part leaves the box, shall not be counted."

Once you know the total number of conductors, refer to *CEC Table 23* and Fig. 5–16 on page 84 to find the box suitable for the conductors.

Use the volume of the box plus the space provided by plaster rings, extension rings, and raised covers to determine the total available volume. If the box contains one or more devices, such as fixture studs or hickeys, the number of conductors permitted in the box shall be one less than shown in *Table 23* for *each type* of device contained in the box.

BONDING OF WALL BOXES

The specifications for the residence state that *all* metal boxes are to be bonded. The bare conductor included in NMSC or armoured cable (BX) is connected to the bonding screw in the box and, for receptacles, carries on from the connection in the box to the bonding screw on the receptacle. This conductor is used only to bond the metal box. This bare grounding conductor must *not* be used as a current-carrying conductor, since severe shocks can result.

According to *Diagram 1* and *Rule 26–700*, grounding-type receptacles must be installed on 15-ampere and 20-ampere branch circuits. Fig. 11–8 shows how to attach the bonding conductor to the proper terminal on the receptacle.

POSITIONING OF SPLIT-CIRCUIT RECEPTACLES

The receptacle outlets shown in the front bedroom and study are split-switched receptacles. The top portion of such a receptacle is hot at all times, and the bottom portion is controlled by the wall switch, Fig. 11–9. The electrician should wire the bottom of the receptacle as the switched section. Then, when the lamp is plugged into the switched bottom portion of the receptacle, the cord does not hang in front of the unswitched section. This unswitched section can be used for a clock, vacuum cleaner, radio, stereo or CD player, personal computer, television, or some other appliance where a switch control is not necessary or desirable.

When mounting split-switched receptacles horizontally, locate the switched portion to the right.

POSITIONING OF RECEPTACLES NEAR ELECTRIC BASEBOARD HEATING

When laying out receptacles in residential applications, ensure that any cords plugged into the receptacles do not pass over heat-generating sources such as baseboard heaters, or hot air, hot water, or steam registers. *Rule 26–712(a), Appendix B.*

Unit 17 discusses installing electric baseboard heating units and their relative positions below receptacle outlets.

FIXTURES IN CLOTHES CLOSETS

Clothing, boxes, and other material normally stored in closets are a potential fire hazard. These items may ignite on contact with the hot surface of exposed lamps. Incandescent lamps have a hotter surface temperature than fluorescent lamps. *Rule 30–204* gives very specific rules about the location and types of lighting fixtures permitted in clothes closets, Figs. 11–10 through 11–14.

Figure 11–15 shows typical bedroom-type lighting fixtures.

LIGHTING BRANCH CIRCUIT A14 FOR MASTER BEDROOM

The residence panel schedules show that the master bedroom is supplied by circuit A14. Because Panel A is located in the southeast corner of the basement workshop, the home run for circuit A14 is brought into the closest outlet to the panel. This would be the receptacle located in the hallway between the front and master bedrooms. Again, it is a matter of studying the circuit to determine the best choice for conservation of cable or conduit in these runs and to economically select the correct size of wall boxes.

Fig. 11–16 on page 210 and Table 11–2 show a sample device and wiring layout in which circuit A14 has four split-switched receptacle outlets in the bedroom, one receptacle in the hall, one outdoor weatherproof light, two recessed closet fixtures each on a separate switch, a paddle fan/light fixture, one telephone outlet, and one television outlet.

In addition, an outdoor GFCI weatherproof receptacle is located adjacent to the sliding door and is connected to circuit A24, which also includes the other outdoor receptacles.

Fig. 11–9   Split-switched wiring for receptacles in the bedroom.

DESCRIPTION	QUANTITY	WATTS	VOLT-AMPERES
Receptacles @ 120 watts each	5	600	600
Outdoor bracket fixture	1	100	100
Closet recessed fixtures One 60-W lamp each	2	120	120
Ceiling outlet	1	100	100
TOTALS	9	920	920

Table 11–2   Master bedroom: outlet count and estimated load, circuit A14.

Fig. 11–10   Typical closet with one shelf and one rod. The shaded area defines storage area. Dimension A is the width of the shelf or 300 mm from wall, whichever is greater. Dimension B is below the rod, 600 mm from the wall.

Fig. 11–11   Large walk-in closet with access to the centre rod from both sides. The shaded area defines storage area. Dimension A is the width of the shelf or 300 mm, whichever is greater. Dimension B is 600 mm from wall.

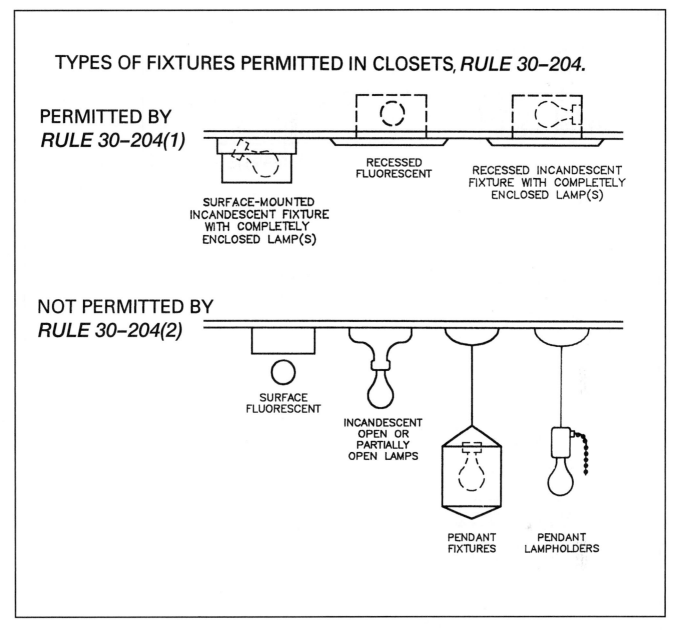

Fig. 11–12 Types of lighting fixtures permitted in clothes closets. The *CEC* does not permit pendant fixtures, pendant lampholders, or luminaires with bare bulbs to be installed in clothes closets.

LOCATION OF FIXTURES IN CLOSETS, *SECTION 30–204*, SHOWING RECOMMENDED CLEARANCE BETWEEN FIXTURES AND STORAGE AREA.

STORAGE AREA

MAY INSTALL SURFACE-MOUNTED INCANDESCENT FIXTURES ON CEILING OR ON WALL ABOVE DOOR.

300 mm

300 mm

STORAGE AREA

MAY INSTALL SURFACE-MOUNTED FLUORESCENT FIXTURES ON CEILING OR ON WALL ABOVE DOOR. FIXTURE MUST BE OF THE ENCLOSED TYPE.

300 mm

300 mm

STORAGE AREA

MAY INSTALL RECESSED INCANDESCENT FIXTURES THAT HAVE THE LAMPS TOTALLY ENCLOSED, OR ENCLOSED RECESSED FLUORESCENT FIXTURES ON CEILING OR ON WALL ABOVE DOOR. IF CEILING IS INSULATED, A FIXTURE TYPE Ic WOULD BE REQUIRED.

150 mm

150 mm

Fig. 11–13   Examples of clearances between fixtures and the storage area.

Fig. 11–14   Recessed incandescent closet lighting fixture.

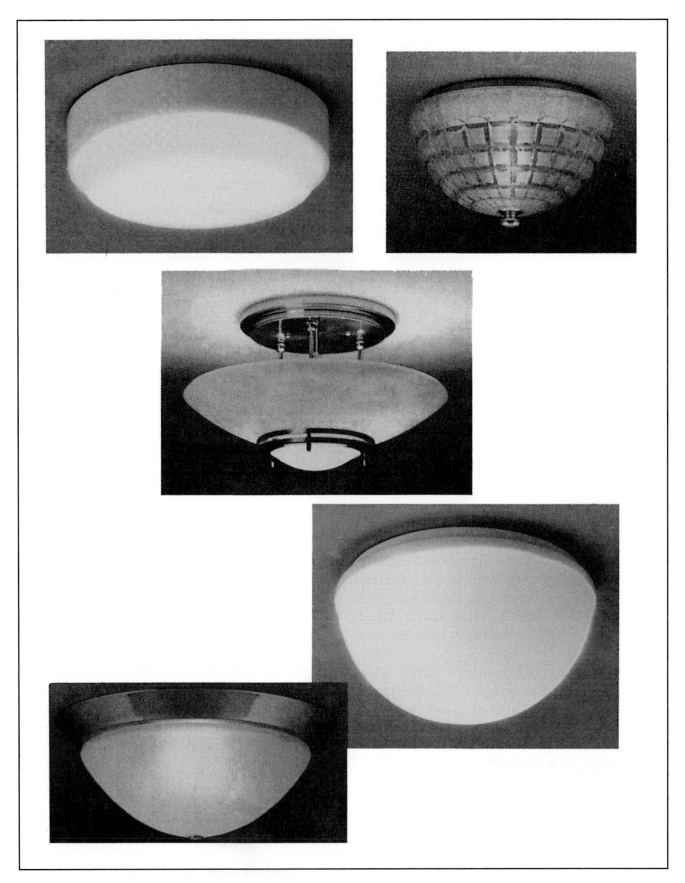

Fig. 11–15   Typical bedroom-type lighting fixtures.

Fig. 11–16 Cable layout for master bedroom, circuit A14.

Because there are two entrances to this room, the split-switched receptacle outlets are controlled by two three-way switches. One switch is located next to the sliding door. The use of split-switched receptacles offers the advantage of having switch control of one of the receptacles at a given outlet, while the other receptacle remains live at all times.

Next to the door to the hall is the other three-way switch plus the single-pole switch for the ceiling outlet.

SLIDING GLASS DOORS

CEC Rule 26–712(c) covers spacing receptacle outlets on walls where sliding glass doors are installed.

SELECTION OF BOXES

In the master bedroom, the electrician may decide to install a two-gang welded box or two sectional device boxes ganged together. The size of box depends on the number of conductors entering the box. The cable layout in Fig. 11–16 shows that three cables enter the box next to the sliding glass door, for a total of 7 conductors, each No. 14 AWG circuit conductors. *Rule 12–3034 (2)(c)* stipulates that the number of conductors shall be reduced by two if the box contains one or more flush devices mounted on a single strap. Therefore, the number of conductors to be included in determining proper box size is 7 circuit conductors plus two conductors for each switch, for a total of 11 wires.

1. Add the circuit conductors
   $2 + 2 + 3 =$ 7
2. Add two for each switch $(2 + 2) = $ $\underline{4}$

                              Total $\overline{11}$

Now look at the Quick-Check Box Selector, Fig. 5–16, and select a box or a combination of gangable device boxes that are permitted to hold

11 conductors. For example, you can gang two 76 × 51 × 64-mm (3 × 2 × 2$\frac{1}{2}$-in) device boxes together (8 × 2 = 16 No. 14 AWG conductors allowed).

LIGHTING CIRCUIT FOR STUDY AND MAIN BATHROOM

The study/bedroom circuit feeds the bedroom, which provides excellent space for a home office or den. Should it become necessary to have a third bedroom, the changeover is easily done. Since this room is considered a bedroom, AFCI protection is required for the receptacles. The circuit also supplies the light and receptacle in the main bathroom.

Circuit A16 originates at main Panel A. This circuit feeds the study/bedroom at the closet switch. From this point, the circuit feeds the closet recessed light and then on to the overhead light. Here it splits to feed the hallway light and smoke detector. The other leg goes to the switches by the door. A three-wire cable runs around the room to feed the other four split-switched AFCI-protected receptacles from A18. Fig. 11-17 shows the cable layout for circuit A16.

There are four split-switched receptacles in the study/bedroom controlled by two three-way switches; one of the receptacles is surge protected. The black and white conductors carry the live circuit, and the red conductor is the switch return for the control of the switched half of the split-circuit receptacles.

A paddle fan/light fixture is mounted on the ceiling. Table 11–3 summarizes the outlets and estimated load for the study/bedroom.

DESCRIPTION	QUANTITY	WATTS	VOLT-AMPERES
Receptacle @ 120 W	1	120	120
Closet recessed fixture	1	60	60
Paddle fan/light Three 50-W lamps Fan motor (0.75 A @ 120V)	1	150 80	150 90
Valance lighting Two 34-W fluorescent lamps	1	68	80
Hall light	1	100	100
Bathroom light	2	100	100
TOTALS	7	678	700

Table 11–3   Study/bedroom: outlet count and estimated load, circuit A16.

VALANCE LIGHTING

Indirect fluorescent lighting is provided above the windows. Fig. 11–18 illustrates one method of installing fluorescent valance lighting.

Fig. 11–19 shows one way to arrange the wiring for the paddle fan/light control, the valance lighting control, and the receptacle outlet control.

Since this room is intended to be used as a study or home office, a personal computer will probably be located here. Transient voltage surge suppressors are covered in Unit 9. The wiring requirements for telephone and data systems are covered in Unit 24.

PADDLE FANS

Paddle fans have become extremely popular in recent years, Fig. 11–20. These fans rotate slowly (60 to 250 rpm for home-type fans). The air currents they create can save energy during the heating season because they destratify the warm air at the ceiling, bringing it down to living levels. In the summer, even with air conditioning, a slight circulation of air creates a wind-chill factor and causes a person to feel cooler.

Residential paddle fans usually extend 305 mm or less below the ceiling; so, for 2.44 m ceilings, the fan blades are about 229 to 254 mm below the ceiling, Fig. 11–21.

Safely supporting a paddle fan involves:

- the actual weight of the fan
- the twisting and turning motion when started
- vibration.

To ensure the safe support of the fan, you should not use outlet boxes as the sole support of ceiling (paddle) fans unless the CSA approves and lists them for this purpose. The CSA also requires that the motor housing have a cable attached to an independent means of support to prevent the fan from falling if it vibrates loose from its mounting screws.

Always check the instructions furnished with the ceiling fan to be sure that the mounting and supporting methods will be safe. Boxes that are permitted to support a fan are marked "acceptable for fan support."

Rule 30–302(5) requires that a lighting fixture be supported independently of the outlet box if the fixture weighs more than 23 kg.

Fig. 11–17   Cable layout for study/bedroom circuit A16. For receptacles in study, see Fig. 11–7.

Fig. 11–18   Fluorescent valance with draperies.

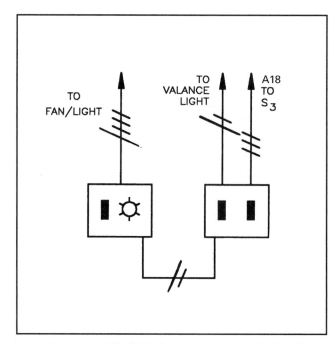

Fig. 11–19   Conceptual view of how the study/bedroom switch arrangement is to be accomplished, circuit A16.

Fig. 11–20   Paddle fan.

Fig. 11–22 shows the wiring for a fan/light combination with the supply at the fan/light unit. Fig. 11–23 shows the fan/light combination with the supply at the switch.

The electrician must install a junction box where the fan or fan/light is to be located. Since the fan is quite heavy, the box must be solidly secured. A mounting kit furnished by one manufacturer of paddle fans is shown in Fig. 11–24.

Fig. 11–21   Typical mounting of a ceiling paddle fan.

Fig. 11–22   Fan/light combination with the supply at the fan/light unit.

Fig. 11–23   Fan/light combination with the supply at the switch.

Fig. 11–25 shows an adjustable-type ceiling fan hanger installed. This type of hanger is useful when there is no access to the ceiling cavity, since it can be installed through the hole where the outlet box will be mounted and adjusted to tighten against the joists.

Fig. 11–26 illustrates a typical combination light switch and three-speed fan control. The electrician must install a deep 102-mm box with a two-gang raised plaster ring for this light/fan control.

A typical home-type paddle fan motor draws 50 to 100 watts (50 to 100 volt-amperes), about 0.4 to 0.8 amperes.

Paddle fan/light combinations increase the load requirements somewhat. Some paddle fan/light units have one lamp socket, whereas others have four or five lamp sockets.

For the paddle fan/light unit in the master bedroom, 240 volt-amperes were included in the load calculations. The current draw is:

$$I = \frac{VA}{V} = \frac{240}{120} = 2 \text{ amperes}$$

Many fans have a pull-chain motor speed control (OFF–HIGH–MED–LOW) on the bottom of the fan housing. Some fans can be used with a solid-state control that is continuously variable.

Some fan motors are of the reversible type to allow proper circulation of air in the room below.

Fig. 11–26   Combination light switch and three-speed fan control.

Fig. 11–24   A mounting kit furnished by one manufacturer of paddle fans.

Fig. 11–25   An adjustable-type ceiling fan hanger that is CSA listed to support 60 kg. The hanger adjusts between 400 and 600 mm. It can be installed in new construction or where an existing light is installed, since it can be installed from below through a hole cut in the ceiling.

REVIEW

Note: Refer to the *CEC* or the plans where necessary.

1. Why do some electricians prefer to have more than one circuit feeding a room? _____

   _____

   _____

2. Is it good practice to have outlets on different floors on the same circuit? Why?

   _____

   _____

3. What usually determines the grouping of outlets into a circuit? _____

   _____

   _____

4. The continuous load on a lighting branch circuit must not exceed _____ percent of the branch circuit rating.

5. To determine the maximum number of outlets in a circuit, _____ amperes per outlet are allowed. For a 15-ampere circuit, this results in a maximum of _____ outlets.

6. For this residence, what are the estimated wattages used in determining the loading of branch circuit A16?

   Receptacles _____ watts (volt-amperes)

   Closet recessed fixture _____ watts (volts-amperes)

7. With respect to *Rule 8–104(1)*, the ampere rating of circuit A16 is _____.

8. What size wire is used for the lighting circuit in the front bedroom? _____

9. How many receptacles are connected to this circuit? _____

   _____

10. What main factor influences the choice of wall boxes? _____

    _____

11. How is a wall box grounded? _____

    _____

12. What is a split-switched receptacle? _____

    _____

13. Is the switched portion of an outlet mounted toward the top or the bottom? Why?

    _____

    _____

14. The following questions pertain to lighting fixtures in clothes closets.

   a. Does the *CEC* allow bare incandescent lamp fixtures such as porcelain keyless or porcelain pull-chain lampholders to be installed? _____

   _____

   b. Does the *CEC* allow bare fluorescent lamp fixtures to be installed? _____

   _____

   c. Does the *CEC* permit pendant fixtures or pendant lampholders to be installed?

   _____

15. How many switches are in the front bedroom circuit, and of what type are they? _____

   _____

   _____

16. The following is a layout of the lighting circuit for the front bedroom. Using the cable layout shown in Fig. 11–7, make a complete wiring diagram of this circuit. Indicate the colour of each conductor.

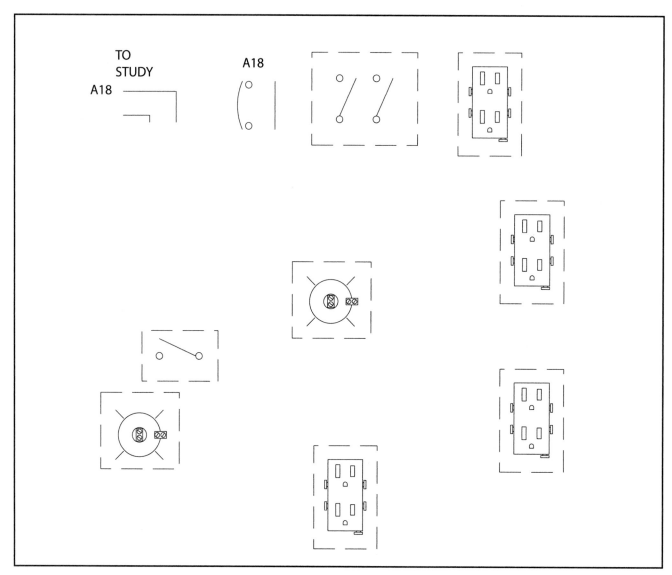

Front Bedroom

17. When planning circuits, how should you divide loads? _____
_____
_____

18. The *CEC* uses the terms *watts, volt-amperes,* and *kW.* Explain their significance in calculating loads. _____
_____
_____
_____
_____
_____

19. How many No. 14 AWG conductors are permitted in a device box that measures $76 \times 51 \times 64$ mm ($3 \times 2 \times 2\frac{1}{2}$ in)? _____

20. A 102×38 mm ($4 \times 1\frac{1}{2}$ in) octagon box has one cable clamp, one fixture stud, and three wire connectors. How many No. 14 AWG conductors are permitted? _____
_____
_____

21. What circuit supplies the master bedroom? _____

22. For the recessed closet fixtures, what wattage was used for calculating their contribution to the load on the circuit? _____

23. What is the estimated load in volt-amperes for the circuit supplying the master bedroom? _____

24. The sliding glass door in the master bedroom could affect the number of required receptacle outlets. Answer the following statements true or false.

   a. Sliding glass panels are considered to be wall space. _____

   b. Fixed panels of glass doors are considered to be wall space. _____

25. What type of receptacle will be installed outdoors, just outside of the master bedroom?
_____
_____

26. Based on the total estimated load calculations, what is the current draw on the study/bedroom lighting circuit? _____

27. The study/bedroom is connected to Circuit _____

28. The conductor for this circuit is _____

29. Is it necessary to install a receptacle in the wall space leading to the bedroom hallway?
_____
_____

30. Show the calculations needed to select a properly sized box for the receptacle outlet in Question 29. Non-metallic-sheathed cable is the wiring method. Refer to the Quick-Check Box Selection Guide, Fig. 5–16 in Unit 5.

_____

_____

_____

_____

31. Using the cable layout shown in Fig. 11–17, make a complete wiring diagram of circuit A16.

Study Receptacle Circuit

Study Lighting Circuit

32. The following is a layout of the lighting circuit for the master bedroom. Using the cable layout shown in Fig. 11–16, make a complete wiring diagram of this circuit. Indicate the colour of each conductor.

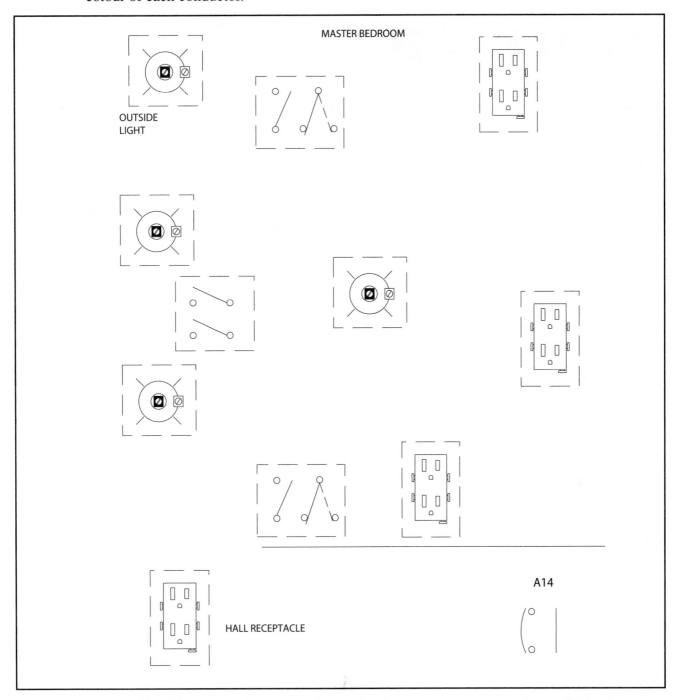

33. Using the electrical plans supplied with this text, lay out the devices and wiring in the master bedroom according to **minimum** *CEC* standards.

UNIT 12

Branch Circuits for the Living Room and Front Entry

OBJECTIVES

After studying this unit, the student will be able to

- understand various types of dimmer controls
- connect dimmers to incandescent lamp loads
- understand the phenomenon of incandescent lamp inrush current
- discuss Class P ballasts and ballast overcurrent protection
- discuss the basics of track lighting
- understand how to install a switch in a doorjamb for automatic ON/OFF when the door is opened or closed
- discuss the type of fixtures recommended for porches and entries
- discuss the advantages of switching outdoor receptacles from indoors
- define wet and damp locations

LIGHTING CIRCUIT OVERVIEW

Fig. 12–1 shows sample device and wiring layout for the living-room area that is connected to circuit B16, a 15-ampere circuit. The home run is brought from Panel B to the split-switched receptacle in the living room next to the sliding door.

A three-wire cable is then carried around the living room, feeding in and out of the eight split-switched receptacles. This three-wire cable carries the black and white circuit conductors plus the red wire, which is the switched conductor. Since these receptacles will provide most of the general lighting for the living room through floor and table lamps, it is a good idea to have some of these lamps controlled by the three-switches (two three-way switches and one four-way switch). Only one half of each receptacle is switched, *Rule 26–710(k)*.

The receptacle below the four-way switch is required because the wall space is more than 900 mm wide, *Rule 26–712(c)*, and there must be a receptacle within 1.8 m of any point along the wall line, *Rule 26–712(a)*. Because it is unlikely that a split-switched receptacle (Fig. 12–2) will be needed in this location, a standard receptacle is used.

Accent lighting above the fireplace is in the form of a recessed fixture controlled by a single-pole dimmer switch. See Unit 10 for installation data and *CEC* requirements for recessed fixtures.

Table 12–1 summarizes the number of outlets and estimated load for the living room circuit.

Unit 6 covers requirements for spacing receptacle outlets and locating lighting outlets as well as spacing receptacles on exterior walls where sliding glass doors are involved, *Rule 26–712(c)*.

Fig. 12–1   Cable layout for the living room, circuit B16.

Fig. 12–2   Split-switched receptacle wiring diagram.

DESCRIPTION	QUANTITY	WATTS	VOLT-AMPERES
Receptacles @ 120 W	9	1080	1080
Fireplace recessed fixtures @ 60 W each	1	60	60
TOTALS	10	1140	1140

Table 12–1   Living room: outlet count and estimated load, circuit B16.

TRACK LIGHTING

The plans show that track lighting is mounted on the ceiling of the living room on the wall opposite the fireplace. Track lighting provides accent lighting for a fireplace, a painting on the wall, or an item (sculpture or collection) that the homeowner wishes to focus attention on. It is also used as task lighting for counters, game tables, or kitchen dining tables.

Unlike recessed fixtures and individual ceiling fixtures that are fixed in place, track lighting offers flexibility because the actual lampholders can be moved and relocated on the track as desired.

Lampholders, selected from hundreds of styles, are inserted into the extruded aluminum track at any point. The circuit conductors are in the track, and the plug-in connector on the lampholder completes the connection. The track is generally fas-tened to the ceiling at the outlet box and with toggle bolts or screws. Various track lighting installations are shown in Figs. 12–3, 12–4, and 12–5.

On the plans for the residence, the living room ceiling has wood beams, a challenge when installing track lighting that is longer than the space between them.

There are a number of things the electrician can do:

1. Install a pendant kit assembly that will allow the track to hang below the beam, Fig. 12–6. This fixture may be supported directly by the outlet box if the weight does not exceed 23 kg, *Rule 30–302(4)*.

2. Install conduit connector fittings on the ends of sections of the track, then drill a hole in the beam, and use 16-mm EMT to join the track sections. Running lighting track through walls or partitions is not recommended.

3. Run the branch circuit wiring concealed above the ceiling to outlet boxes located at the points where the track is to be fed by the branch-circuit wiring, Fig. 12–7.

Where to Mount Track Lighting

To install the lighting track at the proper distance from the wall, you must consider the ceiling height, the aiming angle, the type of lampholder, the type of lamp, and the type of lighting to be achieved.

Fig. 12–3   End-feed track light.

Fig. 12–4   Centre-feed track light.

Fig. 12–5   Plug-in track light.

For example, if you are designing an installation to illuminate an oil painting that is to be centred 1.65 m (average eye level) from the floor, where the ceiling height is 2.75 m, and the lampholder and lamp information tell us that the aiming angle should be 60°, you would find in the manufacturer's installation data that the track should be mounted 600 mm from the wall.

If the ceiling height in this example were 2.4 m, then the distance to the track from the wall would be 450 mm.

Since many variables affect the positioning of the track, it is necessary to consult the track lighting manufacturer's catalogue for recommendations that will achieve the desired results.

The track itself must be properly grounded to the outlet box. The plug-in connector on the move-

able fixture also has a ground contact to ensure that the body of the fixture is safely grounded. Fig. 12–8 shows tracks in cross-sectional views.

Fig. 12–9 shows typical fixtures that attach to track lighting systems.

The track lighting in the living room is turned on and off by a single-pole dimmer switch.

Supporting Track Lighting

Each lighting track must be fastened at a minimum of two places. This provides the secure support required by *Rule 30–302(1)*. If the track is extended, install at least one additional support for every 1.5 m or less of extension.

Load Calculations for Lighting Track

For track lighting installed in homes, it is not necessary to add additional loads for feeder or service entrance calculations. *Rule 8–200(1)(a)* includes lighting track load as part of the regular watts-per-square-metre computations.

For further information on track lighting, refer to lighting manufacturers' catalogues.

DIMMER CONTROLS

Electronic Solid-State Dimmer Controls

According to their CSA listing, solid-state electronic dimmers of the type sold for home use "are

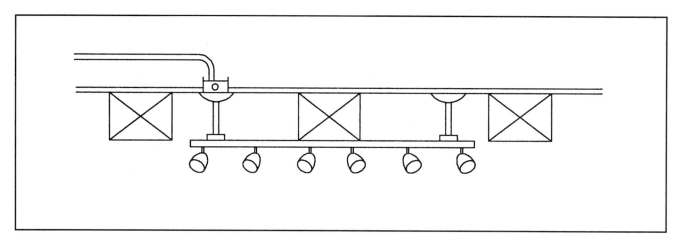

Fig. 12–6   A pendant kit assembly allows the track to hang below the beam.

Fig. 12–7   An outlet box feeds each section of track.

Fig. 12–8   Tracks in cross-sectional views and various track connectors.

Fig. 12–9   Typical fixtures that attach to track lighting systems.

intended only for the control of permanently installed incandescent fixtures," Fig. 12–10. For residential installations, the dimmer commonly used is a 600-watt unit that fits into a standard single-gang wall box. Both single-pole and three-way dimmers are available. The control knob is pushed to turn the lights on or off and is rotated to dim the lights. Compact, 1000-watt dimmer units are available for use in two-gang wall boxes.

These boxes should be No. 1004 boxes, which are 3 inches deep. Fig. 12–11 shows how single-pole and three-way solid-state dimmer controls are used in circuits.

Autotransformer Dimmer Control

This type of dimmer control has a continuously adjustable autotransformer that changes the lighting intensity by varying the voltage to the lamps, Figs. 12–12 and 12–13. When the knob of the transformer is rotated, a brush contact moves over a

Fig. 12–10   Solid-state rotary dimmer control.

bared portion of the transformer winding to vary the lighting intensity from complete darkness to full brightness. This type of control uses only the current required to produce the lighting desired.

Dimmer controls are used on 120-volt, 60-hertz, single-phase AC systems. The operation of such controls depends on transformer action, so they will not operate on DC lines.

The electrician can prevent overloading of the dimmer control transformer by checking the total load in watts to be controlled by the dimmer against the rated capacity of the dimmer. The maximum wattage that can be connected safely is shown on the nameplate of the dimmer. In general, overcurrent protection is built into dimmer controls to prevent burnout due to an overload.

Instructions are furnished with most dimmer controls to help prevent such burnout. These instructions state that the dimmers are for control of incandescent lighting only and are not to be used to control receptacle outlets:

- Serious overloading of the dimmer could result, since it may not be possible to determine or limit what other loads might be plugged into the receptacle outlets.

- Reduced voltage supplying a television, radio, stereo components, home computer, or motors of any kind, such as vacuum cleaners and food processors, can result in costly damage to the appliance.

Because they contain a bulky transformer, autotransformer dimmers are larger than solid-state dimmers. Autotransformers require a special box that is furnished with the dimmer. They are common in commercial establishments, whereas the smaller solid-state dimmer is most commonly used in residential installations.

SINGLE POLE

NOTE:
THIS WHITE
WIRE IS HOT
ONLY WHEN
USED TO
FEED POWER
TO A SWITCH.

SPLICE   WHITE
HOT
NEUTRAL
TO LIGHT

GREEN GROUND LEAD
POWER FED AT LOAD

LOAD BLACK   NEUTRAL WHITE
SPLICE
LINE
GREEN   GROUND LEAD
POWER FED AT SWITCH

POWER FED AT SWITCH

LINE
NEUTRAL   HOT
(WHITE)   (BLACK)

COMMON TERMINAL
BLACK OR
COPPER COLOUR

WHT   BLK   RED

BLACK
RED
WHITE

TWIN LEADS
FROM DIMMER
(2) BLACK

GREEN
GROUND

DIMMER AND SWITCH
MAY BE INTERCHANGED.

POWER FED AT LOAD

NOTE; THIS WHITE WIRE IS HOT
ONLY WHEN USED TO FEED
POWER TO A SWITCH

COMMON TERMINAL
BLACK OR
COPPER COLOUR

HOT(BLACK)   LINE
NEUTRAL(WHITE)

WHT.   BLK.   RED

BLACK
RED
WHITE

TWIN LEADS
FROM DIMMER
(2) BLACK

GREEN
GROUND

DIMMER AND SWITCH MAY BE INTERCHANGED.

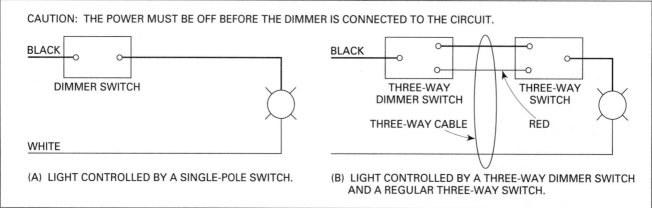

CAUTION: THE POWER MUST BE OFF BEFORE THE DIMMER IS CONNECTED TO THE CIRCUIT.

BLACK
DIMMER SWITCH
WHITE

(A) LIGHT CONTROLLED BY A SINGLE-POLE SWITCH.

BLACK
THREE-WAY
DIMMER SWITCH
THREE-WAY CABLE
THREE-WAY
SWITCH
RED

(B) LIGHT CONTROLLED BY A THREE-WAY DIMMER SWITCH
AND A REGULAR THREE-WAY SWITCH.

Fig. 12–11   Single-pole and three-way solid-state dimmer control circuits.

INCANDESCENT LAMP LOAD INRUSH CURRENTS

An unusual action occurs when a circuit is energized to supply a tungsten filament lamp load. A tungsten filament lamp has a very low resistance when it is cold. When the lamp is connected to the proper voltage, the resistance of the filament increases very quickly, within 1/240 second, or one-quarter of a 60-Hz cycle, after the circuit is energized. During this period, there is an inrush current

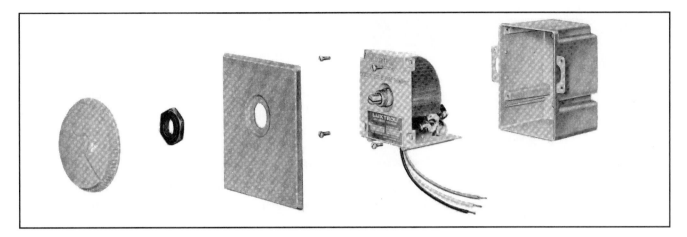

Fig. 12–12   Dimmer control, autotransformer type.

Fig. 12–13   Dimmer control (autotransformer) wiring diagram for incandescent lamps.

that is about 10 to 20 times greater than the normal operating current of the lamp.

"T" rated switches are required for many applications of tungsten-filament lamps. Although "T" rated switches are generally not required in dwelling units, canopy switches are required to be "T" rated and have an ampacity at least three times the connected load if they are used to supply incandescent lighting.

Dimmer Controls for Incandescent Lighting

The wiring diagram in Fig. 12–11 shows the circuit used to control incandescent lamps. The entire load is cut off when the internal switch is in the OFF position. These dimmers are available in single-pole and three-way switches. Fig. 12–14 shows slider-type dimmer switches with a rocker-type ON/OFF switch, which allows the slider to remain at the same place, thus maintaining the same light level.

Lutron manufactures a full line of dimmers ranging from rotary, toggle, and slider up to a designer series with special features such as remote control on one model. The Lutron dimmers, depending on the model, are available for incandescent, fluorescent, and magnetic low-voltage fixtures for loads up to 600 or 1000 watts, with many models having matching paddle fan speed controls. The Maestro series has LEDs to indicate light intensity,

Fig. 12–14 *Leviton* **solid-state slider dimmer control with rocker ON-OFF switch to maintain same light level.**

a soft-glow night-light, and a tap on and tap off with a 10-second fade to off. One of these dimmers may be used in combination with up to nine smart remote modules to provide control from up to 10 locations. The Lutron Faedra dimmers have features similar to the Maestro with a design like a standard toggle switch.

DIMMING FLUORESCENT LAMPS

To dim incandescent lamps, the voltage to the lamp filament is varied (raised or lowered) by the dimmer control. The lower the voltage, the less intense the light. The dimmer is simply connected "in series" with the incandescent lamp load.

In fluorescent lamps, the connection is not a simple "series" circuit. To dim fluorescent lamps, special dimming ballasts and dimmer switches are required. The dimmer control allows the dimming ballast to maintain a voltage to the cathodes that will maintain the cathodes' proper operating temperature and allow the dimming ballast to vary the current flowing in the arc. This varies the intensity of light coming from the fluorescent lamp.

However, some compact fluorescent lamps have an integrated circuit built right into the lamp, allowing it to be controlled by a standard incandescent wall dimmer. These dimmable compact fluorescent lamps can replace standard incandescent lamps of the medium screw shell base type. No additional wiring is necessary. They are designed for use with dimmers, photocells, occupancy sensors, and electronic timers that are marked "Incandescent Only."

Fig. 12–15 illustrates the connections for an autotransformer-type dimmer for the control of fluorescent lamps using a special dimming ballast. The fluorescent lamps are F40T12 rapid-start lamps. With the newer electronic dimming ballasts, 32-watt, rapid-start T8 lamps are generally used.

Fig. 12–16 shows an electronic dimmer for the control of fluorescent lamps using a special dimming ballast.

Fluorescent dimming ballasts and controls are available in many types such as: simple "ON/OFF," 100 to 1 percent variable light output, 100 to 10

Fig. 12–15   Dimmer control wiring diagram for fluorescent lamps.

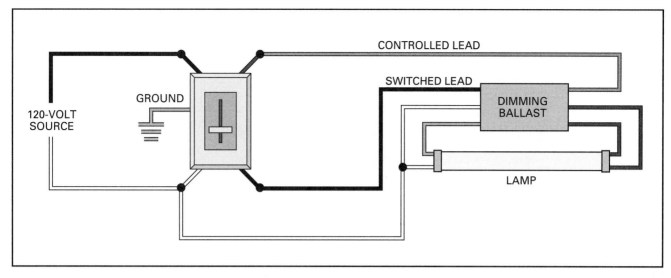

Fig. 12–16 This wiring diagram shows the connections for an electronic dimmer designed for use with a special dimming ballast for rapid-start lamps. Always check the dimmer and ballast manufacturers' wiring diagrams, as the colour coding of the leads and the electrical connections may be different from this. The dimmer shown is a "slider" type, and offers wide-range (20 to 100 percent) adjustment of light output for the fluorescent lamp.

percent variable light output, and two-level 100 percent/50 percent light output. Wireless remote control is also an option.

To ensure proper dimming performance of fluorescent lamps:

- Make sure that the lamp reflector and the ballast case are solidly grounded.

- Do not mix ballasts or lamps from different manufacturers on the same dimmer.

- Do not mix single- and two-lamp ballasts.

- "Age" the lamps for about 100 hours before dimming to allow the gas cathodes in the lamps to stabilize. Without "aging" (sometimes referred to as "seasoning"), it is possible to have unstable dimming, lamp striation (stripes), and/or unbalanced dimming characteristics between lamps.

- Because incandescent lamps and fluorescent lamps have different characteristics, they cannot be controlled simultaneously with a single dimmer control.

- Do not control electronic ballasts and magnetic ballasts simultaneously with a single dimmer control.

Always read and follow the manufacturer's instructions furnished with dimmers and ballasts.

Without special dimming fluorescent ballasts and controls, a simple way to have two light levels is to use four-lamp fluorescent luminaires. Then, control the ballast that supplies the centre two lamps with one wall switch, and control the ballast that supplies the outer two lamps with a second wall switch. This is sometimes referred to as an "inboard/outboard" connection.

BRANCH CIRCUIT FOR FRONT ENTRY, PORCH

Fig. 12–17 shows a possible device and wiring layout for the front entry and porch area. This area is connected to circuit A15. The home run enters the six-gang switch box next to the front door. From this box, the circuit spreads out, feeding the recessed closet light, the porch bracket fixture, the two bracket fixtures on the front of the garage, one receptacle outlet in the entry, and the post light in the front lawn. The outdoor weatherproof GFCI receptacle on the porch is connected to circuit A24 along with the other outdoor receptacles. Table 12–2 summarizes the outlets and estimated load for the entry and outdoor lights at the front of the house.

Fig. 12–17   Cable layout of the entry–porch, circuit A15.

The receptacle on the porch is controlled by a single-pole switch just inside the front door, which gives the homeowner control of decorative lighting plugged into the outdoor receptacle, Fig. 12–18.

Typical ceiling fixtures commonly installed in front entryways, where it is desirable to make a good first impression on guests, are shown in Fig. 12–19. Typical wall-mounted porch and entrance lighting fixtures are shown in Fig. 12–20.

According to *Rule 30–318*, any wiring of light fixtures in damp or wet locations requires fixtures to be approved and marked for use in such locations. For example, the fixtures will be marked "Suitable for wet locations."

The *CEC* defines partially protected areas under roofs, canopies, or open porches as damp locations. Fixtures to be used in these locations must be marked "Suitable for damp locations." Many types of fixtures are available. Always check the label on the fixture for CSA approval or consult with your electrical supplier to determine the suitability of the fixture for a wet or damp location. See Fig. 12–21.

DESCRIPTION	QUANTITY	WATTS	VOLT-AMPERES
Receptacles @ 120 W	1	120	120
Outdoor porch bracket fixture	1	100	100
Outdoor garage bracket fixtures @ 100 W each	2	200	200
Ceiling fixture	1	100	100
Living room track light	1	200	200
Post light	1	100	100
Closet recessed fixture	1	60	60
TOTALS	8	880	880

Table 12–2   Entry and porch: outlet count and estimated load, circuit A15.

Probably the most difficult part of the A15 circuit is planning the connections at the switch location just inside the front door, where there are six toggle switches:

- One single-pole switch for the front entrance ceiling fixture

- One single-pole switch for the light fixtures at the front of the garage
- One single-pole switch for the post light
- One single-pole switch for the porch light
- One three-way switch for the light over the basement stairs
- One single-pole switch for the weatherproof receptacle on the porch (circuit A24).

This location is another challenge for determining the proper size wall box.

1. Count the circuit conductors
   $2 + 2 + 2 + 2 + 2 + 3 + 3$ $= 16$
2. Add the switches
   6 switches $= 12$
3. Add the wire connectors
   2 connectors $= 1$

Total   29

The Quick-check box selection guide (Fig. 5–16) and Tables 5–1 and 5–2 give many possibilities. For example, two sets of three $76 \times 51 \times 64$-mm ($3 \times 2 \times 2\frac{1}{2}$-in) device boxes can be ganged together ($6 \times 8 = 48$ conductors).

LOW-VOLTAGE DECORATIVE LAWN LIGHTING

120- TO 12-VOLT TRANSFORMER

Fig. 12–18   Typical low-voltage decorative lighting for use around shrubbery. Cord plugs into 120-volt receptacle and feeds transformer that reduces 120 volts to a safer 12 volts.

Fig. 12–19   Hall luminaires: ceiling mount and chain mount. Photography courtesy of Progress Lighting/Hubbell.

Fig. 12–20   Outdoor porch and entrance luminaires: wall bracket styles. Photography courtesy of Progress Lighting/Hubbell.

Fig. 12–21   Outdoor areas considered to be *damp* or wet locations.

An interesting possibility presents itself for the front entry closet. Although the plans show that the recessed closet fixture is turned on and off by a standard single-pole switch to the left as you face the closet, a doorjamb switch could have been installed. A doorjamb switch is usually mounted about 1.83 m above the floor, on the inside of the 2 × 4 framing for the closet door. The electrician runs the cable to this point and lets it hang out until the carpenter can cut the proper size opening into the doorjamb for the box. After the finish woodwork is completed, the electrician installs the switch, Fig. 12–22. The plunger on the switch is pushed inward when the edge of the door pushes on it as it closes, shutting off the light. The plunger can be adjusted in or out to make the switch work properly. These doorjamb switches come complete with their own special wall box.

Fig. 12–22   Door switch.

REVIEW

Note: Refer to the *CEC* or the plans where necessary.

1. How many convenience receptacles are connected to the living room circuit? _____

   _____

2. a. How many wires enter the switch box at the four-way switch location? _____

   _____

   b. What type and size of box may be installed at this location? _____

   _____

3. a. What is meant by incandescent lamp inrush current? _____
   _____

   b. Does the *CEC* require "T" rated switches for incandescent lamps in dwelling units?
   _____

4. Complete the wiring diagram for the dimmer and lamp.

5. What is the difference between a split-switched receptacle and a split-wired (three-wire) receptacle? _____
   _____
   _____
   _____

6. How many branch circuits are required for

   a. a split-switched receptacle? _____

   b. a split-wired (three-wire) receptacle? _____

7. Explain why fluorescent lamps having the same wattage can draw different current values. _____
   _____
   _____
   _____
   _____

8. What is the total current consumption of the track lighting?

9. a. How many television outlets are provided in the living room? _____
   _____

   b. What symbol is used on the drawing to indicate the TV outlet? _____
   _____

10. a. Where is the telephone outlet located in the living room? _____
    _____

    b. What symbol is used on the drawing to indicate the telephone outlet? _____
    _____

11. A layout of the outlets, switches, dimmers, track lighting, and recessed fixtures is shown in the following diagram. Using the cable layout of Fig. 12–1, make a complete wiring diagram of this circuit. Use coloured pencils or marking pens to indicate conductors.

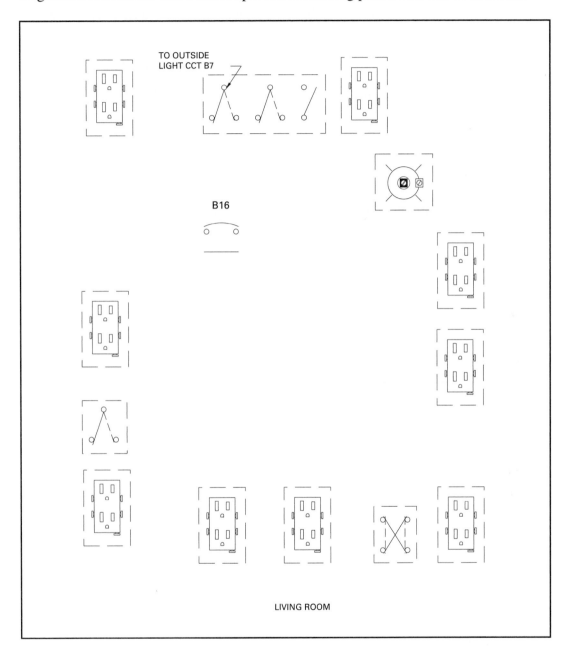

12. Must track lighting always be fed (connected) at one end of the track? _____

_____

13. If there is an a seven-metre piece of track lighting, how many outlets is this considered to have? _____

_____

14. Using the electrical plans provided, lay out devices and wiring in the living room area to *minimum CEC* standards.

15. a. How many circuit wires enter the front entry ceiling box? _____

   b. How many equipment bonding conductors enter the front entry ceiling box? _____

16. How many receptacle outlets and lighting outlets are supplied by circuit A15?

   _____

17. Outdoor fixtures directly exposed to the weather must be marked as:

   a. suitable for damp locations

   b. suitable for dry locations only

   c. suitable for wet locations

   Circle the correct answer.

18. Make a material list of all types of switches and receptacles connected to circuit A15.

   _____

   _____

   _____

19. The following layout is for lighting circuit A15, the entry, porch, and front garage lights. Using the cable layout shown in Fig. 12–17, make a complete wiring diagram of this circuit. Use coloured pencils to indicate the colour of the conductors' insulation.

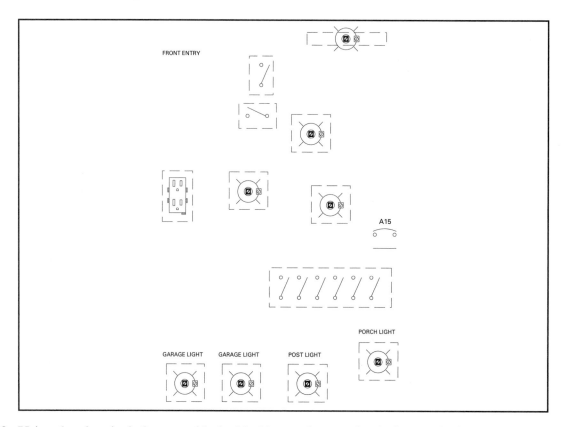

20. Using the electrical plans provided with this text, lay out the devices and wiring, including circuit numbers, for the front porch and entry to *minimum CEC* standards.

UNIT 13

Branch Circuits for Bathrooms

OBJECTIVES

After studying this unit, the student will be able to

- list bonding requirements for bathroom installations
- discuss fundamentals of proper lighting for bathrooms
- explain the operation and switching sequence of the heat/vent/light/night-light
- describe the operation of a humidistat
- install attic exhaust fans with humidistats, both with and without a relay
- list the various methods of controlling exhaust fans
- understand advantages of installing exhaust fans in residences
- understand the electrical circuit and *CEC* requirements for hydromassage bathtubs

Electrical equipment in bathrooms and washrooms requires special installation considerations. For example, the hydromassage bathtub in the en suite bathroom requires ground fault circuit protection. Switches must be properly located. There must be enough circuits to operate all the equipment without the possibility of overloading circuits. It is also important to correctly locate luminaires to achieve proper lighting.

The vanity lighting for the two bathrooms is provided from the circuit that supplies the GFCI-protected receptacle in each bathroom (A20 and A22). Because the receptacles in the main bathroom and en suite bath are likely to be used in the morning at the same time by people using hair dryers that draw as much as a toaster or kettle (1200 W), we have put the receptacles on their own circuit. This is not a *CEC* requirement.

Each bathroom has a ceiling heater/light/night-light/fan that is connected to separate circuits shown as $\blacktriangle_J$ and $\blacktriangle_K$. A hydromassage tub, located in the bathroom serving the master bedroom, is connected to separate circuit $\blacktriangle_A$.

Fig. 13–1 and the electrical plans for this area of the home show that each bathroom has a fixture above the vanity mirror: Some typical fixtures are shown in Fig. 13–2. The homeowner might decide to purchase a medicine cabinet complete with a self-contained lighting fixture. Fig. 13–3 illustrates how to rough-in the wiring for medicine cabinets with or without such a lighting fixture. These fixtures are controlled by single-pole switches at the doors.

A bathroom or powder room should have proper lighting for shaving, combing hair, putting on makeup, and so on. If the face is poorly lit and in shadow, that is what will be reflected in the mirror.

Fig. 13–1   Cable layout for bathrooms.

Fig. 13–2   Vanity luminaires of the side bracket and strip types. Photography courtesy of Progress Lighting/Hubbell.

Fig. 13–3   Two methods most commonly used for roughing-in the wiring for lighting above a vanity.

A lighting fixture directly overhead will light the top of one's head, but will cause shadows on the face. Mirror lighting and/or adequate lighting above and forward of the standing position at the vanity can provide excellent lighting in the bathroom. Figs. 13–4 to 13–6 show pictorial and section views of typical soffit lighting above a bathroom vanity.

The receptacles in each bathroom are located adjacent to the vanity, but must be at least 1 m from the bathtub or shower when practicable, *Rule 26–710(g).* Switches controlling lights must not be within reach of the person in the shower or bathtub. As required by *Rule 26–700(11)* and as discussed in Unit 9, these must be GFCI receptacles. *Section 0* defines a bathroom as "a room containing a wash basin together with bathing or showering facilities," Fig. 13–7.

GENERAL COMMENTS ON LAMPS AND COLOUR

Incandescent lamps (light bulbs) provide pleasant colour tones, bringing out the warm flesh tones similar to those of natural light.

With fluorescent lighting, you should use fluorescent lamps that bring out the warm flesh tones, such as warm white (WW) and warm white deluxe (WWX), which simulate incandescent lamps, or cool white deluxe (CWX), which simulates outdoor daylight on a cloudy day.

White (W) or cool white (CW) fluorescents downplay flesh tones and give the skin a pale appearance. Compact fluorescent lamps provide excellent colour rendition.

WRONG

RIGHT

Fig. 13–4   Positioning bathroom lighting fixtures.

HANGING FIXTURES IN BATHROOMS

Rule 30–320(1) requires any light fixture (luminaire) or lampholder that is located within 2.5 m vertically or 1.5 m horizontally of bathtubs, laundry tubs, or other grounded surfaces to be controlled by a wall switch. The switch must be located at least 1 m from a tub or shower so that a person standing in the tub or shower will not be able to reach the switch. See Figs. 13–8 and 13–9.

HALLWAY LIGHTING

The hallway lighting is provided by one ceiling fixture controlled with two three-way switches located at either end of the hall and supplied from circuit A16.

RECEPTACLE OUTLETS IN HALLWAYS

One receptacle outlet has been provided in the hallway, supplied from circuit A14. *Rule 26–712(f)* requires that in a dwelling unit no point in a hallway shall be more than 4.5 m from a duplex receptacle.

BONDING REQUIREMENTS FOR A BATHROOM CIRCUIT

All exposed metal equipment, including fixtures, electric heaters, faceplates, and similar items, must be bonded. The bonding requirements for bathroom circuits and any other equipment in a residential building are contained in *Rule 10–400*.

In general, all exposed noncurrent-carrying metal parts of electrical equipment must be bonded together and connected to ground:

- if wiring supplying the equipment contains a bonding conductor, *Rule 10–400(b)*
- if the ground or a grounded surface and the electrical equipment can be touched at the same time, *Rule 10–400(d,e)*
- if they are located in wet or damp locations, such as bathrooms, showers, and outdoors, *Rule 10–400(c)*
- if they are in electrical contact with metal, including metal lath and aluminum foil insulation, *Rule 10–400(g)*.

TYPICAL INCANDESCENT RECESSED SOFFIT LIGHTING
OVER BATHROOM VANITY

TYPICAL FLUORESCENT RECESSED SOFFIT LIGHTING
OVER BATHROOM VANITY. NOTE ADDITIONAL
INCANDESCENT "SIDE–OF–MIRROR" LIGHTING.

END CUTAWAY OF SOFFIT SHOWING RECESSED
INCANDESCENT FIXTURES IN TYPICAL SOFFIT ABOVE
BATHROOM VANITY. TWO OR THREE FIXTURES
GENERALLY INSTALLED TO PROVIDE PROPER LIGHTING.

Fig. 13–5   Incandescent soffit lighting.

END CUTAWAY VIEW OF SOFFIT ABOVE BATHROOM
VANITY SHOWING RECESSED FLUORESCENT FIXTURES
CONCEALED ABOVE TRANSLUCENT ACRYLIC LENS

Fig. 13–6   Combination fluorescent and incandescent bathroom lighting.

Equipment is considered bonded to ground when it is properly and permanently connected to a metal raceway, a separate bonding conductor, or the bonding conductor in non-metallic-sheathed cable or armoured cable. The bonding conductor must itself be properly grounded.

According to *Rule 10–408*, residential cord-connected electrical appliances must be grounded.

There are two exceptions to this basic requirement. *Rule 10–408(3)* accepts "double insulating" or equivalent in lieu of grounding. *Rule 10–408(4)* provides an exemption from the bonding requirements of subrule (1) provided the equipment is used only in a location where reliable grounding cannot be obtained and the equipment is supplied from a double-insulated portable Class A GFCI.

(A) THIS IS A WASHROOM.
(BASIN, TOILET)

(B) THIS IS A BATHROOM.
(BASIN, TOILET, TUB)

(C) THIS IS *NOT* A BATHROOM.
(BASIN ONLY) IT IS A WASHROOM.

Fig. 13–7   Definition of a bathroom, *Section 0*.

Double-insulated appliances must be clearly marked ▢ , or with the words "Double insulated," and do not require a separate bonding conductor. These appliances are furnished with a two-wire cord that is designed for a polarized receptacle. The short blade must be connected to the "hot" conductor and the long blade to the neutral or identified conductor, *Rule 10–408(3), Appendix B.*

Residential electrical appliances that must be grounded, unless double insulated, are:

facial saunas

freezers and refrigerators

clothes washers and dryers

dishwashers

aquarium equipment

hand-held motor-operated tools

electric motor-operated hedge trimmers, lawn mowers, and snow blowers

air conditioners

sump pumps

BATHROOM CEILING HEATER CIRCUITS ▲K ▲J

In the ceilings of both bathrooms is a combination heater, light, night-light, and exhaust fan, similar to the one shown in Fig. 13–10. The symbols ▲K and ▲J represent the outlets for these units.

The bathroom ventilation units are Heat/ Vent/Light/Night-light combination units (AK965) supplied by Air King ventilation products. Each unit consists of a 1500-watt heater, separate heater and ventilation motors rated at 0.9 ampere each, a 7-watt night-light, and either a 100-watt incandescent or 26-watt compact fluorescent main lamp. The unit is IC rated for installation in ceilings insulated to R40 and has a sound rating of 3.5 sones. The unit is not designed for mounting over a tub or shower stall or for use with solid state speed control devices on the fans. The unit has been tested and certified by Underwriters Laboratories to CSA standard C22.2 electric air heaters.

According to the manufacturer's specifications and installation instructions, the minimum circuit requirements for the unit are 120 volts, 18.6 amperes, 60 Hz. (The actual full load current based on a 1500-watt heater, two 0.9 FLA motors, a 60-watt main light, and a 7-watt night-light would be 14.9 amperes.) It must be wired on a separate 20-ampere circuit. Install number 12 AWG copper NMD-90 cable from the panel to a 2-gang switch box and from the switch box to the heater box as shown in Fig. 13-11. The switches must be located so they cannot be reached from the tub or shower. For this residence, a two-gang switch box will be used to accommodate the switches for the AK965 unit and a single gang for the vanity light.

Bonding the Combination Heater

To bond the unit, connect the bare bonding conductor of the non-metallic sheathed cable to the bonding terminal or lug on the unit.

Fig. 13–8 The horizontal area around a tub in which luminaires must be controlled by a wall switch, *Rule 30–320.*

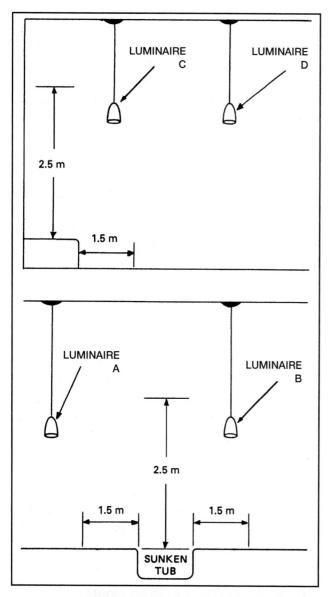

Fig. 13–9 Luminaires B and C must be controlled by a wall switch outside the reach of a person in the tub. A wall switch must be a minimum of 1 m from the tub. Where this is not possible, the spacing may be reduced to 500 mm if the switch is protected by a GFCI, *Rule 30–320.*

Fig. 13–10 Combination heater, ventilation fan, light, and night-light.

Fig. 13–11 Wiring for bathroom combination units.

When connecting ducting to the unit, it is critical to vent the unit to the outdoors. If a unit is vented to the attic, the moisture in the warm exhaust air will condense in the attic during the winter, which can cause a build-up of ice on the inside of the roof resulting in significant damage to the structure of the building during freeze-melt cycles. It also reduces the effectiveness of the ceiling insulation.

HYDROMASSAGE TUB CIRCUIT ▲A

The master bedroom is equipped with a hydromassage bathtub, sometimes referred to as a whirlpool bath. Pools, tubs, and spas are covered by *CEC Section 68.*

Rule 68–050 defines a hydromassage bathtub as a permanently installed bathtub having an integral or remote water pump or air blower, and having a fill and drain water system.

In other words, fill—use—drain!

The significant difference between a hydromassage bathtub and a regular bathtub is the recirculating piping system and the electric pump that circulates the water. Both types of tubs are drained completely after each use.

Spas and hot tubs are intended to be filled, then used. They are *not* drained after each use because they have a filtering and heating system.

Electrical Connections

The hydromassage tub is fed with circuit A9, a separate 15-ampere, 120-volt circuit using No. 14 AWG NMD90.

Table S–1, Schedule of Special-Purpose Outlets, on page 428 of this book's Appendix A, indicates that the hydromassage bathtub in this residence has a $1/2$-hp motor that draws 10 amperes.

Rule 68–302 requires that the circuit supplying a hydromassage bathtub have Class A GFCI protection, as discussed in Unit 9.

Rule 68–304(1) requires that the hydromassage bathtub be controlled by an on/off device. An automatic shut-off timer is recommended, Fig. 13–12.

Rule 68–304(2) requires that the associated controls be located behind a barrier not less than 1 m horizontally from the wall of the hydromassage bathtub.

Spas, hot tubs, and swimming pools introduce additional electrical shock hazards; these issues are covered in Unit 23.

The hydromassage bathtub's electrical control is prewired by the manufacturer. The electrician just has to run the separate 15-ampere, 120-volt GFCI-protected circuit to the end of the tub where the pump and control are located. Generally, the manufacturer will supply a 900-mm length of watertight flexible conduit that contains one black, one white, and one green equipment bonding conductor, Fig. 13–13. Make sure that the equipment bonding conductor of the circuit is properly connected to the green bonding conductor of the hydromassage tub.

Do not connect the green bonding conductor to the grounded (white) circuit conductor.

Make proper electrical connections in the junction box where the branch circuit wiring is to be connected to the hydromassage wires. Because it contains splices, this junction box must be accessible, usually by means of a removable cover on the side of the hydromassage tub.

Fig. 13–12 A 30-minute timer for control of a hydromassage bathtub.

Fig. 13–13 Typical wiring diagram of a hydromassage tub showing the motor, power panel, and electrical supply leads, *Rule 68–300*.

The pump and power panel may also need servicing. Access may be from underneath or the end, whichever is convenient for the installation, Figs. 13–14 and 13–15.

Fig. 13–16 shows a typical hydromassage tub.

Fig. 13–15 Hydromassage bathtub showing motor, pneumatic pushbutton (upper right), and flexible cord.

Fig. 13–14 The basic roughing-in of a hydromassage bathtub is similar to that of a regular bathtub. The electrician runs a separate 15-ampere, 120-volt GFCI-protected circuit to the area where the pump and control are located. Check the manufacturer's specifications for this data. An access panel from the end or from below is necessary to service the wiring, the pump, and the power panel after the installation is complete.

Fig. 13–16 A typical hydromassage bathtub, sometimes referred to as a whirlpool. Courtesy of Robert Nolan/ Shutterstock

REVIEW

Note: Refer to the *CEC* or the plans where necessary.

1. Most appliances of the type commonly used in bathrooms, such as hair dryers, electric shavers, and curling irons, have two-wire cords. These appliances are _____ insulated.

2. The *CEC* requires that all receptacles in bathrooms be _____ protected.

3. Draw the symbol or other markings used to identify double-insulated equipment.

_____

_____

_____

BATHROOM CEILING HEATER CIRCUITS ▲_J ▲_K

1. What is the wattage rating of the heat/vent/light/night-light? _____

2. To what circuits are the heat/vent/lights connected? _____

3. What type of box would you use for the rough-in of the switch assembly for the heat/vent/light? _____

_____

Why? _____

_____

_____

4. a. How many wires are required to connect the control switch and the heat/vent/light/night-light? _____

   b. What size wires are used? _____

5. Can the heating element be energized when the fan is not operating? _____

_____

6. Can the fan be turned on without the heating element? _____

7. What device can be used to provide automatic control of the heating element and the fan of the heat/vent/light? _____

8. Where does the air enter the heat/vent/light? _____

9. Where does the air leave this unit? _____

10. Who is to furnish the heat/vent/light? _____

_____

_____

11. For a ceiling heater rated 1200 watts at 240 volts, what is the current draw?

_____

HYDROMASSAGE BATHTUB CIRCUIT ▲_A

1. What circuit supplies the hydromassage bathtub? _____

2. Should the circuit supplying the hydromassage bathtub have GFCI protection?

_____

3. What conductor size feeds the hydromassage tub? _____

4. What is the fundamental difference between a hydromassage bathtub and a spa?

   _____

   _____

5. What rules of the *CEC* refer to hydromassage bathtubs? _____

   _____

6. Must the metal parts of the pump and power panel of the hydromassage tub be grounded?

   _____

   _____

7. Where should the electric controls be located for the hydromassage tub? _____

   _____

   _____

8. Complete the wiring diagram below.

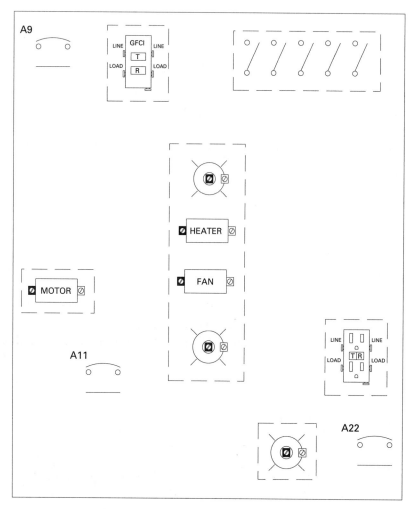

MASTER BATHROOM CCTS A9, 11, 22

UNIT 14

Lighting Branch Circuit and Small Appliance Circuits for the Kitchen

OBJECTIVES

After studying this unit, the student will be able to

- explain the *CEC* requirements for small appliance circuits in kitchens
- understand the *CEC* requirements for receptacles that serve countertops
- discuss split-circuit receptacles
- discuss multiwire circuits
- discuss exhaust fan noise ratings
- discuss typical kitchen, dining area, and undercabinet lighting

LIGHTING CIRCUIT B7

Circuit B7 supplies the lighting in the kitchen and service entrance as well as the outside light between the sliding doors in the kitchen and living room, Fig. 14-1. There are eight outlets on Circuit B7. See Table 14-1.

KITCHEN LIGHTING

The general lighting for the kitchen is provided by two track lights, controlled by three-way switches located next to the doorway leading to the living room and at the rear entrance. All fixtures must be CSA approved.

An adjustable swag fixture (Fig. 14–2A) is mounted on a track to provide lighting above the breakfast nook table. This fixture is controlled by three-way switches: one located adjacent to the sliding door, the other at the doorway leading to the living room. Many fixtures of this type are provided with a dimmer switch to allow one to select the lighting level. Figs. 14–2A, B, and C show other types of ceiling-mounted lighting fixtures used in kitchens, dining areas, and dining rooms.

One goal of good lighting is to reduce shadows in work areas, particularly important in the kitchen. The two ceiling fixtures, the fixture above the sink plus the fixture within the range hood will provide excellent lighting in the work area of the kitchen. Track lighting in the nook offers good lighting for that area.

Undercabinet Lighting

In homes with little natural light, the architect might specify strip fluorescent fixtures under the kitchen cabinets. Several methods may be used to install undercabinet lighting fixtures: They can be fastened

DESCRIPTION	QUANTITY	WATTS	VOLT-AMPS
Track light	2	400	400
Recessed fixture over sink	1	60	60
Outdoor rear bracket	1	100	100
Range hood fan Lamps two × 60W Motor 120 volt, 1.33 amps	1	120 160	120 180
Recessed lights in service entrance	2	120	120
Receptacle in service entrance	1	120	120
Total	8	1080	1100

Table 14–1   Kitchen lighting: outlet count and estimated load, circuit B7.

to the wall just under the upper cabinets, Fig. 14–3(A); installed in a recess that is part of the upper cabinets so that they are hidden from view, Fig. 14–3(B); or fastened under and to the front of the upper cabinets, Fig. 14–3(C). All three require close coordination with the cabinet installer to be sure that the wiring is brought out of the wall at the proper location to con-

nect to the undercabinet fixtures. The installation of extra-low-voltage cabinet lighting systems is covered by *Rule 30–1208.*

Lamp Type

Good colour rendition in a kitchen area is achieved with incandescent lamps or fluorescent fixtures with compact lamps or standard warm white deluxe (WWX), warm white (WW), cool white deluxe (CXW), or Ultralume 3000 lamps.

Fluorescent lamps use up to 80 percent less energy than incandescent ones and last 10 to 20 times as long, saving on the cost of replacement. Fluorescent lighting also produces less heat, thus saving on air conditioning.

For example, two F40 34-watt fluorescent lamps (78 watts including ballast) provide about the same illumination as five 60-watt incandescent lamps (300 watts). When used five hours a day at $0.08/kWH, a luminaire with two fluorescent lamps will save $32.41 annually.

Fig. 14–1   Cable layout for the kitchen lighting, circuit B7. The kitchen receptacles are not shown in this lighting circuit layout.

Fig. 14–2A Fixtures that can be hung either singly or on a track over dinette tables. Photography courtesy of Progress Lighting/Hubbell.

Fig. 14–2B   Typical kitchen ceiling fixture: which might have three 13-watt compact fluorescent lamps or two 34-watt U-shaped lamps.

Fig. 14–2C   Types of fixtures used over dining room tables. Photography courtesy of Progress Lighting/Hubbell.

Fig. 14–3   Methods of installing undercabinet lighting fixtures.

Fan Outlet

The fan outlet is connected to branch lighting circuit B7. Fans can be installed in a kitchen either to exhaust the air to the outside of the building via ducts or vents, or to recirculate the air through various types of filters (e.g., charcoal or foam). Both types of fans can be used in a residence. The ductless hood fan does not exhaust air to the outside and does not remove humidity from the air. Fig. 14–4 shows a typical range hood exhaust fan.

Homes heated electrically with resistance-type baseboard units usually have excess humidity. These heating units neither add nor remove moisture from the air. The proper use of weather stripping, storm doors, storm windows, and a vapour barrier of polyethylene plastic film tends to retain humidity.

Vapour barriers are installed to keep moisture out of the insulation. This protection normally is installed on the warm side of ceilings, walls, and floors; under concrete slabs poured directly on the ground; and as a ground cover in crawl spaces. Insulation must be kept dry to maintain its effectiveness. For example, a one percent increase of moisture in insulating material can reduce its efficiency by five percent. (Humidity control using a humidistat connected to an exhaust fan is covered in Unit 16.) Building inspectors require the vapour barrier to be intact around all outlet boxes on the outside walls of the house (Fig. 5–7B). It is the electrician's responsibility to maintain the integrity of the vapour barrier, *Rule 12–3000, Appendix B*.

An exhaust fan simultaneously removes odours and moisture from the air and exhausts heated air, Fig. 14–4. The electrician, builder, and/or architect may make suggestions to guide the owner in making the choice of type of fan to install.

For the residence in the plans, a separate wall switch is not required, because the speed switch, light, and light switch are integral parts of the fan.

Fan Noise

The fan motor and the air movement through an exhaust fan generate noise. The unit used to define fan noise is the *sone* (rhymes with tone), a subjective unit equal to the loudness of a 1000-hertz tone at 40 decibels as perceived by a person with normal hearing. One sone is about the noise level of an average refrigerator. The lower the sone rating, the quieter the fan. All manufacturers of exhaust fans provide this information in their descriptive literature.

Clock Outlets

A deep sectional box is recommended because a recessed clock outlet takes up considerable room in the box, Fig. 14–5.

Fig. 14–4   Typical range hood exhaust fan. Speed control and light are integral parts of the unit. Courtesy of Broan-NuTone.

Fig. 14–5   Recessed clock receptacle.

Fig. 14–6   Single and duplex receptacles.

Some decorative clocks have the entire motor recessed so that only the hands and numbers are exposed on the surface of the soffit or wall. Some of these clocks require special outlet boxes, usually furnished with the clock. Other clocks require a standard 4-in square outlet box. Accurate measurements must be taken to centre the clock between the ceiling and the bottom edge of the soffit. If the clock is available when the electrician is roughing in the wiring, its dimensions can be checked to help in locating the outlet.

If an electrical clock outlet has not been provided, battery-operated clocks are available.

SMALL APPLIANCE AND BRANCH CIRCUITS FOR CONVENIENCE RECEPTACLES IN KITCHEN

Fig. 14–6 shows single and duplex receptacles. The *CEC* requirements for small appliance circuits and the spacing of receptacles in the kitchens of dwellings are covered in *Rules 26–712, 720, 722, and 724*. This was discussed in detail in Unit 6. Highlights are:

- A separate circuit is required to supply receptacle outlets in the eating area of the kitchen.
- These small appliance circuits shall be assigned a maximum load of 12 amperes each when calculating branch circuit connectors.
- A separate circuit is required for the refrigerator.

Fig. 14–7   A receptacle installed solely to supply and support an electric clock may be connected to a refrigerator receptacle, *Rule 26–720(a).*

- The clock outlet may be connected to the refrigerator circuit only when this outlet is to supply a recessed clock receptacle intended for use with an electric clock, *Rule 26–720(a)* (Fig. 14–7). In the residence in the plans, the clock outlet is connected to the lighting circuit.
- Countertop receptacles must be supplied by at least two 15-ampere, three-wire branch circuits. As an alternative, 5–20 RA receptacles may be used in accordance with *Rule 26–722(b)*, Fig. 14–8.

Fig. 14–8 Layout of receptacles behind the counter in a kitchen. If two or more countertop receptacles are required by *Rule 26–712*, at least two 20-ampere branch circuits or two 15-ampere, three-wire branch circuits are required, *Rule 26–722(b)*.

Fig. 14–9 In kitchens all 125-volt, 15-ampere duplex receptacles located along the wall of counter work surfaces must be of the split-wired type (three-wire circuit) or 20-ampere 125-volt T-slot receptacles must be used. Receptacles must be located so that no point along the wall line is more than 900 mm from a receptacle, *Rule 26–702*.

- There must be a receptacle within 900 mm of any point along the wall behind all counters except those less than 300 mm long, Fig. 14–9.

- Microwave enclosures must be supplied with a receptacle on a separate circuit, *Rule 26–720*. If no cupboard or enclosure is provided, the microwave would be powered by one of the countertop split receptacles.

- All outdoor receptacles must be on a separate GFCI-protected circuit.

- Counter receptacles may be three-conductor split-circuit or 20-ampere T-slot receptacles. Counter receptacles prevent circuit overloading when small appliances such as kettles and toasters are plugged in at the same time.

In the kitchen, split-circuit receptacles and 20-ampere T-slot receptacles prevent circuit overloading from the heavy concentration and use of electrical appliances. For example, one type of cord-connected microwave oven is rated 1440 watts at 120 volts. The current required by this appliance alone is:

$$I = \frac{W}{E} = \frac{1440}{120} = 12.0 \text{ amperes}$$

Fig. 14–10 shows the kitchen and laundry special-purpose receptacle circuiting.

A maximum of two counter receptacles are permitted on one circuit. This provides the best availability of electrical power for the variety of appliances that will be used in the heavily concentrated work areas in the kitchen.

Fig. 14–10 shows the circuiting for the special-purpose outlets and counter receptacles in the kitchen.

A receptacle outlet installed below the sink for easy plug-in connection of a food waste disposal

Fig. 14–10 Circuiting for special-purpose and counter receptacles in the kitchen and laundry.

is *not* included in the *CEC* requirements for split circuits and is not required to be GFCI protected.

SPLIT-CIRCUIT RECEPTACLES AND MULTIWIRE CIRCUITS

Split-wired receptacles may be installed where a heavy concentration of plug-in load is anticipated, in which case each half of the duplex receptacle is connected to a separate circuit, as in Figs. 14–11 and 14–12. Examples of two-pole, 15-ampere breakers that feed three-wire branch circuits for use with kitchen split receptacles are found in Fig. 14–13.

Split or 20-ampere T-slot receptacles are mandatory along the countertop in the kitchen, *Rule 26–712*. These must be arranged so that no more than two receptacles are on a circuit. Any receptacle within 1.5 m of the sink must be a GFCI receptacle or GFCI protected.

When a multiwire branch circuit is used to supply fixed lighting loads and non-split duplex receptacles, the continuity of the identified neutral conductor must be independent of the device con-

nections. Instead of making connections of the neutral at the terminals of the receptacle, the neutral is tailed in the junction box and the tail is brought out to the receptacle, Fig. 14–11.

Rule 14–010(b) does not require all the ungrounded conductors of multiwire branch circuit supplying light fixtures or receptacles (which are connected to the neutral and one ungrounded conductor) to be disconnected simultaneously. Therefore, when the breaker in a panel supplying a receptacle is turned off, there still may be current flowing in the neutral conductor. For that reason, *Rule 4–034(4)* requires that the continuity of an identified conductor be independent of device connections to luminaires and receptacles. The hazards of open neutrals are covered in Unit 19.

When multiwire branch circuits supply more than one receptacle on the same yoke, a means must be provided to simultaneously disconnect all of the hot conductors at the panelboard where the branch circuit originates. Fig. 14–12 shows that although 240 volts is present on the wiring device, only 120 volts is connected to each receptacle on the wiring

Fig. 14–11   Connecting the neutral in a three-wire (multiwire) circuit, *Rule 4–034(4)*.

Fig. 14–12   Split receptacle connected to multiwire circuit.

BQLT 15-215   TWO-POLE FULL MODULE

Fig. 14–13 Examples of two-pole 15-ampere breakers for use in multiwire branch circuits for kitchen split receptacles.

device. The "simultaneous disconnect" rule applies to any receptacles, lampholders, or switches mounted on one yoke, *Rule 14–010(b)*.

In the case of the three-conductor split-duplex receptacles in the kitchen connected to multiwire branch circuits, the two-pole breaker in the panel will disconnect all ungrounded conductors of the circuit simultaneously. The local inspection authority may permit you to make connections of the neutral at the terminals of the receptacle when looping between the counter receptacles in the kitchen.

GENERAL BONDING CONSIDERATIONS

The specifications for the residence require that all outlet boxes, switch boxes, and fixtures be bonded to ground. In other words, the entire wiring installation must be a grounded system.

Rules 10–400 to *414* outlines what types of equipment are required to be bonded together and to ground. Every electrical component located within reach of a grounded surface must be bonded. Post lights, weatherproof receptacles, basement wiring, boxes and fixtures within reach of sinks, ranges, and range hood fans must all be bonded. Hot water heating, hot air heating, and steam heating registers are all grounded surfaces. This means that any electrical equipment, boxes, or fixtures within reach of these surfaces must be bonded. *Rule 10–400(g)* requires that any boxes installed in contact with metal or metal lath, tinfoil, or aluminum insulation also be bonded.

Bonding is accomplished through the proper use of non-metallic-sheathed cable, armoured cable, and metal conduit.

When using non-metallic-sheathed cable, do not confuse the bare *bonding* conductor with the *grounded* circuit conductor. The grounding or bonding conductor is used to ground equipment, whereas the white grounded conductor is one of the branch circuit conductors. They may not be connected together except at the main service entrance equipment where this is achieved through the bonding screw or jumper.

REVIEW

Note: Refer to the *CEC* or the plans where necessary.

1. What provision does the *CEC* require for microwave ovens in enclosures?

   _____

   Rule # _____

2. What circuit feeds the kitchen lighting? What size conductors are used? _____

   _____

   _____

3. How many lighting fixtures are connected to the kitchen lighting circuit? _____

   _____

4. What colour fluorescent lamps are recommended for residential installations? _____

   _____

   _____

5. a. What is the minimum number of 15-ampere split circuits or 20-ampere circuits required for a kitchen according to the *CEC*? _____

   b. How many are there in this kitchen? _____

6. How many duplex receptacle outlets are provided in the kitchen? _____

7. What is meant by the term "split-circuit receptacle"? _____

   _____

   _____

   _____

   _____

   _____

8. A single receptacle connected to a circuit must have a rating _____ the ampere rating of the circuit.

9. In kitchens, receptacles along the countertop must be installed no farther than _____ mm apart.

10. A fundamental rule regarding the grounding of metal boxes, fixtures, and so on is that they must be grounded when "in reach of _____."

11. How many circuit conductors enter the box:

    a. in the track lighting ceiling box over the eating area? _____

    b. at the two-gang switch located to the right of the sliding door? _____

12. How much space is there between the countertop and upper cabinets? _____

13. Where is the speed control for the exhaust fan located? _____

14. Who furnishes the range hood? _____

15. List the appliances in the kitchen that must be connected. _____

    _____

    _____

    _____

    _____

16. Complete the wiring diagram by connecting feedthrough GFCI 1 to protect receptacle 1, both to be supplied by circuit A1. Connect feedthrough GFCI 2 to also protect receptacle 2, both to be supplied by circuit A2. Use coloured pencils or markers to show insulation colours. Assume that the wiring method is EMT, where more freedom in the choice of insulation colours is possible.

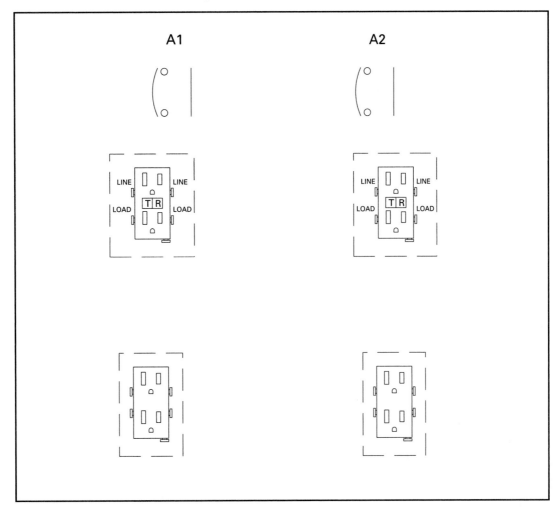

17. Each 15-ampere appliance circuit load demand shall be determined at (1440 watts) (1500 watts) (1800 watts). Choose one.

18. It is permitted to connect an outlet supplying a clock receptacle to the branch circuit dedicated to the refrigerator circuit. (True) (False) Circle one.

19. a. It is permitted to connect a receptacle in the dining area of a kitchen to one of the kitchen split circuits. (True) (False) Circle one.

    What *CEC* rule applies to this situation? _____

    b. Receptacles located above the countertops in the kitchen must be supplied by at least _____ 15-ampere, three-wire circuits.

20. The following is a layout for the lighting circuit for the kitchen. Complete the wiring diagram using coloured pens or pencils to show the conductors' insulation colour.

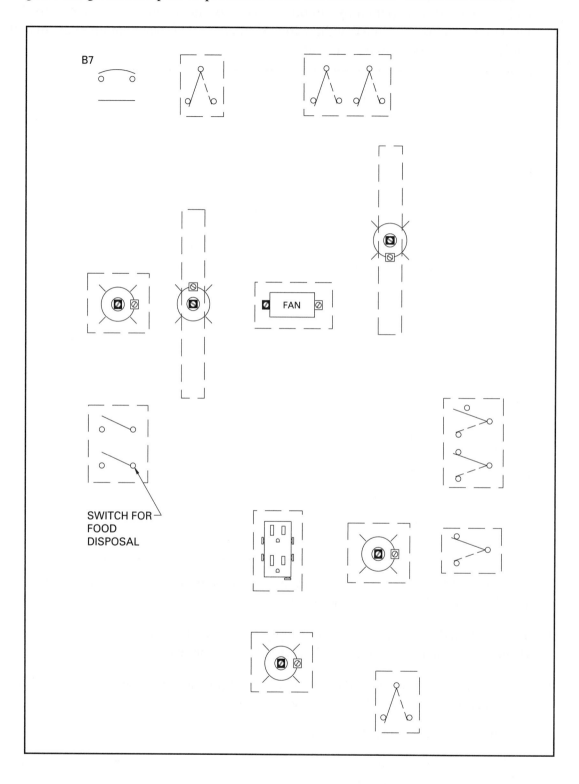

B7

SWITCH FOR
FOOD
DISPOSAL

FAN

21. a. According to *Rule 26–712(a)*, no point along the floor line shall be more than ___ m from a receptacle outlet.

    b. According to *Rule 26–712(d)*, a receptacle must be installed in any wall space ___ m wide or greater.

22. The *CEC* states that in multiwire circuits, the screw terminals of a receptacle must *not* be used to splice the neutral conductors. Why? Quote the rule number.

    _____

    _____

    _____

23. Exhaust fans produce noise. It is possible to compare the noise levels of different fans before installation by comparing their _____ ratings.

24. Is it permitted to connect the white grounded circuit conductor to the grounding terminal of a receptacle? _____

UNIT 15

Special-Purpose Outlets for Ranges, Counter-Mounted Cooking Units ▲_G, Wall-Mounted Ovens ▲_F, Food Waste Disposals ▲_H, and Dishwashers ▲_I

OBJECTIVES

After studying this unit, the student will be able to

- interpret electrical plans to determine special installation requirements for counter-mounted cooking units, wall-mounted ovens, freestanding ranges, dishwashers, and waste disposals
- compute demand factors for ranges, wall-mounted ovens, and counter-mounted cooking units
- select proper conductor sizes for wiring installations based on the ratings of appliances
- ground all appliances properly regardless of the wiring method used
- understand how infinite heat temperature controls operate
- supply counter-mounted cooking units and wall-mounted ovens by connecting them to one feeder using the "tap" rule
- understand how to install a feeder to a load centre, and divide the feeder to individual circuits to supply the appliances
- install circuits for dishwashers and food waste disposals

The residence in this text has one built-in oven and one built-in cooktop. Checking the Schedule of Special-Purpose Outlets in Appendix A, we find that the built-in wall-mounted oven is supplied by circuit B6/B8, and the built-in cooktop is supplied by circuit B2/B4.

Running separate circuits to individual appliances is probably the most common method used to connect appliances. However, according to *Rule 26–746(3)*, two or more separate built-in cooking units shall be counted as one appliance. A single circuit is run from the main panel to a junction box located close to both the oven and the cooktop. From this junction box, "taps" of smaller conductors are made to feed each appliance. These taps should be less than 7.5 m long. Which method to use is governed by local rules and consideration of the cost of installation (time and material).

This is how to use a single circuit for a separate cooktop and oven, and for a freestanding range.

COUNTER-MOUNTED COOKING UNIT CIRCUIT ▲_G

Counter-mounted cooking units are available in many styles. Units may have two, three, or four surface heating elements. The surface heating elements of some cooking units are completely covered by a sheet of high-temperature ceramic. These units have the same controls as standard units and are wired according to the general *CEC* guidelines for cooking units. When the elements of these ceramic-covered units are turned on, the ceramic above each element changes colour slightly. Most of these cooking units have a faint design in the ceramic to locate the heating elements. The ceramic is rugged, attractive, and easy to clean.

In the residence in the plans, a separate circuit is provided for a standard, exposed-element, counter-mounted cooking unit. This unit is rated at 7450 watts and 120/240 volts. The ampere rating is:

$$I = \frac{W}{E} = \frac{7450}{240} \text{ amperes} = 31.04 \text{ amperes}$$

The cooking unit outlet is shown by the symbol ▲_G.

Rule 8–300 indicates how to calculate the demand on the conductors feeding the range top and cooking unit.

Installing the Cooking Unit

CEC Table 2 shows that No. 8 AWG NMD90 may be used as the circuit conductors for the 31.04-ampere cooking unit installation. These circuit conductors are connected to circuit B2/B4, a two-pole, 40-ampere circuit in Panel B.

The electrician must obtain the roughing-in dimensions for the counter-mounted cooking unit so that the cable can be brought out of the wall at the proper height. In this way, the cable is hidden from view. To simplify both servicing and installation, a plug and receptacle combination may be used in

the supply line.

Most cooktop manufacturers provide a junction box on their units: The electrician can terminate the circuit conductors of the appliance in this box and make the proper connections to the residence wiring. Or the manufacturer may connect a short length of flexible conduit to the appliance: This conduit contains the conductors to which the electrician will connect the supply conductors in a junction box furnished by either the electrician or the manufacturer.

In this residence, the cooktop range is installed in the island. (See plans for details.) The feed will be a three-conductor non-metallic-sheathed cable with three No. 8 AWG copper conductors plus an equipment bonding conductor. The cable runs out of the top of Panel B, through the ceiling joists in the recreation room, then upward into the base of the island cabinet. Because the recreation room has a drywall ceiling, the cable is run to a point just below the intended location of the kitchen island, leaving ample cable length hanging in the basement. Before the gypsum board in the basement is installed the cable must be run up into the kitchen. Later when the kitchen cabinets and the island are installed, and other finishing touches are being done in the kitchen, the built-in appliances are set into place. Following this, the electrician completes the final hookup of the built-in appliances.

The plans show that a receptacle outlet is to be installed on the side of the island containing the cooktop. This outlet is supplied by Circuit B15, Fig. 14–10.

Temperature Effects on Conductors

Before beginning the installation, always check the appliance instructions provided by the manufacturer. The supply conductors may require an insulation rated for temperatures higher than the standard rating of 60°C.

If such information is not provided, it is safe to assume that the conductors used may have insulation rated for 60°C. *Table 19* gives the conductor insulation temperature limitations and the accepted applications for conductors.

General Wiring Methods

You may use any of the standard wiring methods to connect counter-mounted cooking units and wall-mounted ovens. Freestanding ranges and clothes dryers must be connected using receptacles and cord-connected plugs, according to *Rule 26–744*. Otherwise, armoured or non-metallic-sheathed cable may be used.

Verify any local restrictions on the use of the wiring methods listed before proceeding.

Bonding Requirements

CEC Section 10 requires electric ranges, oven cooktop units, and electric clothes dryers to be bonded to ground, with a separate bonding conductor sized according to *Table 16*.

Surface Heating Elements for Cooking Units

The heating elements used in surface-type cooking units are manufactured in several steps. First, spiral-wound nichrome resistance wire is impacted in magnesium oxide (a white, chalklike powder). The wire is then encased in a nickel–steel alloy sheath, flattened under very high pressure, and formed into coils. The flattened surface of the coil makes good contact with the bottoms of cooking utensils. Thus, efficient heat transfer takes place between the elements and the utensils.

Cooktops with a smooth ceramic surface offer the user ease in cleaning and a neater, more modern appearance. The electric heating elements are in direct contact with the underside of the ceramic top.

Figs. 15–1 and 15–2 illustrate typical electric range surface units.

TEMPERATURE CONTROL

Infinite-Position Temperature Controls

Modern electric ranges are equipped with "infinite"-position temperature knobs (controls), Fig. 15–3. Older-style heat controls connect the heating elements of a surface unit to 120 volts or 240 volts in series or in parallel to attain a certain number of specific heat levels. Newer switches have a rotating cam that provides an infinite number of heat positions.

The contacts open and close as a result of a heater-bimetal device that is an integral part of the rotary control. When the control is turned on, a separate set of pilot light contacts closes and remains closed, while the contacts for the heating element open and close to attain and maintain the desired temperature of the surface heating unit, Fig. 15–4.

Inside these infinite position temperature controls, contacts open and close repetitively. The temperature of the surface heating element is regulated by the ratio of the time the contacts are in the closed position versus the open position. This is termed *input percentage*.

The contacts in a temperature control of this type have an expected life of over 250 000 automatic cycles. The knob rotation expected life exceeds 30 000 operations.

Fig. 15–1   Typical 240-volt electric range surface heating element of the type used with infinite heat controls. Courtesy of Emerson Heating Products, Emerson Electric Company.

Fig. 15–2   Typical electric range surface unit of the type used with seven-position controls, as discussed in this text. Courtesy of Emerson Heating Products, Emerson Electric Company.

Heat Generation by Surface Heating Elements

The surface heating elements generate a large amount of radiant heat. For example, a 1000-watt heating unit generates about 3415 Btu of heat per hour. A *British thermal unit* (*Btu*) is the amount of heat required to raise the temperature of one pound of water by one degree Fahrenheit.

WALL-MOUNTED OVEN CIRCUIT ▲_F

A separate circuit is provided for the wall-mounted oven. The circuit is connected to circuit B6/B8, a two-pole, 40-ampere circuit in Panel B and wired using three #8 AWG NMD90. The receptacle for the oven is shown by the symbol ▲_F.

The wall-mounted oven installed in the residence is rated at 6.6 kW or 6600 watts at 120/240 volts. The current rating is:

$$I = \frac{W}{E} \text{ amperes} = \frac{6600}{240} \text{ amperes} = 27.5 \text{ amperes}$$

Fig. 15–3   Infinite heat control switch.

After determining the current rating (I = W/E), refer to *Table 2*. You find No. 10 AWG copper NMD90 cable may be used to supply this 27.5-ampere wall-mounted oven. This 27.5-ampere load will exceed 80 percent of the rating of the overcurrent device. Therefore, A 40-ampere circuit breaker and #8 AWG NMD90 will be used. See branch-circuit requirements for ranges, ovens, and counter-top cooking units later in this unit.

Self-Cleaning Oven

Self-cleaning ovens are popular appliances because they automatically remove cooking spills from the inside of the oven. Lined with high-temperature material, a self-cleaning oven can be set for a cleaning temperature that is much higher than baking and broiling temperatures. When the self-clean control is turned on, oven temperatures will approximate 427°C. At this temperature, all residue in the oven, such as grease and drippings, is burned until it is fine ash that can be removed easily. During the self-clean cycle, for safety reasons, a special latching mechanism makes it impossible to open the oven door until a predetermined safe temperature is reached. Self-cleaning ovens do not require special wiring and are electrically connected in the same manner as standard electric ovens.

Operation of the Oven

The oven heating elements are similar to the surface elements. The oven elements are mounted in removable frames. A thermostat controls their temperature. Any oven temperature can be obtained by setting

Fig. 15–4   Typical internal wiring of an infinite heating surface control.

the thermostat knob to the proper point on the dial. The thermostat controls both the baking element and the broiling element. These elements can be used together to preheat the oven.

A combination clock and timer is used on most ovens. The timer can be preset so that the oven turns on automatically at a preset time, heats for a preset duration, and then turns off.

Ovens are insulated with fibreglass or polyurethane foam insulation placed within the oven walls to prevent excessive heat leakage. Fig. 15–5 shows a typical wiring diagram for a standard oven unit.

Microwave Oven

The electrician's primary concern is how to calculate the microwave oven load and how to install the circuit for microwave appliances to conform to *Rule 26–720*. The same rules apply for both electrical and electronic microwave cooking appliances.

BRANCH-CIRCUIT REQUIREMENTS FOR RANGES, OVENS, AND COUNTERTOP COOKING UNITS

The requirements for branch circuits supplying ranges, ovens, and countertop cooking units are covered in *Rules 8–300* and *26–742 to 26–746*.

Fig. 15–5 Wiring diagram for an oven unit.

Some key requirements are:

- Branch-circuit conductors for electric ranges, countertop cooking units, and built-in ovens are based on the demand factors of *Rule 8–300*.

- When sizing the conductors for a range, the demand is considered to be 8 kW if the range is 12 kW or less. If the range has a rating of more than 12 kW, the demand is considered to be 8kW + 40% of the amount by which the range exceeds 12 kW, *Rule 8–300*.

- Two or more separate built-in units may be considered as a single range, *Rule 8–300*.

- When a single branch circuit is used to supply separate built-in units, tap conductors to an individual unit must have an ampacity equal to the rating of the unit they supply, *Rule 26–742*.

- The demand factors of *Rule 8–300* may not be applied to cord-connected appliances such as hot plates and rangettes.

- When a gas range is installed, a receptacle will be required in the wall behind the range. The receptacle should be mounted as close to the centreline of the range as possible and not more than 130 mm above the finished floor, *Rule 26–712(d)*.

FREESTANDING RANGE

The demand calculations for a single freestanding range are simple. For example, a single electric range to be installed has a rating of 14 050 watts (exactly the same as the combined built-in units):

1. 14 050 watts – 12 000 watts = 2050 watts or 2 kW

2. According to *Rule 8–300(1)(b),* the conductors of a branch circuit supplying a range shall be considered to have a demand of 8 kW plus 40 percent of the amount that the rating of the range exceeds 12 kW.

   For example,

   2050 watts × 40% = 820 watts or 0.8 kW

3. Calculated load = 8 kW + 0.8 kW = 8.8 kW

4. In amperes, the calculated load is:

$$I = \frac{W}{E} = \frac{8820}{240} \text{ amperes} = 36.75 \text{ amperes}$$

No. 8 NMD90 conductors may be used for the branch circuit to the range. The size of the neutral is the same as that of the line conductors even though the load is minimal for the element pilot lights and clock. If a range is rated at not over 12 kW, the maximum power demand is based on *Rule 8–300(1)(a)*. For example, if the range is rated at 11.4 kW, the maximum demand is based on 8 kW:

$$I = \frac{W}{E} = \frac{8000}{240} \text{ amperes} = 33.3 \text{ amperes}$$

Therefore, No. 8 AWG three-wire NMD90 cable can be installed for this range.

The disconnecting means for a freestanding range having a demand of 50 amperes or less shall be a cord and plug, Fig. 15–6. For electric ranges, these cords and plugs are rated at 50 amperes and must have an attachment plug of CSA configuration 14-50R. Fig. 15–7 shows the correct wiring to the terminals on the receptacle. The receptacle for this plug must be flush mounted, *Rule 26–744(8)*, and be installed according to *Rule 26–744(5)*:

- centre of the receptacle not more than 130 mm above the finished floor (A.F.F.)
- as near as possible to the centreline of the space the range will occupy
- flush with the wall when practical
- with the U-ground oriented to either side

BRANCH CIRCUITS SUPPLYING SEPARATE BUILT-IN COOKING UNITS

The branch circuit that supplies a separate countertop cooking unit and a built-in oven may be treated the same as a circuit for a single range. The demand on the conductors is considered to be 8 kW for ranges up to 12 kW and 8 kW + 40 percent of the amount by which the cooking units exceed 12 kW when the combined load of the units exceed 12 kW.

Fig. 15–8 shows a countertop cooking unit and a built-in oven connected to a single branch circuit. The calculation of the demand load is:

counter-mounted cooking unit	7 450 watts
built-in oven	6 600 watts
total connected load	14 050 watts

The demand on the conductors will be considered to be:

8 kW + 40% of the amount the units exceed 12 000 watts. The units exceed 12 000 watts by: 14 050 watts – 12 000 = 2050 watts

The demand on the conductors is then:

(8000 + 40% of 2050) watts = 8820 watts

8820/240 amperes = 36.8 amperes

Therefore, number 8 AWG copper NMD90 cable may be used as the branch-circuit conductors for the cooking unit and built-in oven.

The tap conductors to the cooking unit and built-in oven must have a current-carrying capacity that is equal to the ratings of the units they supply.

Counter-mounted cooking unit:
7450/240 amperes = 31.04 amperes
8 AWG copper NMD90 (*Table 2*)

Built-in oven:
6600/240 amperes = 27.50 amperes
#10 AWG copper NMD90 (*Table 2*)

The tap conductors should not be longer than necessary. If you are in doubt about the length of these conductors, consult with your electrical inspector; otherwise consider 7.5 metres to be the limit.

At one time, a separate panel was required to provide overcurrent protection for a countertop cooking unit and built-in oven. This requirement has been deleted from the *CEC*.

FOOD WASTE DISPOSAL ⒶH

The food waste disposal outlet is shown on the plans by the symbol Ⓐ H. The food waste disposal is rated at 7.2 amperes. This appliance is connected to circuit B19, a separate 15-ampere, 120-volt branch circuit in Panel B. No. 14 NMD90 conductors are used to supply the unit. The conductors terminate in a junction box provided on the disposal.

Fig. 15–6   Range receptacle and cord set rated at 50 amperes, 125/250 volts. Be sure to mount the 119-mm box so that the ground pin will be to either side.

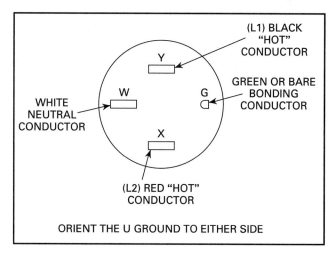

Fig. 15–7   Correct wiring connections for the terminals on the range receptacle rated at 50 amperes, 125/250 volts, CSA configuration 14–50R, *Diagram 1*.

Fig. 15–8   Counter-mounted cooktop and wall-mounted oven connected to one circuit.

Overcurrent Protection

The food waste disposal is a motor-operated appliance. Food waste disposals normally are driven by a ¹/₄- or ¹/₃-hp, split-phase, 120-volt motor. Therefore, running overcurrent protection must be provided. Most food waste disposal manufacturers install a built-in thermal protector to meet *CEC* requirements. Either a manual reset or an automatic reset thermal protector may be used.

Disconnecting Means

All electrical appliances must be provided with some means of disconnecting them, *Rule 14–010*. The disconnecting means may be a separate ON/OFF switch (Fig. 15–9), a cord connection (Fig. 15–10), or a branch-circuit switch or circuit breaker, such as that supplying the installation in Fig. 15–11.

Some local codes may require that food waste disposals, dishwashers, and trash compactors be

Fig. 15–9 **Wiring for a food waste disposal operated by a separate switch located above the countertop near the sink.**

cord-connected (Fig. 15–10) to make it easier to disconnect the unit, replace it, service it, and reduce noise and vibration. The appliance shall be intended or identified for flexible cord connection.

Turning the Food Waste Disposal On and Off

Separate ON/OFF Switch: When a separate ON/OFF switch is used, a simple circuit arrangement can be made by running a two-wire supply

cable to a switch box located next to the sink at a convenient location above the countertop (Fig. 15–9). Not only is this convenient, but it positions the switch out of the reach of children. A second cable is run from the switch box to the junction box of the food waste disposal. A single-pole switch in the switch box provides the ON/OFF control of the disposal.

In the kitchen of this residence, the food waste disposal is controlled by a single-pole switch located to the left of the kitchen sink above the countertop.

Another possibility, but not as convenient, is to install the switch for the disposal inside the cabinet space directly under the sink near the food waste disposal.

Integral ON/OFF Switch: A food waste disposal may be equipped with an integral prewired control switch. This integral control starts and stops the disposal when the user twists the drain cover into place. An extra ON/OFF switch is not required. To connect the disposal, the electrician runs the supply conductors directly to the junction box on the disposal and makes the proper connections, Fig. 15–11.

Disposal with Flow Switch in Cold Water Line: Some manufacturers of food waste disposals recommend that a *flow switch* be installed in the cold water line under the sink, Fig. 15–12. This switch is connected in series with the disposal motor. The flow switch prevents the disposal from operating

Fig. 15–10   **Typical cord connection for a food waste disposal,** *Rule 26–744(7).*

Fig. 15–11 Wiring for a food waste disposal with an integral ON/OFF switch.

until a predetermined quantity of water is flowing through it. The cold water helps prevent clogged drains by solidifying any grease in the disposal. Thus, the addition of a flow switch means that the disposal cannot be operated without water.

Fig. 15–12 shows one method of installing a food waste disposal with a flow switch in the cold water line.

BONDING REQUIREMENTS

Rule 10–400 states that all exposed, noncurrent-carrying parts are to be bonded to ground. The presence of any of the conditions outlined in *Rules 10–400* to *10–404* requires that all electrical appliances in the dwelling be grounded. Food waste disposals can be grounded using any of the methods covered in the previous units, such as the separate bonding conductor of non-metallic-sheathed cable.

DISHWASHER ▲ᵢ

The dishwasher is supplied by a separate 15-ampere circuit connected to circuit B5. No. 14 NMD90 conductors are used to connect the appliance. The dishwasher has a $^1/_4$-hp motor rated at 5.8 amperes and 120 volts. The outlet for the dishwasher is shown by the symbol ▲ᵢ. During the drying cycle of the dishwasher, a thermostatically controlled, 750-watt electric heating element turns on.

Actual connected load		Circuit calculation
Motor	5.8 A	5.8 A × 1.25 = 7.25 A
Heater	6.25 A	6.25 A
Total	12.05 A	13.5 A

Use a 15-ampere circuit.

For most dishwashers, the motor does not run during the drying cycle. Thus, the actual maximum demand on the branch circuit would only be the larger of the two loads—the 750-watt heating element. In some dishwashers, there is a fan or blower to speed up the drying time. In either case, for the dryer example shown, a 15-ampere circuit is more than adequate.

Energy-saving dishwashers are equipped with a built-in booster water heater that allows the homeowner to adjust the regular water heater temperature to 50°C or lower. The booster heater of the dishwasher then raises the water temperature from 50°C to a temperature not less than 60°C for the wash cycle, and 80°C for the rinse cycle, the temperature considered necessary to sanitize dishes.

Fig. 15–12 Disposal with a flow switch in the cold water line.

Wiring the Dishwasher

The dishwasher manufacturer normally supplies a terminal or junction box on the appliance. The electrician connects the supply conductors in this box. The electrician must verify the dimensions of the dishwasher so that the supply conductors are brought through the wall at the proper point.

The dishwasher is a motor-operated appliance and requires running overcurrent protection. Such protection prevents the motor from burning out if it becomes overloaded or stalled. Normally, the required protection is supplied by the manufacturer as an integral part of the dishwasher motor. If the dishwasher does not have integral protection, the electrician must provide it as part of the installation of the unit.

Grounding

The dishwasher must be bonded to ground. The grounding of appliances is described fully in previous units and in the discussion of the food waste disposal.

PORTABLE DISHWASHERS

Portable dishwashers have one hose that is attached to the water faucet and a water drainage hose that hangs in the sink. The obvious location for a portable dishwasher is near the sink. Thus, the dishwasher probably will be plugged into the outlet nearest the sink.

Portable dishwashers are supplied with a three-wire cord containing two circuit conductors and one bonding conductor and a three-wire grounding-type attachment plugcap. If the three-wire plugcap is plugged into the three-wire grounding-type receptacle, the dishwasher is adequately grounded. Whenever there is a chance that the user may touch the appliance and a grounded surface such as the water pipe or faucet at the same time, **the equipment must be bonded to ground to reduce the shock hazard**. The exceptions to *Rule 10–408* are *Subrules (3)* and *(4)*, which permit the double insulation of appliances and portable tools instead of grounding. Double insulation means that the appliance or tool has two separate insulations between the hot conductor and the person using the device. Although refrigerators, cooking units, water heaters, and other large appliances are not double insulated, many portable hand-held appliances and tools have double insulation.

All double-insulated tools and appliances must be marked by the manufacturer to indicate this feature. These tools and appliances must be marked with the words "Double Insulated" or the symbol $\Box$, as specified in *CSA Standard C22.2 No. 0.1–M1985*.

CORD CONNECTION OF FIXED APPLIANCES

As with the previously discussed food waste disposal, residential built-in dishwashers are permitted to be cord-connected, Fig. 15–10.

Some electrical inspectors feel that using the cord-connection method to connect these appliances is better than direct-connection (sometimes referred to as "hard-wired") because it allows them to be easily removed for maintenance and repairs.

REVIEW

Note: Refer to the *CEC* or the plans where necessary.

COUNTER-MOUNTED COOKING UNIT CIRCUIT ▲_G

1. a. What circuit supplies the counter-mounted cooking unit in this residence? _____

_____

   b. What is the rating of this circuit? _____

2. What methods may be used to connect counter-mounted cooking units? _____

_____

_____

_____

_____

_____

_____

_____

3. Is it permissible to use standard 90°C insulated conductors to connect all counter-mounted cooking units? Why? _____

_____

_____

_____

4. Indicate the maximum operating temperature (in degrees Celsius) for the following.

   a. Type TW conductors _____

   b. Type TWN75 conductors _____

   c. Type R90 conductors _____

5. May NMWU cable be used to connect counter-mounted cooking units? _____

_____

6. a. What *CEC* rule applies to bonding a counter-mounted cooking unit?

   _____

   b. How is the counter-mounted cooking unit bonded to ground? _____

   _____

   _____

   _____

   _____

7. a. What is the mounting height of the receptacle supplying a range? _____

   b. In which direction should the U-ground slot be oriented? _____

   _____

   c. Give the *CEC* section number and rule. _____

   _____

   _____

8. When the voltage to an element is doubled, the wattage (circle one)

   a. increases.   b. decreases.

9. One kilowatt equals _____ Btu per hour.

10. How much heat will a 1000-watt heating element produce if operated continuously for one hour? _____

WALL-MOUNTED OVEN CIRCUIT ⒜ F

1. To what circuit is the wall-mounted oven connected? _____

2. An oven is rated at 7.5 kW. This is equal to

   a. _____ watts.   b. _____ amperes at 240 volts.

3. a. What section of the *CEC* governs the bonding of a wall-mounted oven?

     _____

   b. How are wall-mounted ovens bonded to ground? _____

     _____

     _____

     _____

4. What is the type and rating of the overcurrent device protecting the wall-mounted oven?

   _____

   _____

   _____

5. How many metres of cable are required to connect the oven in the residence?

   _____

6. When connecting a wall-mounted oven and a counter-mounted cooking unit to one feeder, how long are the taps to the individual appliances? _____

   _____

   _____

   _____

7. A 6-kW counter-mounted cooking unit and a 4-kW wall-mounted oven are to be installed in a residence. Calculate the demand according to *Rule 8–300(1)*. Show all calculations.

8. What size conductors will feed

    a. from the panel to the junction box? _____

    b. each of the individual units in Question 7? _____ and _____

9. a. A freestanding range is rated at 11.8 kW, 240 volts. According to *Rule 8–300(1)*, what is the demand?_____

    b. What size wire (Type NMD90) is required? _____

    c. What type of receptacle is required for the range? _____

FOOD WASTE DISPOSAL CIRCUIT ▲ H

1. How many amperes does the food waste disposal draw? _____

2. a. To what circuit is the food waste disposal connected? _____

    b. What size wire is used to connect the food waste disposal? _____

3. Means must be provided to disconnect the food waste disposal. The homeowner need not be involved in electrical connections when servicing the disposal if the disconnecting means is _____

    _____

4. How is running overcurrent protection provided in most waste disposal units?

    _____

    _____

5. Why are flow switches sometimes installed on food waste disposals? _____

    _____

    _____

6. What *CEC* sections relate to the grounding of appliances? _____

    _____

7. Do the plans show a wall switch for controlling the food waste disposal? _____

    _____

8. A separate circuit supplies the food waste disposal in this residence. How many metres of cable will be required to connect the disposal? _____

_____

DISHWASHER CIRCUIT ▲₁

1. a. To what circuit is the dishwasher in this residence connected? _____

   _____

   b. What size wire is used to connect the dishwasher? _____

2. The motor on the dishwasher is (circle one)

   a. ¼ hp.   b. ⅓ hp.   c. ½ hp.

3. The heating element is rated at (circle one)

   a. 750 watts.   b. 1000 watts.   c. 1250 watts.

4. How many amperes at 120 volts do the following heating elements draw?

   a. 750 watts _____

   b. 1000 watts _____

   c. 1250 watts _____

5. How is the dishwasher in this residence grounded? _____

   _____

   _____

   _____

6. What type of cord is used on most portable dishwashers? _____

   _____

7. How is a portable dishwasher grounded? _____

   _____

8. a. What is double insulation? _____

      _____

      _____

   b. What is the symbol for double insulation? _____

9. Who is to furnish the dishwasher? _____

   _____

10. Does the *CEC* require a separate circuit for the dishwasher?

   _____

   _____

UNIT 16

Branch Circuits for the Laundry, Washroom, and Attic

OBJECTIVES

After studying this unit, the student will be able to

- make the proper wiring and grounding connections for large appliances, using various wiring methods
- understand the *CEC* requirements governing the receptacle outlet(s) for laundry areas
- discuss the *CEC* rules pertaining to wiring methods in attics
- demonstrate the proper way to connect pilot lights and pilot light switches

DRYER CIRCUIT ▲D

A separate circuit is provided in the laundry room for the electric clothes dryer (Fig. 16–1). This appliance demands a large amount of power. The special circuit provided for the dryer is indicated on the plans by the symbol ▲D. The dryer is connected to circuit B1/B3.

Clothes Dryer

Clothes dryer manufacturers make many dryer models with different wiring arrangements and connection provisions. All electric dryers have electric heating units and a motor-operated drum, which tumbles the clothes as the heat evaporates the moisture. Dryers also have thermostats to regulate the temperature of the air inside the dryer. Timers regulate the lengths of the various drying cycles. Drying time, determined by the type of fabric, can be set to a maximum of about 85 minutes. Dampness con-

trols are often provided to stop the drying process at a preselected stage of dampness so that it is easier to iron the clothes.

Clothes dryers create a lot of humid air. Therefore, be sure the dryer is vented to the outside. This residence also has an exhaust fan located in the laundry room.

Table 16–1 shows the estimated loads for the various receptacles and fixtures in the laundry, washroom, and attic.

Connection of Electric Clothes Dryer

The connection of electric clothes dryers is governed by *Rule 26–744(2,3)*. Fig. 16–2 shows the internal components and wiring of a typical electric clothes dryer.

A separate receptacle, as shown in CSA configuration 14–30R *CEC Diagram 1,* should be installed on the wall adjacent to the dryer. *Rule 26–744(3)*

Fig. 16–1 Cable layout for circuits B9, B11, and B20.

GARAGE

LAUNDRY, WASHROOM
AND ATTIC LIGHT CIRCUITS

SERVICE
ENTRANCE

TO ATTIC
LAMPHOLDERS

DISP.
ATTIC
STAIRS

requires that a cord set as shown in CSA configuration 14–30P be used to connect the dryer to the receptacle, Fig. 16–3. A disconnecting means is not required for this appliance, since the attachment plug and receptacle may be used instead of a switch, *Rule 26–746(2)(a)*.

The receptacle can be wired with NMD90 cable if concealed wiring is desired. If the NMD90 is run exposed, it must be protected from mechanical injury. In the residence, the wiring method for the dryer is non-metallic-sheathed cable, which runs from Panel B to a dryer receptacle on the wall behind the dryer. A cord set is attached to the dryer and is plugged into this receptacle. The receptacle and cord have a current-carrying capacity that is rated at 125 percent greater than the dryer rating.

A standard dryer receptacle and cord set, as shown in Fig. 16–4, are usually rated at 30 amperes, 125/250 volts. Fig. 16–5 shows the correct wiring connections to the terminals on the receptacle. Cords and receptacles rated for a higher current are available. The L-shaped neutral slot of the outlet does not accept a 50-ampere cord set.

The receptacle must be flush mounted, according to *Rule 26–744(8)*. The outlet will fit a 119 × 54-mm square box. Make sure that the box is mounted in the proper orientation to ensure that the appliance cord hangs down and reduces strain on the receptacle.

In this residence, the dryer is installed in the laundry room. The schedule of special-purpose outlets in the specifications shows that the dryer is rated at 5700 watts and 120/240 volts. The schematic wiring diagram, Fig. 16–2, shows that the heating elements are connected to the 240-volt terminals of the terminal block. The dryer motor and light are connected between the hot wire and the neutral terminal, 120 volts.

The motor of the dryer has integral thermal protection. This prevents the motor from reaching dangerous temperatures as the result of an overload or failure to start. *Rule 14–610* specifies the overcurrent protection requirements for electric clothes dryers, and states that, when fuses are used, they must be time delay (TD) or low melting point fuses (P-rated).

DESCRIPTION	QUANTITY	WATTS	VOLT-AMPERES
Washroom receptacles (GFCI)	1	120	120
Washroom vanity fixture	1	200	200
Laundry room fluorescent fixtures @ 32 W	2	64	69
Attic lights @ 60 W each	4	240	240
Laundry exhaust fan	1	80	90
Washroom exhaust fan	1	80	90
TOTALS	10	784	809

Table 16–1 Laundry–washroom–attic, circuit B20.

Fig. 16–2   Laundry dryer: wiring and components.

Fig. 16–3   Dryer connection using a cord set.

Fig. 16–4   Typical dryer receptacle and cord.

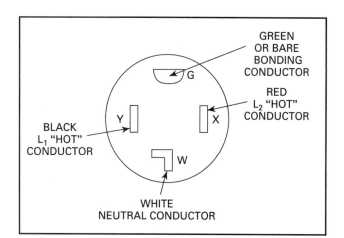

Fig. 16–5   Correct wiring for the terminals on the dryer receptacle rated at 30 amperes, 125/250 volts, CSA configuration 14–30R. Refer to *CEC Diagram 1*.

To determine the feeder rating for the house, the *CEC* requires that the load included for a dryer be 25 percent of the nameplate rating, assuming that an electric range has been included.

The branch circuit ampacity is based on the nameplate rating of the dryer. If it is not clear what the load of an appliance is, always base it on the nameplate rating.

Conductor Size

Rule 26–746(1) states that any appliance rated at more than 1500 watts shall be supplied from a branch circuit used solely for one appliance.

The electric clothes dryer in this residence has a nameplate rating of 5700 watts. Thus:

$$I = \frac{W}{E} = \frac{5700}{240} = 23.75 \text{ amperes}$$

If we apply *Rule 8–104*, that is, to multiply the current rating times 125 percent to find the required circuit ampacity, we find that:

23.75 amperes $\times$ 1.25 = 29.7 (30) amperes

Accordingly, the dryer circuit B1/B3 is a three-wire, 30-ampere, 240-volt circuit. The conductors in the non-metallic-sheathed cable are No. 10 AWG.

By *CEC* definition, a branch circuit refers to the circuit conductors between the final overcurrent device and the outlet. Therefore, the conductors between the overcurrent device in Panel B and the dryer outlet constitute the dryer's branch circuit.

Overcurrent Protection

Rule 14–104 states that the branch circuit overcurrent device rating is not to exceed the allowable ampacity of the conductors. The size of the fuse or circuit breaker may be obtained from *Table 13*. Here, again, the manufacturer of the appliance will mark the circuit overcurrent device rating. If the appliance is not marked, the overcurrent protection shall be in accordance with *Rule 4–004* for the conductors—30 amperes for our example, as discussed above.

Bonding Frames of Ranges and Dryers

Rule 10–408(1) requires that the exposed non-current-carrying metal parts of portable equipment such as clothes dryers or ranges be bonded to ground.

The branch-circuit conductors for the dryer are connected to Panel B, circuit B1/B3. This is a double-pole circuit that breaks only the two ungrounded conductors. The neutral conductor remains solid and unbroken.

RECEPTACLE OUTLETS—LAUNDRY

Rule 26–710 requires that at least two receptacles be installed in a laundry room—one for a washing machine and at least one more. *Rule 26–720* requires a minimum of one branch circuit to be installed solely for receptacles in the laundry room.

In the laundry room in this residence, the 15-ampere circuit B9 supplies the receptacle for the washer. Circuit B9 serves no other outlets. The receptacle is not required to be GFCI protected. Fig. 16–1 shows the cable layout for the laundry room.

Circuit B11 connects to the two receptacles in the laundry, which will serve an iron, a sewing machine, or other small portable appliances.

Circuits B18 and B20 have already been included in the watts per square metre calculations for service entrance and feeder conductor sizing, *Rule 8–200(1)(a)(i,ii)*. See the complete calculations in Unit 3.

LIGHTING CIRCUIT

The general lighting circuit for the laundry and washroom is supplied by circuit B20.

The types of vanity lights, ceiling fixtures, and standard wall receptacle outlets, as well as the circuitry and switching arrangements, are similar to those discussed in other units of this text.

Exhaust Fans

Ceiling exhaust fans are installed in the laundry and washroom and connected to circuit B20.

The exhaust fan in the laundry will remove excess moisture resulting from the use of the clothes washer.

Exhaust fans may be installed in walls or ceilings, Fig. 16–6. The wall-mounted fan can be adjusted to fit the thickness of the wall. If a ceiling-mounted fan is used, sheet-metal duct is recommended between the fan unit and the outside of the house.

Any exhaust ducting located in an attic space must be insulated. During the cold season, any moisture that is removed by an exhaust fan will condense in an uninsulated duct. This will lead to premature rusting of the exhaust duct and fan. The moisture can also leak out of the fan or duct, causing serious water damage.

The fan unit terminates in a metal hood or grille on the exterior of the house. The fan has a shutter that opens as the fan starts up and closes as the fan stops. The fan may have an integral pull-chain switch for starting and stopping, or it may be used with a separate wall switch. In either case, single-speed or

WALL TYPE

CEILING TYPE

Fig. 16–6   Exhaust fans. Courtesy of Broan-NuTone.

multispeed control is available. The fan in use has a very small power demand, 90 volt-amperes.

HUMIDITY CONTROL

An electrically heated dwelling can experience a problem with excess humidity due to the "tightness" of the house, because of the care taken in the installation of the insulation and vapour barriers. High humidity is uncomfortable. It promotes the growth of mould and the deterioration of fabrics and floor coverings. In addition, the framing members, wall panels, and plaster or drywall of a dwelling may deteriorate because of the humidity. Insulation must be kept dry, or its efficiency decreases. A low humidity level can be maintained by automatically controlling the exhaust fan.

One type of automatic control is the *humidistat*. This device starts the exhaust fan when the relative humidity reaches a high level. The fan exhausts air until the relative humidity drops to a comfortable level. The comfort level is about 50 percent relative humidity. Adjustable settings are from 0 to 90 percent relative humidity.

The electrician must check the maximum current and voltage ratings of the humidistat before the device is installed. Some humidistats are extra-low-voltage devices and require a relay. Other humidistats are rated at line voltage and can be used to switch the motor directly (a relay is not required). However, a relay must be installed on the line-voltage humidistat if the connected load exceeds the maximum allowable current rating of the humidistat.

The switching mechanism of a humidistat is controlled by a nylon element, Fig. 16–7, which is very sensitive to changes in humidity. A bimetallic element cannot be used because it reacts to temperature changes only.

Wiring

The humidistat and relay in the dwelling are installed using No. 14 NMD90. The cable runs from circuit A10 to a 4-in square $1\frac{1}{2}$-in deep outlet box located in the attic near the fan, Fig. 16–8(B). A motor circuit switch is mounted on this outlet box. This switch unit serves as a disconnecting means within sight of the motor, as required by

SET POINT ADJUSTING KNOB   2 X 3 IN. SWITCH BOX

WALL PLATE

SWITCH   NYLON ELEMENT

Fig. 16–7   Details of a humidity control used with an exhaust fan.

Rules 28–600 and *28–602(3)(e)*. Motor overload protection is provided by the heater unit installed in the motor-rated switch. As an alternative, an AC general purpose switch rated at 125 percent of the full load amperes (FLA) of the motor may be used as the disconnecting means.

Mounted next to the switch unit is a relay that can carry the full-load current of the motor. Extra-low-voltage thermostat wire runs from this relay to the humidistat. (The humidistat is located in the hall between the bedrooms.) The thermostat cable is a two-conductor cable since the humidistat has a single-pole, single-throw switching action. The line-voltage side of the relay provides single-pole switching to the motor. Normally the white grounded conductor is not broken.

Although the example in this residence utilizes a relay to use extra-low-voltage wiring, it is also proper to install line-voltage wiring and use line-voltage thermostats and humidity controls. These are becoming very popular, with the switches actually being solid-state speed controllers, permitting operation of the fan at an infinite number of speeds between high and low settings.

ATTIC LIGHTING AND PILOT LIGHT SWITCHES

It is always good practice to install at least one lighting outlet in an attic and to control it by a switch located near to the access point of the attic. The light should have a guard to protect it from injury, or a short flexible drop light may be used, *Rule 30–314(2)*. Many attics are used for storage. They may also contain equipment that might need to be serviced, so install a lighting outlet near it.

Fig. 16–8   Wiring for the humidity control.

Ambient temperatures in attics can exceed the ratings of some equipment. Any circuit and equipment wiring calculations and requirements must take this into account, *Rule 4–004(8)*. Table 16–2, which reproduces *Table 5A,* shows the derating factors for ambient temperatures greater than 30°C. These must be applied for wiring in attics where the ambient temperature could easily exceed 30°C.

The four keyless lampholders in the attic are turned on and off by a single-pole switch on the garage wall close to the attic storable stairway. Associated with this single-pole switch is a pilot light. The pilot light may be located in the handle of the switch or separately mounted. Fig. 16–9 shows how pilot lamps are connected in circuits containing either single-pole or three-way switches.

If a neon pilot lamp in the handle (toggle) of a switch does not have a separate grounded conductor connection, it will glow only when the switch is in the OFF position, because the neon lamp will then be in series with the lamp load, Fig. 16–10. The voltage across the load lamp is virtually zero, so it does not light, and the voltage across the neon lamp is 120 volts, allowing it to glow. When the switch is turned on, the neon lamp is bypassed (shunted), causing it to turn off and the lamp load to turn on.

Use this type of switch when it is desirable to have a switch glow in the dark to make it easy to locate.

Installation of Cable in Attics

The wiring in the attic is to be done in cable and must meet the requirements of *Rules 12–500* to *12–526* for *non-metallic-sheathed* cable. This rule describes how the cable is to be protected, Fig. 16–11.

In accessible attics, Fig. 16–11(A), cables must *not* be:

- run across the top of floor or ceiling joists ①
- run across the face of studs ② or rafters ③ that have a vertical distance between the floor or floor joists exceeding 1 metre.

Guard strips are not required if the cable is run along the sides of rafters, studs, or floor joists ④ provided *Rule 12–516(1,2)* is met.

CEC **TABLE 5A**
CORRECTION FACTORS APPLYING TO TABLES 1, 2, 3, AND 4
(AMPACITY CORRECTION FACTORS FOR AMBIENT TEMPERATURES ABOVE 30°C)
(See Rules 4–004(8) and 12–2210 and Tables 1, 2, 3, 4, 57, and 58.)

AMBIENT TEMPERATURE, °C	CORRECTION FACTOR					
	60°C TYPE TW	75°C TYPES RW75, TW75	85–90°C TYPES R90, RW90, T90 NYLON	110°C SEE NOTE (2)	125°C SEE NOTE (2)	200°C SEE NOTE (2)
40	0.82	0.88	0.90	0.94	0.95	1.00
45	0.71	0.82	0.85	0.90	0.92	1.00
50	0.58	0.75	0.80	0.87	0.89	1.00
55	0.41	0.65	0.74	0.83	0.86	1.00
60	—	0.58	0.67	0.79	0.83	0.91
70	—	0.35	0.52	0.71	0.76	0.87
75	—	—	0.43	0.66	0.72	0.86
80	—	—	0.30	0.61	0.69	0.84
90	—	—	—	0.50	0.61	0.80
100	—	—	—	—	0.51	0.77
120	—	—	—	—	—	0.69
140	—	—	—	—	—	0.59
	Col. 2	Col. 3	Col. 4	Col. 5	Col. 6	Col. 7

(The left column is labelled Col. 1)

NOTES: (1) These correction factors apply, column for column, to Tables 1, 2, 3, and 4. The correction factors in Column 2 also apply to Table 57.
    (2) The ampacity of a given conductor type at these higher ambient temperatures is obtained by multiplying the appropriate value from Table 1, 2, 3, or 4 by the correction factor for that higher temperature.
    (3) These ampacities are applicable only under special circumstances where the use of insulated conductors having this temperature rating is acceptable.

Table 16–2 *CEC Table 5A. See Rules 4–004(8) and 12–2212 and Tables 1, 2, 3, 4, 5D, 58, and D3.*

Fig. 16–9   Pilot lamp connections.

Fig. 16–10   Neon pilot lamp in the handle of the toggle switch.

Fig. 16–11(C) illustrates a cable installation that most electrical inspectors consider to be safe. Because the cables are installed close to the point where the ceiling joists and the roof rafters meet, they are protected from physical damage. It would be very difficult for a person to crawl into this space, or to store cartons in an area with a clearance of less than 1 metre, Fig. 16–11(B). Although the plans for this residence do not show a 600-mm-wide catwalk in the attic, it is installed. In the future the owner may decide to install additional flooring in the attic to obtain more storage space. Because of the large number of cables required to complete the circuits, it would interfere with flooring to install guard strips wherever the cables run across the tops of the joists, Fig. 16–12. However, the cables can be run through holes bored in the joists and along the sides of the joists and rafters. In this way, the cables do not interfere with the flooring. **Most building codes will not allow manufactured roof systems to be drilled.** This includes roof trusses.

When running cables through framing members, be careful to maintain at least 32 mm between the cable and the edge of the framing member. This minimizes the possibility of driving nails into the cable.

ATTIC EXHAUST FAN CIRCUIT ▲L

An exhaust fan is mounted in the roof of the house, Fig. 16–13(A). When running, this fan removes hot, stagnant, or humid air from the attic. It will also draw in fresh air through louvres located in the end gables. Such an attic exhaust fan can be used to lower the indoor temperature of the house by as much as 5 to 10°C.

The exhaust fan in this residence is connected to circuit A10, a 120-volt, 15-ampere circuit in the main panel utilizing No. 14 NMD90 wire. The derating factors in *Table 5A* must be applied because of the high ambient temperature in the attic.

Exhaust fans can also be installed in a gable of a house, Fig. 16–13(B).

On hot days, the air in the attic can reach a temperature of 65°C or more. Thus, it is desirable to provide a means of taking this attic air to the outside. The heat from the attic can radiate through the ceiling into the living areas, raising their temperature and increasing the air-conditioning load. Personal discomfort increases as well. These problems are minimized by properly vented attic exhaust fans, Fig. 16–14.

Fig. 16–11   Protection of cable in attic.

Fan Operation

The exhaust fan installed in this residence has a $^1/_4$-hp direct-drive motor. Direct-drive means that the fan blade is attached directly to the shaft of the motor. It is rated 120 volts, 60 hertz (60 cycles per second), 5.8 amperes, 696 watts, and operates at a maximum speed of 1050 rpm. The motor is protected against overload by internal thermal protection. The motor is mounted on rubber cushions for quiet operation.

When the fan is not running, the louvre remains in the closed position. When the fan is turned on, the louvre opens. When the fan is turned off, the louvre automatically closes to prevent the escape of air.

Fan Control

Typical residential exhaust fans can be controlled by a choice of switches.

Some electricians, architects, and home designers prefer to have fan switches (controls),

Fig. 16–12 Methods of protecting cable installations in attics.

Figs. 16–15 and 16–16, mounted 1.83 m above the floor so that they will not be confused with other wall switches. This is a matter of personal preference and should be verified with the owner.

Fig. 16–17 shows some of the options available for the control of an exhaust fan.

a. A simple ON/OFF switch (A).

b. A speed control switch that allows multiple and/or infinite number of speeds, (B). (The exhaust fan in this residence has this type of control.) See also Fig. 16–15.

c. A timer switch that allows the user to select how long the exhaust fan is to run, up to 12 hours (C). See also Fig. 16–16.

d. A humidity control switch that senses moisture build-ups (D). See the section on "Humidity Control" in this unit for further details on this type of switch.

e. Exhaust fans mounted into end gables or roofs of residences are available with an adjustable temperature control on the frame of the fan. This thermostat generally has a start range of 21°C to 54°C, and will automatically stop at a temperature 5°C below the start setting, thus providing totally automatic ON/OFF control of the exhaust fan (E).

Fig. 16–13 Exhaust fan installation in an attic.

Fig. 16–14   Ceiling exhaust fan viewed from the roof. Courtesy of NuTone.

Fig. 16–15   Speed control switch. Courtesy of NuTone.

Fig. 16–16   Timer switch with speed control. Courtesy of NuTone.

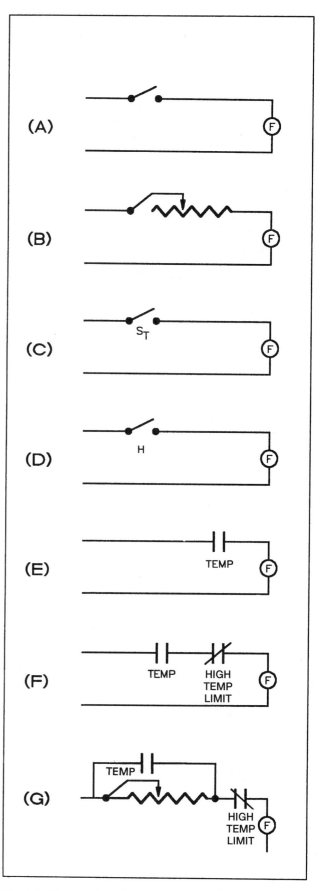

Fig. 16–17   Options for the control of exhaust fans.

f. A high-temperature automatic heat sensor that will shut off the fan motor when the temperature reaches 93°C (F). This is a safety feature so that the fan will not spread a fire. Connect in series with other control switches.

g. A combination of controls, this circuit shows an exhaust fan controlled by an infinite-speed switch, a temperature control, and a high-temperature heat sensor (G). The speed control and the temperature control are connected in parallel so that either can start the fan. The high-temperature heat sensor is in series so that it will shut off the power to the fan when it senses 93°C, even if the speed control or temperature control is in the ON position.

Overload Protection for the Fan Motor

There are several ways to provide running overload protection for the $\frac{1}{4}$-hp attic fan motor. For example, an overload device can be built into the motor, or a combination overload tripping device and thermal element can be added to the switch assembly. Some manufacturers of motor controls also make switches with overcurrent devices. Such a switch may be installed in any standard flush switch box opening 3×2 in (76×51 mm). A pilot light may be used to indicate that the fan is running. Thus a two-gang switch box or $4 \times 1\frac{1}{2}$-in (102×38-mm) square outlet box with a two-gang raised plaster cover may be used for both the switch and the pilot light.

Rule 28–308 states that an automatically started motor having a rating of 1 hp or less does not need to have overload protection if the motor on the unit has protection features that will safely protect the motor from damage if it stalls or the rotor locks. This must be indicated on a nameplate located so as to be visible after installation. This would be indicated on this particular fan motor since it does have integral overload protection.

Branch Circuit Short-Circuit Protection for the Fan Motor

Table D16 specifies short-circuit and overload protection for motors; *Table 45* lists the full-load current in amperes for single-phase AC motors. For a full-load current rating of 5.8 amperes, *Table D16* shows that the rating of the branch circuit overcurrent device shall not exceed 15 amperes if time-delay or non-time-delay fuses are used and 15 amperes if standard thermal-magnetic circuit breakers are used.

The exhaust fan is connected to circuit A10, which is a 15-ampere, 120-volt circuit. This circuit does not feed any other loads. The circuit supplies the exhaust fan only.

Time-delay fuses sized at 115 to 125 percent of the full-load ampere rating of the motor will provide both running overload protection and branch-circuit short-circuit protection. If the motor is equipped with inherent, built-in overload protection, the time-delay fuse in the circuit provides secondary backup protection, providing double protection against possible motor burnout.

REVIEW

Note: Refer to the *CEC* or the plans where necessary.

DRYER CIRCUIT ▲D AND LAUNDRY ROOM CIRCUITS

1. List the switches, receptacles, and other wiring devices that are connected to circuit B20. _____

   _____

   _____

   _____

2. What *CEC* rule(s) refers (refer) to the receptacle required for the laundry equipment? Briefly state the requirements. _____

_____

_____

_____

_____

_____

_____

3. The receptacle outlets are connected to branch circuits that have been calculated to have a maximum load of _____ watts per circuit. Circle one: 1200, 1440, 1500 watts.

   a. What regulates the temperature in the dryer? _____

   _____

   b. What regulates the drying time? _____

   _____

4. What is the method of connection to an electric clothes dryer? _____

   _____

5. What is the unique shape of the neutral blade of a 30-ampere dryer cord set? _____

   _____

   _____

6. a. What is the minimum number of receptacle outlets that must be installed in a laundry room? _____

   b. Is a separate branch circuit required? _____

   c. Quote the *CEC* rule numbers. _____

7. What is the maximum permitted current rating of a portable appliance on a 30-ampere branch circuit? _____

   _____

   _____

8. What is the current draw of the exhaust fan in the laundry? _____

   _____

   _____

   _____

   _____

9. What special type of switch is controlling the attic lights? _____

   _____

10. If an attic exhaust fan draws 12 amperes at 120 volts and the ambient temperature reaches 50°C, what size and type of wire will be required to feed the fan? Refer to *CEC, Table 5A.* _____

_____

11. When installing cables in an attic along the top of the floor joists _____ m clearance must not be exceeded.

12. The total estimated watts for a circuit has been calculated to be 1300 volt-amperes. How many amperes is this at 120 volts? _____

13. In what circumstances would lighting and/or power be provided in an attic? _____

_____

14. The following is a layout of the lighting circuit for the laundry, washroom, rear entrance hall, and attic. Complete the wiring diagram using coloured pens or pencils to indicate conductor insulation colour.

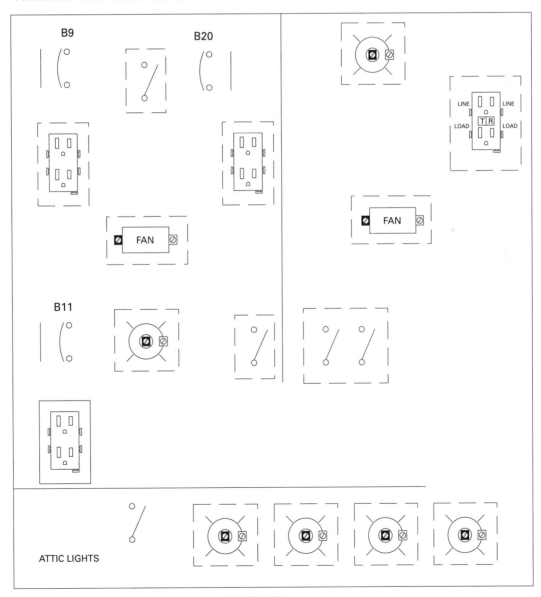

LAUNDRY AND
WASHROOM

ATTIC EXHAUST FAN CIRCUIT ▲L

1. What is the purpose of the attic exhaust fan? _____

    _____

2. At what voltage does the fan operate? _____

3. What is the horsepower rating of the fan motor? _____

4. Is the fan direct or belt driven? _____

5. How is the fan controlled? _____

6. What is the setting of the running overload device? _____

7. What is the rating of the running overload protection if the motor is rated at 10 amperes?

    _____

8. What is the basic difference between a thermostat and a humidistat? _____

    _____

    _____

9. What size conductors are to be used for this circuit? _____

10. How many metres of cable are required to complete the wiring for the attic exhaust fan circuit? _____

11. May the metal frame of the fan be grounded to the grounded circuit conductor?

    _____

12. What *CEC* rule prohibits grounding equipment to a grounded circuit conductor?

    _____

UNIT 17

Electric Heating and Air Conditioning

OBJECTIVES

After studying this unit, the student will be able to

- list the advantages of electric heating
- describe the components and operation of electric heating systems (baseboard heater, cable, furnace)
- describe thermostat control systems for electric heating units
- install electric heaters with appropriate temperature control according to *CEC* rules
- discuss the *CEC* requirements and electrical connections for air conditioners and heat pumps
- explain how heating and cooling may be connected to the same circuit

GENERAL DISCUSSION

Many types of electric heat are available for heating homes—heating cable, unit heaters, boilers, electric furnaces, duct heaters, baseboard heaters, and radiant heating panels. The residence discussed in this text is heated by an electric furnace located in the workshop.

Detailed requirements are found in *Section 62,* which covers fixed electric space and surface heating systems.

This text cannot cover in detail the methods used to calculate heat loss and the wattage required to provide a comfortable level of heat in the building. For this residence, the total estimated wattage is 13 000 watts (13 kW). The heating load depends on the location of the residence—the West Coast, the Prairies, Central Canada, the Maritime Provinces, or the Far North. Each region has a different number of degree days, which affects the heat requirements to maintain a comfortable living temperature.

Compared with other types of heating systems, electric heating is not widely used. However, it does have a number of advantages over other systems. Electric heating is flexible when baseboard heating is used, because each room can have its own thermostat. Thus, one room can be kept cool while an adjoining room is warm. This type of zone control for an electric, gas-, or oil-fired central heating system is more complex and more expensive.

Electric heating is safer than heating with fuels. The system does not require storage space, tanks, or chimneys. Electric heating is quiet and does not add to or remove anything from the air. As a result, electric heat is cleaner. This type of heating is considered to be healthier than fuel heating systems, which remove oxygen from the air. The only moving part of an electric baseboard heating system is the thermostat. This means that there is a minimum of maintenance.

If an electric heating system is used, adequate insulation must be provided to minimize electric bills. Insulation also helps to keep the residence cool during the hot summer months. The cost of extra insulation is offset through the years by the decreased burden on the air-conditioning and heating equipment. Energy conservation measures require the installation of proper and adequate insulation as well as a good-quality vapour barrier.

TYPES OF ELECTRIC HEATING SYSTEMS

Electric heating units are available in baseboard, wall-mounted, and floor-mounted styles. These units may have built-in thermal overload protection. The type of unit to be installed depends on structural conditions and the purpose for which the room is to be used.

Another method of providing electric heating is to embed resistance-type cables in the plaster of ceilings or between two layers of drywall on the ceiling, *Rules 62–214* through *62–222*. Resistance-type heating cables may also be used for melting snow and ice on roof overhangs, eavestroughs, walkways, and driveways, *Rule 62–300*.

Electric heat can also be supplied by an electric furnace and a duct system similar to the type used on conventional hot-air central heating systems. The heat is supplied by electric heating elements rather than the burning of fuel. Air conditioning, humidity control, air circulation, electronic air cleaning, and zone control can be provided on an electric furnace system.

Many homes are now built or renovated with ceramic tile flooring. Often, heating cable is installed under the tile to eliminate cold flooring. It is very important to follow the manufacturer's installation instructions. Although some heat will radiate from the tile into the room, underfloor heating cable sets are not intended to be the sole source of heat for the room. The wattage of the heating cable set is usually quite low, as it is meant to heat only the floor.

SEPARATE CIRCUIT REQUIRED

Rules 62–108(1)(a) and *26–806(1)* require that central heating equipment be supplied by a separate branch circuit. This includes electric, gas, and oil central furnaces and heat pumps. The exception to *Rules 62–108(1)* and *26–806(1)* is equipment directly associated with central heating equipment, such as humidifiers and electrostatic air cleaners, which are permitted to be connected to any branch circuit, *Rule 26–806(2)*.

This "separate circuit" rule is due to the possibility that if the central heating equipment were connected to another circuit, such as a lighting circuit, a fault on that circuit would shut off the power to the heating system. This could cause freezing of water pipes. Refer to Fig. 17–1.

Rule 26–804, Appendix B, gives excerpts from *CSA Standard C22.2 No. 3* entitled *Electrical Features of Fuel-Burning Equipment.* Rule 4.8.8 of this standard states that a safety control shall interrupt the current in the ungrounded conductor of the circuit between the overcurrent protection and the load. This must be provided for in the control circuit connected to the Class 2 transformer.

CONTROL OF ELECTRIC HEATING SYSTEMS

Line-voltage thermostats can be used to control the heat load for most baseboard electric heating systems, Fig. 17–2 (page 294). Common wattage limits for line-voltage thermostats are 2500 watts, 3000 watts, and 5000 watts. (Additional ratings are available.) The electrician must check the nameplate ratings of the heating unit and the thermostat to ensure that the total connected load does not exceed the thermostat rating.

The wattage limit of a thermostat is found using the amperage formula. For example, if a thermostat has a rating of 20 amperes at 125 or 250 volts, the wattage value is:

$$W = E \times I = 125 \times 20 \text{ watts} = 2500 \text{ watts}$$

$$or$$

$$W = 250 \times 20 \text{ watts} = 5000 \text{ watts}$$

The total connected load for this thermostat must not exceed 20 amperes.

THERMOSTAT

FIELD WIRING OF LOW-VOLTAGE, CLASS 2 CONTROL CIRCUIT CONDUCTORS SHALL NOT BE PLACED IN THE SAME RACEWAY, BOX, OR ENCLOSURE WITH POWER CONDUCTORS EXCEPT WHERE THEY ARE SEPARATED BY AN ACCEPTABLE BARRIER, *RULE 16–212(3).*

TYPICAL ELECTRIC FURNACE

DISCONNECTING MEANS:

• SUITABLE DISCONNECTING MEANS SHALL BE PROVIDED FOR THE BRANCH CIRCUIT, *RULE 26–806(5).*

• MAY BE A BRANCH CIRCUIT BREAKER AT THE DISTRIBUTION PANELBOARD, PROVIDED THE PANELBOARD IS LOCATED BETWEEN THE FURNACE AND THE POINT OF ENTRY TO THE AREA WHERE THE FURNACE IS LOCATED, *RULE 26–806(6).*

• MAY NOT BE LOCATED ON THE FURNACE AND MUST BE MARKED TO INDICATE THE EQUIPMENT THAT IT CONTROLS, *RULE 26–806(7a, b).*

MUST BE SEPARATE CIRCUIT, *RULE 62–108(1)(a).*

BRANCH CIRCUIT CONDUCTORS:

• THESE CONDUCTORS ARE REQUIRED TO HAVE AN AMPACITY NOT LESS THAN THE CURRENT RATINGS OF ALL THE EQUIPMENT THEY SUPPLY, THAT IS, THEY MUST BE ABLE TO CARRY THE LOAD, *RULES 62–108(1)(b), 62–114(7)(a), (b),* AND *62–116(1).*

NAMEPLATE WILL SHOW MANUFACTURER'S NAME, VOLTS AND AMPERES, OR VOLTS AND WATTS, OR VOLTS AND KILOWATTS.

EXAMPLE: WHAT SIZE COPPER CONDUCTORS (NMD90), FUSES, AND DISCONNECT SWITCH ARE REQUIRED FOR A FURNACE MARKED 79 AMPERES, 240 VOLT, SINGLE PHASE, 60 CYCLES? TERMINALS ON FURNACE AND SWITCH MARKED 75°C.

ANSWER: CONDUCTOR SIZE: FROM TABLE 2, SELECT NO. 4 NMD90. (85 AMPERES AT 90°C.) *RULE 62–116(i)*

SIZE OF OVERCURRENT DEVICE: INSTALL 100-AMPERE FUSE OR CIRCUIT BREAKER. *SECTION 62–114(7b)*

SWITCH: 100-AMPERE SWITCH 79 × 1.25 = 98.75 AMPERE FUSES.

Fig. 17–1   Basic *CEC* requirements for an electric furnace. These "package" units have all of the internal components, such as limit switches, relays, motors, and contractors, prewired. The electrician provides both the branch circuit and the control circuit wiring together with the thermostat wiring. See *Section 62* for additional information.

When the connected load is larger than the rating of a line-voltage thermostat, as in the case of an electric furnace, low-voltage thermostats may be used. In this case, a relay must be connected between the thermostat and load. The relay is an integral component of the furnace. The low-voltage contacts of the relay are connected to the thermostat. The line-voltage contacts of the relay are used to switch the actual heater load. Low-voltage ther-

mostat cable is run between the low-voltage terminals of the relay and the thermostat, Fig. 17–3.

When the connected load is larger than the maximum current rating of a thermostat and relay combination, the thermostat can be used to control a heavy-duty relay or contactor, Fig. 17–4. An example of such a relay is the magnetic switch used for motor controls. The load side of the magnetic switch feeds a distribution panel containing as many

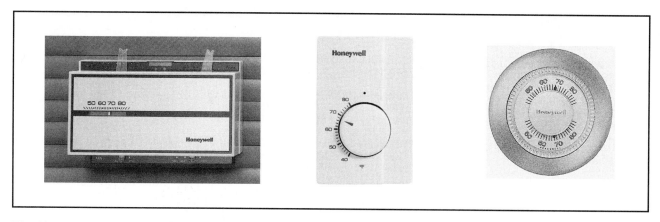

Fig. 17–2   Thermostats for electric heating systems. Courtesy of Honeywell Inc.

Fig. 17–3   Wiring for a baseboard electric heating unit having a low-voltage thermostat and relay. *Note*: **If using black and white wires in a two-wire NMSC or AC cable, tape or paint the white wire red,** *Rule 4–036(3)*. **When using non-metallic sheathed cable, many inspection authorities require the use of special cable, which contains one black and one red conductor.**

Fig. 17–4   Relay for controlling baseboard heating.

15- or 20-ampere circuits as needed for the connected load. A 40-ampere load may be divided into three 15-ampere circuits and still be controlled by one thermostat. The proper overcurrent protection is obtained by dividing the 40-ampere load into several circuits having lower ratings.

CIRCUIT REQUIREMENTS FOR BASEBOARD UNITS

Fig. 17–5 shows typical baseboard electric heating units. These units are rated at 240 volts, but are also available in 120-volt ratings.

The wiring for an individual baseboard heating unit or group of units is shown in Figs. 17–6 and 17–7. A two-wire cable (armoured cable or non-metallic-sheathed cable with ground) would be run from a 240-volt, two-pole circuit in the main panel to the outlet box or switch box installed at the thermostat location. A second two-wire cable runs from the thermostat to the junction box on the heater unit. The proper connections are made in this junction box. Most heating unit manufacturers provide knockouts at the rear and on the bottom of the junction box. The supply conductors can be run through these knockouts using proper connectors for their wiring method.

Fig. 17–5 Baseboard electric heating systems.

Fig. 17–6 Wiring for a single baseboard electric heating unit. See note in caption of Fig. 17–3.

Fig. 17–7 Wiring for baseboard electric heating units at different locations within a room. See note in caption of Fig. 17–3.

Most baseboard units also have a channel or wiring space running the full length of the unit, usually at the bottom. When two or more heating units are joined together, the conductors are run in this wiring channel, *Rule 62–112(2)*. Most manufacturers indicate the type of wire required for these units because of conductor temperature limitations.

A variety of fittings, such as internal and external elbows (for turning corners) and blank sections, are available from the manufacturer.

Most wall and baseboard heating units are available with built-in thermostats (see Fig. 17–5). The supply cable for such a unit runs from the main panel to the junction box on the unit.

Manufacturers of electric baseboard heaters list "blank" sections ranging in length from 0.5 m to 2 m. These blank sections give the installer the flexibility needed to spread out the heater sections, and to install these blanks where wall receptacle outlets are encountered, Fig. 17–8.

LOCATION OF ELECTRIC BASEBOARD HEATERS IN RELATION TO RECEPTACLE OUTLETS

Approved electric baseboard heaters shall not be installed below wall receptacle outlets unless the instructions furnished with the heaters indicate that

Fig. 17–8 Use of blank baseboard heating sections.

they may be installed there, *Rule 26–712(a), Appendix B*. The reason for this is the possible fire and shock hazard if a cord hangs over and touches the heated electric baseboard unit. The insulation on the cord can melt from the heat. See Figs. 17–9 and 17–10.

The electrician must pay close attention to the location of receptacle outlets and the electric baseboard heaters in the plans and specifications.

It may be necessary to carefully select the location of the receptacle outlets and the electric baseboard heaters. Blank spacer sections may be installed, following the *CEC* requirements for spacing receptacle outlets and just how much (wattage and length) baseboard heating is required.

The CSA approves the use of factory-installed or factory-furnished receptacle outlet assemblies as part of permanently installed electric baseboard heaters. These receptacle outlets shall not be connected to the heater circuit, Fig. 17–11.

ELECTRIC BASEBOARD HEATER

Fig. 17–9 Code violation. Unless the instructions furnished with the electric baseboard heater specifically state that the unit is listed for installation below a receptacle outlet, this installation is a violation of *Rule 26–712(a), Appendix B*. In this type of installation, cords attached to the receptacle outlet would hang over the heater, creating a fire hazard and possible shock hazard should the insulation on the cord melt.

CIRCUIT REQUIREMENTS FOR ELECTRIC FURNACES

In an electric furnace, the source of heat is the resistance heating element(s). The blower motor assembly, filter section, condensate coil, refrigerant line connection, fan motor speed control, high-temperature limit controls, relays, and similar components are similar to those found in gas furnaces.

In most installations, the supply conductors must be copper conductors having a minimum 90°C temperature rating, such as type T90. The instructions furnished with the unit will have information relative to supply conductor size and type.

Since proper supply voltage is critical to ensure full output of the heating elements, voltage must be maintained at *not less* than 98 percent of the unit's rated voltage to provide its rated wattage. Therefore, the voltage drop should not be allowed to exceed 2 percent of the unit's rated voltage to achieve its rated output. The *CEC* allows a maximum voltage drop of 3 percent on a branch circuit, *Rule 8–102(1)(c)*.

Many electric furnaces furnish the wattage (or kilowatts) output for more than one voltage level. For every 1 percent drop in voltage, there will be an almost 2 percent drop in wattage output. For instance, if a 240-volt rated heater is connected to 208 volts, its actual output of heat is less than its rated watts, Table 17–1.

The actual voltage at the supply terminals of the electric furnace or baseboard heating unit determines the true wattage output of the heating elements. The effect that different voltages have on electric heating elements was covered in Unit 15.

The demand factors for service conductors used to supply heating equipment are set out in

WATTS @ 240 VOLTS	AMPERES @ 240 VOLTS	WATTS @ 208 VOLTS	AMPERES @ 208 VOLTS
500	2.08	375.6	1.81
750	3.13	563.3	2.71
1000	4.17	751.1	3.61
1250	5.21	938.9	4.51
1500	6.25	1126.7	5.42
1750	7.29	1314.4	6.32
2000	8.33	1502.2	7.22
2500	10.42	1877.8	9.03

Table 17–1 Comparison of wattages and currents for electric heating elements at 240 and 208 volts.

(A) POSITION ELECTRICAL BASEBOARD HEATING UNITS SO THEY WILL *NOT* BE DIRECTLY BELOW A WALL RECEPTACLE OUTLET.

(B) IF INSTALLED AS SHOWN, ELECTRICAL CORDS COULD COME IN CONTACT WITH THE BASEBOARD UNIT, SUBJECTING THIS CORD TO RUBBING (ABRASION) AND HEAT, WHICH MIGHT RESULT IN FAILURE OF THE INSULATION OF THE CORD, A POTENTIAL FIRE AND SHOCK HAZARD.

BASEBOARD HEATER

BASEBOARD HEATER

RECEPTACLE OUTLET

(C) AN EXAMPLE OF HOW THE RECEPTACLE OUTLETS IN THE WINDOW CORNER OF BEDROOM #1 MIGHT BE INSTALLED.

Fig. 17–10 Position of electric baseboard heaters.

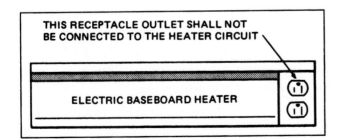

THIS RECEPTACLE OUTLET SHALL NOT BE CONNECTED TO THE HEATER CIRCUIT

ELECTRIC BASEBOARD HEATER

Fig. 17–11 Factory-installed receptacle outlets or receptacle outlet assemblies provided by the manufacturer for use with its electric baseboard heaters may be counted as the required receptacle outlet for the space occupied by a permanently installed heater, *Rule 26–712(a)*.

Rule 62–116, which requires service conductors to have an ampacity not less than the current ratings of all the equipment supplied. In the case of an electric furnace, this would be 100 percent of the total connected load.

If baseboard heating is being used, *Rule 62–116(2)* allows the ampacity of the service conductors to be reduced. If each room or heated area is thermostatically controlled, the demand factor is 75 percent for each kW above 10 kW. Thus, if the total connected load was 20 kW for baseboard heating with thermostats in each room, the demand load must be calculated as:

1st 10 kW @ 100% =	10	kW
2nd 10 kW @ 75% =	7.5	kW
Total	17.5	kW

Total connected load = 20 kW
Total demand load = 17.5 kW

These demand watts would be applied to the service conductors.

The branch-circuit overcurrent devices, for unit equipment such as baseboards, must not be rated at more than 30 amperes, *Rule 62–114(2)*. Therefore, the maximum load on each branch circuit is 80 percent or 24 amperes or 5760 watts at 240 volts.

The electric furnace must be supplied by an individual branch circuit, *Rule 26–806(1)*. The "separate circuit" requirement pertains to any type of central heating, including oil, gas, electric, and heat pump systems.

HEAT PUMPS

A heat pump cools and dehumidifies on hot days and heats on cool days. To provide heat, the pump picks up heat from the outside air, which it adds to the heat created by the compression of the refrigerant. This heat is used to warm the air flowing through the unit, which is then delivered to the inside of the building.

The heat pump can also be used to cool the building when the outside air temperature is warm. To achieve this, the direction of flow of the refrigerant is reversed. Warm air is then delivered to the outside, and the house is cooled.

Heat pumps are available as self-contained packages that provide both cooling and heating. In the Canadian climate, the heat pump alone may not provide enough heat. Therefore, the heat pump may use supplementary heating elements contained in the air-handling unit, or be used in combination with a forced-air gas furnace.

BONDING

Bonding of electrical baseboard heating units and other electrical equipment and appliances has been covered previously in this text. Refer to *Rule 10–400*.

MARKING THE CONDUCTORS OF CABLES

Two-wire cable contains one white conductor, one black conductor, and one bonding conductor. Using two-wire cable to supply electric baseboard heaters appears to violate the *CEC*, since *Rule 4–036(3)* shows that one black and one red conductor are required.

According to *Rule 4–034(1)*, two-wire cable may be used for the 240-volt heaters if the white conductor is permanently marked as ungrounded by paint, coloured tape, or other effective means. This is necessary, since people may assume that an unmarked white conductor is a grounded circuit conductor having no voltage to ground. However, the white wire in the heater circuit is connected to a "hot" phase and has 120 volts to ground. **Therefore, a person can be subject to a harmful shock by touching this wire and the grounded baseboard heater (or any other grounded object) at the same time.** Most electrical inspectors accept black paint or tape as a means of changing the colour of the white conductor. The white conductor must be made permanently unidentified at the electric heater terminals and at the panels where these cables originate. A more commonly used method is to install two-wire cable with black and red conductors, *Rule 4–036(3)*. One such cable is Heatex™, a NMSC cable manufactured by Canada Wire Inc., specifically designed for heating circuits. Reread the sections on "Colour Coding" in Unit 8.

ROOM AIR CONDITIONERS

For homes without central air conditioning, window or through-the-wall air conditioners may be installed. These room air conditioners are available in both 120-volt and 240-volt ratings. Because room air conditioners are plug-and-cord connected, the receptacle outlet and the circuit capacity must be selected and installed according to the *CEC*.

The basic *CEC* rules for installing these units and their receptacle outlets are:

- the air conditioners must be grounded;
- the air conditioners must be connected using a cord and attachment plug;
- the rating of the branch-circuit overcurrent device must not exceed the branch-circuit conductor rating or the receptacle rating, whichever is less;

- the air-conditioner load shall not exceed 80 percent of the branch-circuit ampacity if no other loads are served;

- the attachment plug cap may serve as the disconnecting means, *Rule 28–602(3)(c)*.

RECEPTACLES FOR AIR CONDITIONERS

Fig. 17–12 shows three types of receptacles that can be used for air-conditioning installations: (A) shows a 20-ampere, 250-volt receptacle for use on a 240-volt installation; (B) a 20-ampere, 125-volt receptacle for use with a 120-volt air conditioner circuit; and (C) a 15-ampere, 250-volt receptacle for use

Fig. 17–12 Types of receptacles used for air conditioners, *Diagram 1*.

with a 240-volt installation. Refer to the *CEC Diagram 1*. CSA configuration 5–20RA may also be used.

CENTRAL HEATING AND AIR CONDITIONING

The residence in this text has central electric heating and air conditioning, consisting of an electric furnace and a central air-conditioning unit.

The wiring for central heating and cooling systems is shown in Fig. 17–13. One feeder runs to the electric furnace, and another runs to the air conditioner or heat pump outside the dwelling. Extra-low-voltage wiring is used between the inside and outside units to provide control of the systems. The extra-low-voltage Class 2 circuit wiring must not be run in the same raceway as the power conductors, *Rule 16–212(1)*.

SPECIAL TERMINOLOGY

These terms pertain to air conditioners, heat pumps, and other hermetically sealed motor-compressors.

Rated-load current is the current drawn when the unit is operating at rated load, at rated voltage, and at rated frequency.

Branch-circuit conductor ampacity is based on the rated-load current, which must be marked on the label, *Rule 2–100(1)(d)*. This will determine

Fig. 17–13 Connection diagram showing typical electrical furnace and air-conditioner/heat pump installation.

conductor size, disconnect switch size, controller size, and short-circuit and ground-fault protective device ratings, Fig. 17–14.

Rule 2–100(4) requires that approved equipment be used only in accordance with instructions included in the original approval of the equipment. For

example, the nameplate of an air-conditioning unit reads "Maximum size fuse 40 amperes." Thus, only 40-ampere fuses can be used. Forty-ampere circuit breakers are not permitted. If the nameplate called for "Maximum overcurrent protection," either fuses or circuit breakers could be used. See Fig. 17–15.

BRANCH CIRCUIT OVERCURRENT PROTECTION:
- MUST BE ABLE TO CARRY STARTING CURRENT
- SIZE ACCORDING TO DATA ON THE EQUIPMENT LABEL
- MUST BE FUSES UNLESS LABEL ON EQUIPMENT SHOWS THAT CIRCUIT BREAKERS ARE PERMITTED

DISCONNECTING MEANS:
- SELECT SIZE BASED ON NAMEPLATE RATED-LOAD CURRENT OR BRANCH CIRCUIT SELECTION CURRENT— WHICHEVER IS GREATER—AND LOCKED ROTOR CURRENT
- AMPERE RATING OF SWITCH MUST BE AT LEAST 115% OF NAMEPLATE RATED-LOAD CURRENT OR BRANCH CIRCUIT SELECTION CURRENT, *RULE 28–714(1)*
- MUST ALSO BE HORSEPOWER RATED, *RULE 28–704.*
- MUST BE WITHIN SIGHT OF EQUIPMENT

BRANCH CIRCUIT CONDUCTORS:
- THE CONDUCTOR AMPACITY RATING REQUIRED FOR THE AIR-CONDITIONING UNIT IS FOUND ON THE LABEL. THIS HAS BEEN DETERMINED BY THE MANUFACTURER, TAKING INTO CONSIDERATION THE COMPRESSOR MOTOR CURRENT, FAN MOTOR CURRENT, AND HEATER CURRENT. THIS IS GENERALLY 125% OF THE LARGEST MOTOR PLUS THE FULL-LOAD RATING OF THE REST OF THE EQUIPMENT'S LOADS, SUCH AS FANS AND HEATERS, *RULE 28–108(1).*

TYPICAL AIR-CONDITIONING UNIT

OVERLOAD PROTECTION
- THIS IS USUALLY AN INTEGRAL PART OF THE EQUIPMENT, SUPPLIED BY THE MANUFACTURER.

FAN MOTOR

HERMETICALLY SEALED MOTOR

LABEL:
- MANUFACTURER'S NAME
- VOLTAGE
- FREQUENCY
- PHASES
- MINIMUM SUPPLY CIRCUIT CONDUCTOR AMPACITY
- MAXIMUM RATING OF BRANCH CIRCUIT, SHORT-CIRCUIT, AND GROUND-FAULT PROTECTIVE DEVICE
- WILL STATE "MAXIMUM SIZE FUSE" *OR* "MAXIMUM SIZE FUSE OR CIRCUIT BREAKER"

Fig. 17–14 **Basic circuit requirements for a typical residential-type air conditioner or heat pump. Reading the label is important as the manufacturer has determined the minimum conductor size and branch circuit fuse size.**

The air conditioner requires a motor rated disconnect to be installed within three metres and within sight of the unit, *Rule 28–604(5)*, Fig. 17–16.

LOADS NOT USED SIMULTANEOUSLY

Heating and air-conditioning loads are not likely to operate at the same time. Therefore, when calculating feeder sizes to an electric furnace and an air conditioner, you need to consider only the larger load. In *Rule 8–106(4)*, the words "where it is known" refer to the fact that common sense would dictate that heating and air conditioning would not be in use at the same time. The larger connected load is used to calculate the demand on the service feeders.

If the air conditioning has the larger demand, this demand is used at 100 percent. If the heating demand is greater, this is used to calculate the demand watts according to *Rule 62–116*.

Fig. 17–15   This installation conforms to *Rule 28–604*. The panel contains the 40-ampere circuit breaker called for on the air conditioner nameplate as the branch circuit protection. The disconnect must be capable of being locked off, *Rule 28–604(1)(b)(ii)*.

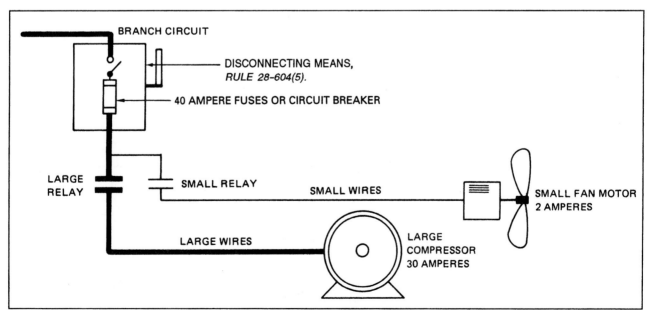

Fig. 17–16   In a typical air-conditioner unit, the branch-circuit protective device is called upon to protect the large components (wire, relay, compressor) and the small components (wire, relay, fan motor).

REVIEW

Note: Refer to the *CEC* or the plans where necessary.

ELECTRIC HEAT

1. a. What is the allowance in watts made for electric heat in this residence? _____

   _____

   b. What is the value in amperes of this load? _____

   _____

2. What are some of the advantages of electric heating? _____

   _____

   _____

   _____

   _____

   _____

   _____

   _____

3. List the different types of electric heating system installations. _____

   _____

   _____

   _____

   _____

   _____

   _____

   _____

   _____

   _____

4. There are two basic voltage classifications for thermostats. What are they? _____

   _____

   _____

5. What device is required when the total connected load exceeds the maximum rating of a thermostat? _____

   _____

6. The electric heat in this residence is provided by what type of equipment? _____

   _____

7. At what voltage does the electric furnace operate? _____

   _____

8. A certain type of control connects electric heating units to a 120-volt supply or a 240-volt supply, depending on the amount of the temperature drop in a room. These controls are supplied from a 120/240-volt, three-wire, single-phase source. Assuming that this type of device controls a 240-volt, 2000-watt heating unit, what is the wattage produced when the control supplies 120 volts to the heating unit? Show all calculations.

9. What advantages does a 240-volt heating unit have over a 120-volt heating unit?

_____

_____

_____

_____

_____

_____

_____

10. a. The white wire of a cable may be used to connect to a hot circuit conductor only if

_____

_____

   b. Quote the rule number. _____

11. Electric baseboard heaters shall not be installed beneath wall receptacle outlets. Explain and quote the *CEC* rule. _____

_____

_____

_____

_____

_____

12. The branch circuit supplying a heater shall have an ampacity of

   a. not less than the connected load supplied.

   b. 125 percent of the current rating of the heater.

   Circle the correct answer.

13. Compute the current draw of the following electric furnaces. The furnaces are all rated 240 volts.

    a. 7.5 kW

    b. 15 kW

    c. 20 kW

14. The wattage output of a 240-volt electric furnace connected to a 208-volt supply will be about 75 percent of the wattage output had the furnace been connected to a 240-volt supply. Referring to Question 13, calculate the wattage output of (a), (b), and (c) if connected to 208 volts. _____

    _____

    _____

    _____

    _____

15. A central electric furnace heating system is installed in a home. Should a separate branch circuit supply this furnace? Explain. _____

    _____

    _____

    _____

    _____

16. What *CEC* rule provides the answer to Question 15?

    _____

AIR CONDITIONING

1. When calculating air-conditioner load requirements and electric heating load requirements, is it necessary to add the two loads together to determine the combined load on the service? Explain your answer. _____

   _____

   _____

   _____

   _____

   _____

   _____

2. The total load of an air conditioner shall not exceed what percentage of a separate branch circuit? (circle one)

   a. 75%   b. 80%   c. 125%

3. Do window air conditioners count as a connected load when calculating the demand watts on a service? _____

4. Must an air conditioner installed in a window opening be bonded to ground?

   _____

5. A 120-volt air conditioner draws 12 amperes. What size is the circuit to which the air conditioner will be connected? _____

6. When a central air-conditioning unit is installed and the label states "Maximum size fuse 50 amperes," is it permissible to connect the unit to a 50-ampere circuit breaker?

   _____

7. What *CEC* rule prohibits installing Class 2 control circuit conductors in the same raceway as the power conductors? _____

   _____

UNIT 18

Oil and Gas Heating Systems

OBJECTIVES

After studying this unit, the student will be able to

- understand the basics of gas and oil heating systems
- understand some of the more important components of gas and oil heating systems
- interpret basic schematic wiring diagrams
- explain the principles of the thermocouple and the thermopile in a self-generating system
- understand and apply the requirements for control circuit wiring, *Section 16*

Unit 17 discussed the circuit requirements for electric heating (furnace, baseboard, ceiling cable). Unit 18 discusses typical residential oil and gas heating systems.

Heating a home with gas or oil is possible with:

- warm air—fan-forced warm air or gravity (hot air rises naturally)
- Hot water—hot water circulated through the piping system by a circulating pump; the system uses tubing embedded in the floor or radiators that look very much like electric baseboard heating units.

PRINCIPLES OF OPERATION

Most residential gas or oil heating systems operate as follows. A room thermostat is connected to the proper electrical terminals on the furnace or boiler, where the controls and valves are interconnected to provide safe and adequate heat to the home. Some typical wiring diagrams are shown in Figs. 18–1, 18–2, and 18–3A and B.

Most residential gas or oil heating systems are "packaged units" in which the safety controls, valves, fan controls, limit switches, and so on are integral, preassembled prewired parts of the furnace or boiler.

For the electrician, the "field wiring" generally consists of installing and connecting the wire to the thermostat, and installing and connecting the power supply to the furnace or boiler. Refer to Fig. 18–4.

MAJOR COMPONENTS

Gas- and oil-fired warm air systems and hot water boilers contain many components.

Aquastat: This direct immersion water temperature thermostat regulates boiler or tank temperature in hydronic heating systems, ensuring that the circulating water maintains proper, satisfactory temperature, Fig. 18–5.

Cad cell: This is a cadmium sulfide sensor for sensing flame. Cad cells have generally replaced the older style stack-mounted bimetal switches.

Circulating pump: This pump circulates hot water in a hydronic central heating system.

Combustion chamber: The combustion chamber surrounds the flame and radiates heat back into the flame to aid in combustion.

Fig. 18–1 Simplified wiring diagram for a gas burner, forced hot water system.

Fig. 18–2 Simplified wiring diagram for a gas burner, forced warm air system.

Combustion head: This creates a specific pattern of air at the end of the air tube. The air is directed to force oxygen into the oil spray so the oil can burn. A combustion head is also referred to as the turbulator, fire ring, retention ring, or end cone.

Control circuit: This is any electric circuit that controls any other circuit through a relay or an equivalent device; also referred to as a remote-control circuit.

Draft regulator: This counterweighted swinging door opens and closes to help maintain a constant level of draft over the fire.

Fan control: This controls the blower fan on forced warm air systems. On newer furnaces, the fan control is an electronic timer that starts the fan motor a given time in seconds after the main burner has come on. This timing ensures that the blower will blow warm air. Older adjustable temperature fan controls still exist. See Fig. 18–6.

Fig. 18–3A Simplified wiring diagram of an oil-burning heating installation.

Fig. 18–3B An oil burner featuring an electronic microprocessor that controls all facets of the burner operation. Most components are integral to the unit, thus keeping field wiring to a minimum. All manufacturers provide detailed wiring diagrams for their products. Always consult the manufacturer's wiring diagrams for a particular model heating unit.

Fig. 18–4   Diagram showing thermostat, power supply wiring, and disconnect switch.

Fig. 18–5   Aquastat, or liquid immersion controller.

Fan motor: This electric motor forces warm air through the heating ducts. In typical residential heating systems, the motor is provided with integral overload protection. The motor may be single, multiple, or variable speed.

Flow valve: This prevents gravity circulation of water through the piping system, Fig 18–7.

Fig. 18–6   Fan/limit control.

Fig. 18–7   Flow valve.

Flue: A flue is a channel in a chimney or a pipe for conveying flame and smoke to the outer air.

Heat exchanger: This transfers the heat energy from the combustion gases to the air in the furnace or to the water in a boiler.

High-temperature limit control: This safety device limits the temperature in the plenum of the furnace to a safe value predetermined by the manufacturer. When this temperature is reached, the control shuts off the power to the burner. In newer furnaces, the control is part of an integrated circuit board and is communicated to by a sensor strategically located in the plenum. Some older furnaces have individual high-temperature limit controls and a separate fan control; others have combination high-temperature limit/fan controls.

Hot surface igniter: In furnaces without a standing constant pilot light, the hot surface igniter heats to a cherry red/orange colour. When this reaches the proper temperature, the main gas valve is allowed to open. If the hot surface igniter does not come on, the main gas valve will not open. Most energy-efficient furnaces use this kind of ignition.

Hydronic system: This system of heating or cooling involves transfer of heat by circulating fluid such as water in a closed system of pipes.

Ignition:

    Constant: The igniter is designed to stay on continuously.

    Intermittent duty: Defined by UL 296 as "ignition by an energy source that is continuously maintained throughout the time the burner is firing"; the igniter is on the entire time the burner is firing.

    Interrupted duty: Defined by UL 296 as "an ignition system that is energized each time the main burner is to be fired and de-energized at the end of a timed trial for ignition period or after the main flame is proven to be established"; the igniter comes on to light the flame. After the flame is established, the igniter is turned off and the main flame keeps burning.

Induced draft blower: When the thermostat calls for heat, this blower first expels any gases remaining in the combustion chamber from a previous burning cycle. It continues to run, pulling hot combustion gases through the heat exchangers, then vents the gases to the outdoors.

Integrated control: This printed circuit board contains many electronic components. When the thermostat calls for heat, the integrated circuit board takes over to manage the sequence of events that allow the burner to operate safely.

Low-water control: Also called a low-water cutoff, this device senses low-water levels in a boiler. When a low-water situation occurs, this control shuts off the electrical power to the burner.

Main gas valve: This valve allows the gas to flow to the main burner of a gas-fired furnace, Fig. 18–8. It may also function as a pressure regulator, a safety shut-off, and the pilot and main gas valve. Should there be a "flame-out," if the pilot light fails to operate, or in the event of a power failure, this valve will shut off the flow of gas to the main burner.

Fig. 18–8   Main gas valve.

Nozzle: A nozzle produces the desired spray pattern for the particular appliance in which the burner is used.

Oil burner: This burner breaks fuel oil into small droplets, mixes the droplets with air, and ignites the resulting spray to form a flame.

Primary control: This controls the oil burner motor, ignition, and oil valve in response to commands from a thermostat, Fig. 18–9. If the oil fails to ignite, the controller shuts down the oil burner.

Pump and zone controls: These controls regulate the flow of water or steam in boiler systems to specific "zones" in the building.

Safety controls: Safety controls such as pressure relief valves, high-temperature limit controls, low-water cutoffs, and burner primary controls protect against appliance malfunction.

Solenoid: This is an electrically operated device, which, when the valve coil is energized, a magnetic field is developed, causing a spring-loaded

Fig. 18–9   Oil burner primary control.

steel valve piston to overcome the resistance of the spring and immediately pull the piston into the stem. The valve is now in the open position. When de-energized, the solenoid coil magnetic field instantly dissipates, and the spring-loaded valve piston snaps closed, stopping oil flow or gas flow to the nozzle. The flame is extinguished, allowing fuel to flow. Usually a spring returns the valve back to the closed position.

Spark ignition: In spark ignition systems, the spark comes on while pilot gas flows to the pilot orifice. Once the pilot flame is "proven" through an electronic sensing circuit, the main gas valve opens. Once ignition takes place, the sensor will monitor and prove existence of a main flame. If the main flame goes off, the furnace shuts down. Energy-efficient furnaces may use this for ignition.

Switching relay: This is used between the low-voltage wiring and line-voltage wiring, Fig. 18–10. When a low-voltage thermostat calls for heat, the relay "pulls in," closing the line-voltage contacts on the relay, which turns on the main power to the furnace or boiler. Usually they have a built-in transformer that provides the low-voltage Class 2 control circuit.

Thermocouple: In systems having a standing pilot, a thermocouple holds open a gas valve pilot solenoid magnet. A thermocouple consists of two dissimilar metals connected together to form a circuit. The metals might be iron and copper, copper and iron constantan, copper-nickel alloy and chrome-iron alloy, platinum and platinum-rhodium alloy, or chromel and alumel. The types

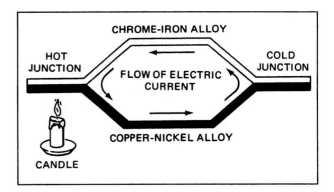

Fig. 18–11   Principle of a thermocouple.

of metals used depend on the temperatures involved. When one of the junctions is heated, an electrical current flows, Fig. 18–11. To operate, there must be a temperature difference between the metal junctions. In a gas burner, the source of heat for the thermocouple is the pilot light. The cold junction of the thermocouple remains open and is connected to the pilot safety shutoff gas valve circuit. A single thermocouple develops a voltage of about 25 to 30 DC millivolts. Because of this extremely low voltage, circuit resistance is kept to a very low value.

Thermopile: More than one thermocouple connected in series is a thermopile, Fig. 18–12. The power output of a thermopile is greater than that of a single thermocouple—typically either 250 or 750 DC millivolts. For example, ten thermocouples (25 DC millivolts each) connected in series result in a 250 DC millivolt thermopile. Twenty-six thermocouples connected in series results in a 750 DC millivolt thermopile.

Thermostat: A thermostat turns the heating/cooling system on and off. Programmable thermostats might have day/night setback settings, a clock, a digital thermometer, humidity control, ventilation, and filtration control. Thermostats

Fig. 18–10   Switching relay.

Fig. 18–12   Principle of a thermopile.

are usually mounted about 1.3 m above the finished floor. Do not mount thermostats in areas influenced by drafts or air currents from hot or cold air registers, near fireplaces, near concealed hot or cold water pipes or ducts, in the direct rays of the sun, or on outside walls, Fig. 18–13. Some thermostats contain a small vial of mercury that "tips," so the mercury closes the circuit by bridging the gap between the electrical contacts inside the vial. These thermostats must be kept perfectly level to ensure accuracy. **WARNING: Do not throw mercury thermostats into the trash. Mercury is a hazardous waste material, to be returned to a waste management organization for proper disposal.**

Transformer: This converts (transforms) the branch circuit voltage to low voltage, Fig. 18–14. For residential applications, these are Class 2 transformers, 120 volts to 24 volts. The low-voltage wiring on the secondary of this transformer is Class 2 wiring.

Water circulating pump: In hydronic (wet) systems, a circulating pump circulates hot water through the piping system. Multizone systems have more than one pump, each controlled by a

Fig. 18–13   A digital electronic programmable thermostat provides proper cycling of the heating system, resulting in comfort and energy savings. This type of thermostat could be programmed to control temperature, time of operation, humidity, ventilation, filtration, circulation of air, and zone control. Installation and operational instructions are furnished with the thermostat. Courtesy of Honeywell, Inc.

Fig. 18–14   Transformer.

thermostat for a particular zone. In multizone systems, a high-temperature control (aquastat) might be set to maintain the boiler water temperature to a specific temperature. Some multizone systems have one circulating pump and use electrically operated valves that open and close on command from the thermostat(s) located in a particular zone(s). See Fig. 18–15.

SELF-GENERATING (MILLIVOLT) SYSTEM

Several manufacturers of gas burners provide self-generating systems on decorative appliances such as gas fireplaces. These systems do not require an outside power supply. Fig. 18–16 shows the wiring diagram of a typical self-generating system. The components are connected in series. The small amount of energy required by the gas valve is supplied by a thermopile. Many of these burners are prewired at the factory. Thus, the only wiring at the site is done by the electrician, who runs a low-voltage cable to the thermostat. This system is not affected by a power failure because there is no outside power source.

Self-generating systems operate in the millivolt

Fig. 18–15   Water circulator.

Fig. 18–16 Wiring diagram of a self-generating system.

range, typically 250 or 750 mV. A special thermostat is required to function at this low voltage level. All other components of the system are matched to operate at the low voltage.

SUPPLY CIRCUIT WIRING

The wiring for the branch circuit for heating equipment that uses solid, liquid, or gaseous fuel is covered in *CEC Rules 26–800 to 808*.

The branch circuit wiring must be adequate for the power requirements of the motors, fans, relays, pumps, ignition, transformers, and other power-consuming components of the furnace or boiler. The sizing of conductors, overcurrent protection, conduit fill, and so on has been covered in other units of this text.

The basic requirements for the branch circuit are:

- The power for the heating unit and associated equipment must be supplied from a single circuit. For exceptions, see *Rule 26–806(2)(3)*.

- A suitable disconnecting means is required. The branch circuit breaker in the panel may be used as the disconnecting means, if no one has to walk past the furnace to reach the breaker.

- If a separate disconnecting means is required, it cannot be mounted on the furnace or in a location that would require a person to walk past the furnace to operate it.

- The switch must be marked to indicate that it controls the furnace.

CONTROL CIRCUIT WIRING

CEC Section 16 covers four types of control circuits:

- Class 1 and 2 remote control circuits
- Class 1 and 2 signal circuits
- Class 1 extra-low-voltage power circuits
- Class 2 low-energy power circuits

As soon as the conductors for remote-control, signaling, or power-limited circuits leave the equipment, they are classified as Class 1 or Class 2 circuits, Fig. 18–17, *Rule 16–002*. There are different requirements for Class 1 and Class 2 circuits regarding wire sizes, derating factors, overcurrent protection, insulation, wiring methods, and materials that "differentiate" control circuit wiring from regular (normal) light and power circuits, *Sections 4 and 12*.

In Fig. 18–17(A), the control circuitry is contained completely within the equipment, and, as such, none of the control wiring is governed by *Section 16*. In Fig. 18–17(B), the control wires leave the equipment. These conductors are classified as Class 1 or Class 2, depending on voltage, current, and other issues discussed in *Section 16*. Motor control circuits may be protected by control fuses in the controller.

When #16 and #18 conductors of a Class 1 circuit are run between pieces of equipment, they must be protected by overcurrent devices rated or set at not more than 10 and 5 amperes, respectively.

Fig. 18–18 shows a Class 1 circuit. Where the failure of equipment or wiring, such as high-limit and low-water controls, would "introduce a direct fire or life hazard," *Rule 16–010*, the circuit is deemed to be a Class 1 circuit. Because of this potential hazard, *Rule 16–116* requires that the wiring be installed in rigid or flexible metal conduit, EMT, or armoured cable, or be otherwise suitably protected from physical damage. See *Rules 16–100 to 16–118* for requirements applicable to Class 1 circuits.

Class 1 remote-control and signaling circuits may be supplied from a source having a maximum voltage of 600 volts. Class 1 extra-low-voltage power circuits are limited to 30V and 1000 VA, *Rule 16–100*.

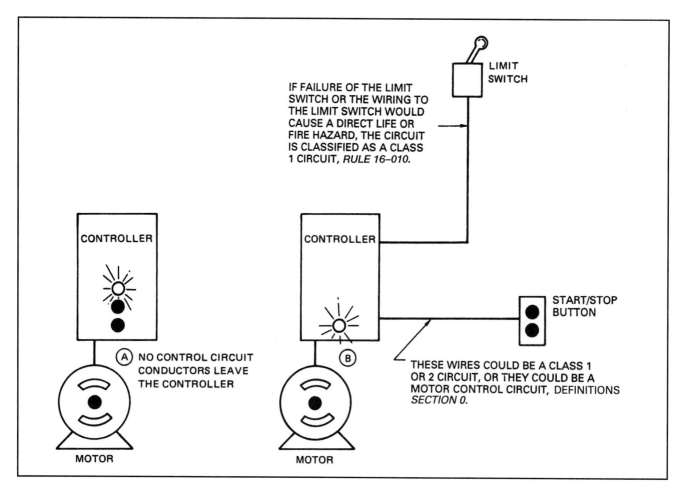

Fig. 18–17 Diagram showing Class 1 and 2 circuits.

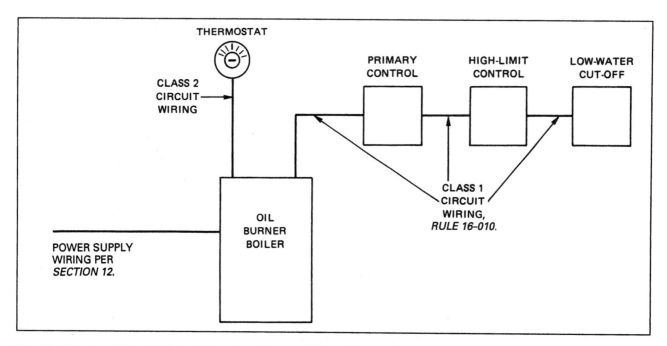

Fig. 18–18 Any failure in Class 1 circuit wiring would introduce a direct fire or life hazard. A failure in the Class 2 circuit wiring would *not* introduce a direct fire or life hazard. See *Rule 16–010.*

The conductors of Class 1 circuits and power conductors may be run in the same raceway **only** if they are connected to the same equipment and the Class 1 conductors have the same voltage rating as the highest rated wire in the raceway, *Rule 16–114*. Most wiring diagrams for residential air-conditioning units indicate that their extra-low-voltage circuity is Class 2, in which case the field wiring of the power and extra-low-voltage circuits must be run separately. See Figs. 18–19A and B.

Fig. 18–20 shows a Class 2 circuit, where the wiring runs to the thermostat. This will be done with LVT wire as listed in *Table 19*. Type ELC cable is not approved for heating control circuits, *Rule 16–210*, but would be suitable for door chimes.

Table 57 lists the allowable ampacities for Class 2 copper conductors. Note that the derating factors for more than three wires or an ambient temperature above 30°C are the same as in *Table 5A* and *Table 5C*.

For example, a No. 16/4 LVT cable at 35°C has the following ampacity:

Temperature derating from *Table 5A* Column 2:
$$10 \text{ A} \times 0.82 = 8.2 \text{ A per conductor}$$

Table 57 correction factor for 4 to 6 conductors:
$$8.2 \text{ A} \times 0.8 = 6.56 \text{ A per conductor}$$

Class 1, Class 2, and communication cables include:

- CIC, ACIC, control and instrumentation cable
- LVT and ELC Class 2 remote-control, signaling, and power-limited cables
- type FAS, FAS 90 power-limited fire protective signaling cables
- type MP multipurpose cables.

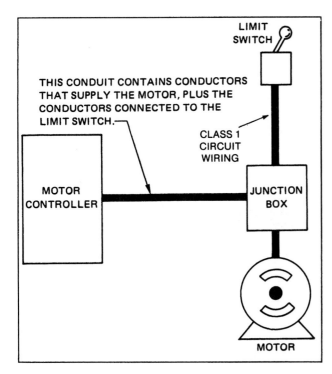

Fig. 18–19A   The power conductors supplying the motor and the Class 1 circuit remote-control conductors to the limit switch and controller may be run in the same conduit when the Class 1 conductors are insulated for the maximum voltage of any other conductors in the raceway.

Fig. 18–19B   The conduit running out of the basement wall connecting the air-conditioner unit is permitted to contain the power conductors and the Class 1 control circuit conductors because it is connected to the same equipment. All conductors must be insulated for the maximum voltage of any conductors in the raceway, *Rule 16–114(2)*.

Other communication cables are listed in *CEC Table 19*.

Conductors for Class 2 circuits are **not** permitted to be run in the same raceway, enclosure, box compartment, or fitting with the regular lighting circuits, power circuits, or Class 1 circuit wiring, *Rule 16–212(1,3)*.

Sometimes it is tempting to cut corners by installing Class 2 conductors in the same raceway as the power conductors. Even if the Class 2 conductors were insulated with 600-volt insulation, the same as the power conductors, this is a violation of *Rule 16–212(3)*, Fig. 18–20.

It is permitted to fasten a Class 2 extra-low-voltage control cable to a conduit containing the power conductors that supply a furnace or similar equipment, Fig. 18–20.

Type ELC conductors in Class 2 wiring are not permitted for heating control circuits, *Rule 16–210(3)*.

Conductor sizing and overcurrent protection requirements are found in *Subsection 16–100* for Class 1 circuits and *Subsection 16–200* for Class 2 circuits.

Transformers that are intended to supply Class 2 circuits are listed by UL and approved by the CSA and are marked "Class 2 Transformer." Fig. 18–21 illustrates the definitions of remote control, signaling, and power-limited terminology.

Fig. 18–20 *Rule 16–212* **prohibits installing the extra-low-voltage Class 2 conductor in the same raceway as the power conductors even if the Class 2 conductor has 600-volt insulation.**

A. A REMOTE-CONTROL, SIGNALING, OR POWER-LIMITED CIRCUIT IS THAT PORTION OF
   THE WIRING SYSTEM BETWEEN THE LOAD SIDE OF THE OVERCURRENT DEVICE OR THE
   POWER-LIMITED SUPPLY AND ALL CONNECTED EQUIPMENT, *CEC SECTION 0* and *RULE 16–002.*

B. A REMOTE-CONTROL CIRCUIT IS ANY ELECTRICAL CIRCUIT THAT CONTROLS ANY OTHER
   CIRCUIT THROUGH A RELAY OR EQUIVALENT DEVICE, *CEC SECTION 0, DEFINITIONS.*

C. A SIGNAL CIRCUIT IS ANY ELECTRICAL CIRCUIT THAT ENERGIZES SIGNALING
   EQUIPMENT, SUCH AS DOOR BELLS AND SIGNAL LIGHTS, *CEC SECTION 0, DEFINITIONS.*

Fig. 18–21   A remote-control circuit and a signaling circuit.

REVIEW

Note: Refer to the *CEC* or the plans where necessary.

1. How is the residence in the text heated? _____

2. Where does the *CEC* require that a disconnecting means be installed for a furnace or boiler? _____
   _____

3. What device in an oil burning system will shut off the burner in the event of a flame-out?
   _____

4. Name the control that limits dangerous water temperatures in a boiler. _____
   _____

5. Name the control that limits the temperature in the plenum of a furnace. _____
   _____

6. In the wiring diagrams in this unit, are the safety controls connected in series or parallel?
   _____

7. What is meant by the term "self-generating"? _____
   _____
   _____
   _____

8. Because a self-generating unit is not connected to an outside source of power, the electrician need not use care in selecting the conductor size or making electrical connections. Is this statement true or false? Explain. _____
   _____
   _____
   _____
   _____
   _____
   _____

9. Explain what a thermocouple is and how it operates. _____
   _____
   _____
   _____
   _____
   _____
   _____

10. Explain what a thermopile is. _____

    _____

    _____

    _____

11. If Class 2 extra-low-voltage conductors are insulated for 32 volts, and the power con-
    ductors are insulated for 600 volts, does the *CEC* permit pulling these conductors
    through the same raceway? _____

12. If Class 2 extra-low-voltage conductors are used on a 24-volt system but are actually insu-
    lated for 600 volts (T90 nylon No. 14, for example), does the *CEC* permit running these
    conductors together in the same raceway as the 600-volt power conductors?

    _____

    _____

13. Under what conditions may Class 1 conductors and power conductors be installed in the
    same raceway?_____

    _____

    _____

14. What is the *CEC* section and rule for Question 13? _____

    _____

    _____

    _____

UNIT 19

Recreation Room

OBJECTIVES

After studying this unit, the student will be able to

- understand three-wire (multiwire) circuits: the advantages and the cautions
- understand how to install lay-in fixtures
- calculate watts loss and voltage drop in two-wire and three-wire circuits
- understand the term *fixture drops*
- understand the advantages of installing multiwire branch circuits
- understand problems that can be encountered on multiwire branch circuits as a result of open neutrals

RECREATION ROOM LIGHTING

The recreation room is well lighted by six lay-in 600-mm × 1200-mm fluorescent fixtures, installed into 600-mm by 1200-mm plaster frames that are fastened to the ceiling's framing. The fixtures rest inside of the plaster frames. Fig. 19–1. Fig. 19–2 shows a device and wiring layout. When conduit is used, junction boxes are mounted above the dropped ceiling, usually within half a metre of the intended fixture location. One to two metres of AC90 armoured cable are installed between these junction boxes and the fixtures. These short cables are commonly called *fixture drops,* Fig. 19–3. They must contain the correct type and size of conductor for the *CEC* temperature ratings and load requirements for recessed fixtures. In houses wired with NMSC, the luminaire may be wired directly using NMD90.

Fluorescent fixtures of the type shown in Fig. 19–3 might bear the label "Recessed fluorescent fixture," which might also state "Suspended ceiling fluorescent fixture." Although they might look the same, there is a difference.

Recessed fluorescent fixtures are intended for installation in cavities in ceilings and walls and are to be wired according to *Rules 30–900* to *30–912.* These fixtures may also be installed in suspended ceilings if they have the necessary mounting hardware.

Suspended fluorescent fixtures are intended only for installation in suspended ceilings where the acoustical tiles, lay-in panels, and suspended grid are not part of the actual building structure.

The building code and most electrical inspection departments require fixtures to be chained. This provides an independent means of support so, if the ceiling collapses, the fixtures remain supported.

CSA Standard C22.2 E598 covers recessed and suspended ceiling fixtures in detail.

The fluorescent fixtures in the recreation room are four-lamp fixtures having warm white deluxe (WWX) or cool white deluxe (CWX) lamps installed in them. They are controlled by a single-pole switch located at the bottom of the stairs.

Circuit B14 feeds the fluorescent fixtures in the recreation room, Fig. 19–2. Table 19–1 summarizes the outlets and estimated load for the recreation room.

Fig. 19–1   Lay-in fluorescent fixture commonly used in conjunction with dropped acoustical ceilings.

Fig. 19–2   Cable layout for recreation room.

Fig. 19–3   Recessed lay-in fluorescent fixture showing a flexible connection to the fixture. Conductors in fixture drops must have temperature rating as required on the label in the fixture: "For supply use 90°C conductors."

DESCRIPTION	QUANTITY	WATTS	VOLT-AMPERES
Circuit B10			
Wet bar receptacles (GFCI protected)	2	240	240
Total	2	240	240
Circuit B12			
Receptacles	5	600	600
Recessed lights	4	240	240
Fan	1	160	300
Total	10	1000	1140
Circuit B14			
Receptacles	4	480	480
Recessed lights	2	120	120
Lay-in fluorescent lights	6	336	366
Total	12	936	966

Table 19–1   Recreation room: outlets and estimated load (three circuits).

How to Connect Recessed Lay-in Fixtures

Fig. 19–3 shows how to connect these fixtures. A junction box is located above the ceiling near the fixture. Fixture drops are usually made up of No. 14/2 AC90 cable if non-combustible construction is used. This cable must be supported a maximum of 300 mm from the box, *Rule 12–618.* NMD90 cable can also be installed directly into fixtures and looped between fixtures if combustible construction is used.

Rule 12–1004(a) permits 12-mm flexible metal conduit in lengths not to exceed 1.5 m to make the connection between the junction box and the fixture. Bonding of the fixture is done according to *Rule 10–400* when 12-mm flexible metal conduit is used.

A fixture drop of flexible metal conduit is not required to be supported if the drop is no more than 900 mm, *Rule 12–1010(3).* Yet, many electricians secure the flex within 300 mm of the outlet box to avoid pulling the flex out of the connector while they are making up the electrical connections and installing the fixture in the lay-in ceiling.

RECEPTACLES AND WET BAR

The circuitry for the recreation room wall receptacles and the wet bar lighting area introduces a new type of circuit: a *multiwire*, or *three-wire*, *circuit*.

Using a multiwire circuit can save money since one three-wire circuit will do the job of two two-wire circuits. Also, if the loads are nearly balanced in a three-wire circuit, the neutral conductor carries only the unbalanced current. This results in less voltage drop and watts loss for a three-wire circuit compared to similar loads connected to two separate two-wire circuits.

Figs. 19–4 and 19–5 illustrate the benefits of a multiwire circuit for watts loss and voltage drop. The example is a purely resistive circuit with equal loads.

The distance from the load to the source is 15 m. The conductor size is No. 14 AWG copper. The

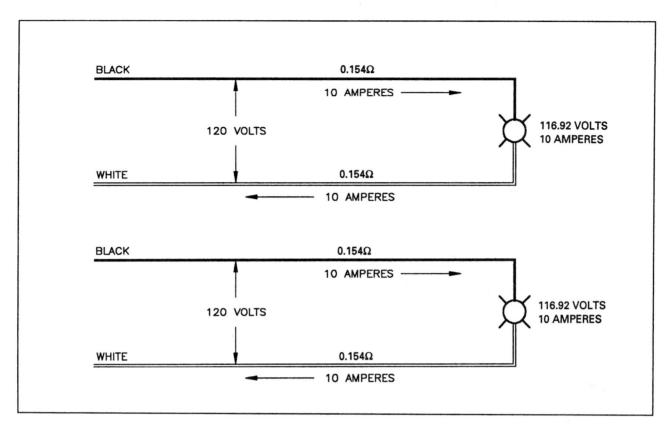

Fig. 19–4   Diagram of Example 1.

Fig. 19–5   Diagram of Example 2.

resistance of each 15-m length of No. 14 AWG is about 0.154 ohm.

The maximum distance that a No. 14/2 NMD90 copper wire, based on a 3 percent voltage drop, may be run is 18.3 m, using *Table D3* at a 10-ampere load. Therefore, for runs longer than 20 m in residential applications, *Table D3* should be used to ensure that the wire is not too small, *Rule 8–102* and *Appendix D.*

EXAMPLE 1: TWO TWO-WIRE CIRCUITS
(Fig. 19–4)

Watts loss in each current-carrying conductor is:

$$\text{Watts} = I^2R = 10 \times 10 \times 0.154 \text{ or } 15.4 \text{ watts}$$

Watts loss in all four current-carrying conductors is:

$$15.4 \text{ watts} \times 4 = 61.6 \text{ watts}$$

Voltage drop in each current-carrying conductor is:

$$\begin{aligned}
E_d &= IR \\
&= 10 \times 0.154 \text{ volts} \\
&= 1.54 \text{ volts}
\end{aligned}$$

E_d for both current-carrying conductors in each circuit is:

$$2 \times 1.54 \text{ volts} = 3.08 \text{ volts}$$

Voltage available at load is:

$$(120 - 3.08) \text{ volts} = 116.92 \text{ volts}$$

EXAMPLE 2: ONE THREE-WIRE CIRCUIT
(Fig. 19–5)

Watts loss in each current-carrying conductor is:

$$\text{Watts} = I^2R = 10 \times 10 \times 0.154 \text{ watts} = 15.4 \text{ watts}$$

Watts loss in both current-carrying conductors is:

$$15.4 \text{ watts} \times 2 = 30.8 \text{ watts}$$

Voltage drop in each current-carrying conductor is:

$$\begin{aligned}
E_d &= IR \\
&= 10 \times 0.154 \text{ volts} \\
&= 1.54 \text{ volts}
\end{aligned}$$

E_d for both current-carrying conductors is:

$$2 \times 1.54 \text{ volts} = 3.08 \text{ volts}$$

Voltage available at loads is:

$$(240 - 3.08) \text{ volts} = 236.92 \text{ volts}$$

Voltage available at each load is:

$$\frac{236.92}{2} \text{ volts} = 118.46 \text{ volts}$$

It is apparent that a multiwire circuit can greatly reduce watts loss and voltage drop. This is the same reasoning that is used for the three-wire service-entrance conductors and the three-wire feeder to Panel B in this residence.

In the recreation room, a three-wire cable carrying circuits B10 and B12 feeds from Panel B to the closest receptacle. This three-wire cable continues to each receptacle wall box along the same wall. Circuit B10 feeds the two counter receptacles at the wet bar, and circuit B12 feeds the wall receptacles, fan, and pot lights over the wet bar counter.

The white conductor of the three-wire cable is common to both the receptacle circuit and the wet bar circuits. The black conductor feeds the receptacles. The red conductor is spliced straight through the boxes and feeds the wet bar area.

It is important to connect the neutral according to *Rule 4–034(4)*, which states that the continuity of the identified (white) conductor must be independent of the device connections. This means that a tail must be left at receptacles and lampholders, Fig. 11–8.

Take care when connecting a three-wire circuit to the panel. The black and red conductors *must* be connected to the opposite phases in the panel to prevent heavy overloading of the neutral (white) conductor. On some panels, you can do this by installing the black and the red wires to adjacent circuit breakers. It is essential to verify that the circuit breakers are on alternate phases.

The neutral conductor of the three-wire cable carries the unbalanced current. This current is the difference between the current in the black wire and the current in the red wire. For example, if one load is 12 amperes and the other is 10 amperes, the neutral current is the difference between these loads, 2 amperes, Fig. 19–6.

If the black and red conductors of the three-wire cable are connected to the same phase in Panel A, Fig. 19–7, the neutral conductor must carry the total current of both the red and black conductors rather than the unbalanced current. As a result, the neutral conductor will be overloaded. All single-phase, 120/240-volt panels are clearly marked to help prevent an error in phase wiring. The electrician must check all panels for the proper wiring diagrams before beginning an installation.

Fig. 19–6   Correct wiring connections for a three-wire (multiwire) circuit.

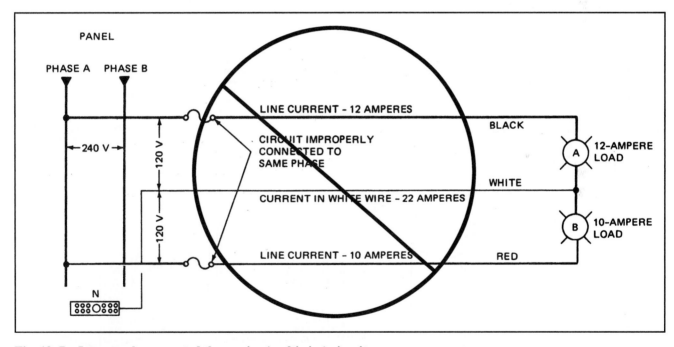

Fig. 19–7   Improperly connected three-wire (multiwire) circuit.

Fig. 19–7 shows how an improperly connected three-wire circuit results in an overloaded neutral conductor. If an open neutral occurs on a three-wire circuit, some electrical appliances in operation may experience voltages higher than normal for the circuit.

Fig. 19–8 shows that for an open neutral condition, the voltage across load A decreases, and the voltage across load B increases. If the load on each circuit changes, the voltage on each circuit also changes.

According to Ohm's law, the voltage drop across any device in a series circuit is directly proportional to the resistance of that device. Therefore, if load B has twice the resistance of load A, then load B will be subjected to twice the voltage of load A for an open neutral condition. To ensure the proper connection, take care when splicing the conductors.

Fig. 19–9 shows what can occur should the neutral of a three-wire multiwire circuit open. Trace the

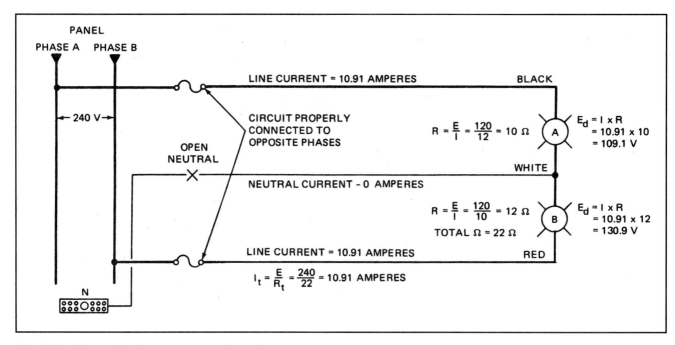

Fig. 19–8   Example of an open neutral conductor.

Fig. 19–9   Problems that can occur with an open neutral on a three-wire (multiwire) circuit.

flow of current from phase A through the television set, then through the toaster, then back to phase B, thus completing the circuit. The following simple calculations show why the television set (or stereo or home computer) can be expected to burn up.

With the neutral broken, the TV and the toaster are connected in series across 240 volts. The total resistance (R_t) is now the sum of these two loads:

$$R_t = (8.45 + 80) \text{ ohms} = 88.45 \text{ ohms}$$

The total current of the circuit is the available voltage (240 volts) divided by the total resistance (88.45 ohms), therefore:

$$I = \frac{E}{R} = \frac{240}{88.45} \text{ amperes} = 2.71 \text{ amperes}$$

Voltage appearing across the toaster is:

$$E = I \times R = 2.71 \times 8.45 \text{ volts} = 22.9 \text{ volts}$$

Voltage appearing across the television is:

$$E = I \times R = 2.71 \times 80 \text{ volts} = 216.8 \text{ volts}$$

The toaster that is rated at 1700 watts is producing only 62 watts, whereas the TV that is rated at 180 watts is producing 587 watts. Therefore, the toaster will not operate, and the TV will burn up. This example illustrates the problems that can arise with an open neutral on a three-wire, 120/240-volt multiwire branch circuit.

The same problem can arise when the neutral of the utility company's incoming service entrance conductors (underground or overhead) opens. The problem is minimized because the neutral of the service is solidly grounded to the water piping system within the building. However, there are cases where poor service grounding has resulted in serious and expensive damage to appliances within the home because of an open neutral in the incoming service entrance conductors.

Three-wire circuits are used occasionally in residential installations. The two hot conductors of three-wire cable are connected to opposite phases. One white neutral (grounded) conductor is common to both hot conductors. All three-wire cables supplying multiwire circuits must be connected to a common-trip, two-pole breaker unless they feed fixed lighting loads or non-split receptacles, *Rule 14–010(b)*.

The two receptacle outlets above the wet bar in the recreation room are GFCI protected as indicated on the plans. All receptacles within 1.5 m of a sink require GFCI protection, *Rule 26–700(11)*.

A single-pole switch to the right of the wet bar controls the four recessed fixtures above it. See Unit 10 for information on recessed fixtures.

An exhaust fan similar to the one in the laundry room is installed in the ceiling of the recreation room to exhaust stale, stagnant, and smoky air from the room.

The basic switch, receptacle, and fixture connections are similar to those discussed in previous units.

The selection of wall boxes for the receptacles is based on the measurements of the furring strips on the walls. Two-by-two furring strips will require the use of shallow device boxes. If the walls are furred with two-by-fours, regular sectional device boxes could be used. Select a box that can contain the number of conductors, devices, and connectors to meet code, as discussed in this text. After all, the installation must be safe.

REVIEW

Note: Refer to the *CEC* or the plans where necessary.

1. What is the total current draw when all six fluorescent fixtures are turned on?

   _____

   _____

2. How many No. 14 AWG conductors will enter the junction box that will be installed above the dropped ceiling near the fluorescent fixtures closest to the stairway?

   _____

3. Why is it important that the hot conductors in a three-wire circuit be properly connected to opposite phases in a panel? _____

_____

_____

4. In the diagram, load A is rated at 10 amperes, 120 volts. Load B is rated at 5 amperes, 120 volts.

a. When connected to the three-wire circuit as indicated, how much current will flow in the neutral? _____

b. If the neutral should open, to what voltage would each load be subjected, assuming both loads were operating at the time the neutral opened? Show all calculations.

5. Calculate the watts loss and voltage drop in each conductor in the following circuit.

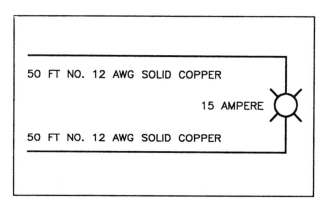

6. Unless specially designed, all recessed incandescent fixtures must be provided with factory-installed _____.

7. If the fluorescent fixtures in the recreation room were to be mounted on the ceiling, what sort of marking would you look for on the label of the fixture? The ceiling is low-density cellulose fibreboard. _____

_____

_____

8. What is the current draw of the recessed fixtures above the bar? _____

_____

_____

_____

9. Calculate the total current draw for circuits B10, 12, and 14.

10. Complete the wiring diagram for the recreation room. Follow the suggested cable layout in Fig 19–2. Use coloured pencils or pens to identify the various colours of the conductor insulation.

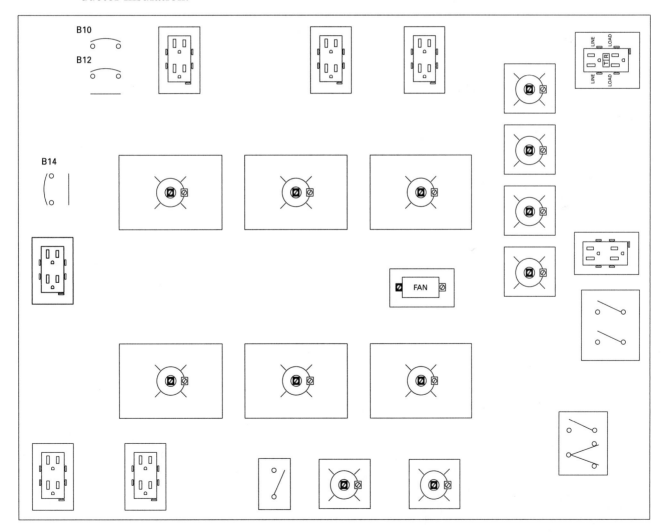

11. a. May a fluorescent fixture that is marked "Recessed fluorescent fixture" be installed in a suspended ceiling? _____

   b. May a fluorescent fixture that is marked "Suspended ceiling fluorescent fixture" be installed in a recessed cavity of a ceiling? _____

   _____

   c. What *CEC, Part 1* section covers the installation of recessed fixtures?

   _____

   _____

12. Using *Table D3* in *Appendix D*, calculate the maximum distance for a No. 12/2 NMD90 wire carrying 10 amperes at 240 volts if a 3% voltage drop is on the cable. Answer in metres. Show calculations. _____

   _____

   _____

13. Using the plans provided with this text, complete a device and wiring layout for the recreation room area to *minimum CEC* standards.

UNIT 20

Branch Circuits for Workshop and Utility Area

OBJECTIVES

After studying this unit, the student will be able to

- understand the meaning, use, and installation of multi-outlet assemblies
- list the requirements for a deep-well jet pump installation
- calculate (given the rating of the motor) the conductor size, conduit size, and overcurrent protection required for the pump circuit
- list the electrical circuits used for connecting electric water heaters
- describe the basic operation of the water heater
- make the proper grounding connections for the water heater

WORKSHOP

The workshop area is supplied by four circuits:

A12   separate circuit for freezer

A17   lighting

A19   plug-in strip (multi-outlet assembly)

A21   two receptacles on window wall

A single-pole switch at the entry controls all four ceiling porcelain lampholders. However, note on the plans (Fig. 20–1) that two of these lampholders have pull-chains to allow the homeowner to turn these lampholders on and off as needed to save energy, Fig. 20–2.

In addition to supplying the lighting, circuit A17 feeds the chime transformers and ceiling exhaust fan. The current draw of the chime transformer is extremely small. Table 20–1 shows the estimated load for lighting; Table 20–2 the load for the other three 15-ampere circuits. See also *Rules 30–104(1)* and *14–600*.

WORKBENCH LIGHTING

Two two-lamp, 34-watt fluorescent fixtures are mounted above the workbench to reduce shadows over the work area. The electrical plans show that these fixtures are controlled by a single-pole wall switch. Either a junction box is mounted immediately above or adjacent to the fluorescent fixture so that the connections can be made readily, or the fixture may be directly wired with NMD90.

NMD90 can be used to connect the fixture to the junction box. Flexible cord (Type S, SJ, or equivalent) cannot be used to connect the fixture to the junction box since it cannot be used as a substitute for the fixed wiring of structures, *Rule 4–010(3)*. The fluorescent fixture must be grounded.

RECEPTACLE OUTLETS

Rule 26–710(e) stipulates that at least one duplex receptacle shall be provided in any unfinished

Fig. 20–1 Cabling layout for the workshop.

DESCRIPTION	QUANTITY	WATTS	VOLT-AMPERES
Ceiling lights @ 100 W each	4	400	400
Fluorescent fixtures Two 34-W lamps each	2	136	156
Chime transformer	1	8	10
Exhaust fan	1	80	90
TOTALS	8	624	656

Table 20–1   Workshop: outlet count and estimated lighting load.

basement area. *Rule 26–710(e)* requires that at least one duplex receptacle be provided for a central vacuum system, where the duct-work is already installed. In this residence, the outlet for the central vacuum is in the garage. In some jurisdictions, the installation of a sump pump in the basement is mandatory. A separate circuit is recommended for the receptacle feeding the sump pump and does not have to be GFCI protected.

DESCRIPTION	QUANTITY	WATTS	VOLT-AMPERES
15-A circuit, A19			
Receptacle next to multi-outlet assembly @ 1A	1	120	120
A six-receptacle plug-in multi-outlet assembly at 1 A per outlet (120 VA)	1	720	720
TOTALS	2	840	840
15-A circuit, A21			
Receptacles on window wall @ 120 W each	2	240	240
TOTALS	2	240	240
15-A circuit, A12			
Single receptacle for freezer (*not* GFCI)	1	360	696
TOTAL of 3 Circuits	5	1440	1776

Table 20–2   Workshop: outlet count and estimated load for three 15-ampere receptacle circuits (wire with conduit). *Rule 8–304(4)* counts each 300 mm of multi-outlet assembly as one outlet for areas where a number of appliances will be used simultaneously.

A. KEYLESS   B. PULL CHAIN

Fig. 20–2   Porcelain lampholders.

CABLE INSTALLATION IN BASEMENTS

Unit 7 covers the installation of non-metallic-sheathed cable (Romex) and armoured cable (BX).

Running non-metallic-sheathed cable is covered by the *12–500* series of rules, and armoured cable (BX) installations by the *12–600* series of rules. When laying out the basement, consider:

- Cables cannot be run across the lower face of basement joists unless they are protected from mechanical injury.
- When run through holes in the joists, cables are considered to be supported.
- Cables for concealed installations must be kept at least 32 mm from the edge of the joist.
- Check with your local inspection authority to see how many cables may be run through holes in the joists before *Rule 4–004(11)* is applied.

- When cables are run near heating ducts, they must be 25 mm from the duct to prevent heat transfer from the duct to the cable. If it is not possible to obtain a 25-mm separation, a thermal barrier must be installed. Running the cable below the ducts whenever possible will also help prevent heat transfer from the ducts.
- If the cable is run near a concrete or masonry chimney, a distance of 50 mm must be maintained between the conductor and a chimney, and 150 mm must be maintained between the conductor and a chimney cleanout.
- Some installations will require drilling of the basement joists. To make the cable easier to pull, start at one end of the basement and drill every second joist. At the end of the run turn around and drill the remaining joists from the opposite direction. The cable will be much easier to pull through holes drilled this way.

MULTI-OUTLET ASSEMBLY

A multi-outlet assembly, Fig. 20–3, has been installed above the workbench. For safety and adequate wiring, any workshop receptacles should be connected to a separate circuit, Fig. 20–4. In the event of a malfunction in any power tool commonly used in the home workshop (saw, planer, lathe, or drill), only receptacle circuit A19 is affected. The lighting in the workshop is not affected by a power outage on the receptacle circuit.

The installation of multi-outlet assemblies must conform to the requirements of *Rule 12–3028*.

Fig. 20–3   Multi-outlet assembly showing track, approvals, voltage, and current ratings.

Fig. 20–4   Detail of how the workbench receptacle plug-in strip is connected through the receptacle located adjacent to the workbench. This is circuit A19.

Load Considerations for Multi-Outlet Assemblies

A maximum of 12 outlets are permitted on any two-wire circuit. These outlets have a load of one ampere each. The maximum load, therefore, is 12 amperes, which is 80 percent of the 15-ampere circuit breaker that feeds the circuit A19. Where a number of appliances may be used simultaneously, each 300 mm of multi-outlet assembly is considered to be one outlet, *Rule 8–304(4)*.

Because the multi-outlet assembly, Fig. 20–3, has been provided in the workshop for plugging in portable tools, a separate circuit A19 is provided. This circuit has been included in the service entrance calculations as part of the general lighting and power load requirements for the residence.

Fig. 20–5   One type of multi-outlet assembly showing track, receptacle, and track cover.

Many types of convenience receptacles are available for various sizes of multi-outlet assemblies, including duplex, single-circuit grounding, split-circuit grounding, and duplex split-circuit receptacles.

Wiring the Multi-Outlet Assembly

The electrician first attaches the base of the multi-outlet assembly to the wall or baseboard. The receptacles are snapped into the base, Fig. 20–5. Covers are then cut to the lengths required to fill the spaces between the receptacles. Manufacturers of these assemblies can supply prefabricated covers to be used when evenly spaced receptacles are installed. These receptacles are usually installed 305 mm or 450 mm apart.

In general, multi-outlet installations are simple because of the large number of fittings and accessories available. Connectors, couplings, ground clamps, blank end fittings, elbows for turning corners, and end entrance fittings can all be used to simplify the installation.

EMPTY CONDUITS

Although not truly part of the workshop wiring, Note 8 to the first-floor electrical plans indicates that two empty 16-mm EMT raceways are to be installed, running from the workshop to the attic. This is unusual but certainly handy when at some later date, additional wiring might be required. The empty raceways will provide the necessary route from the basement to the attic, thus eliminating the need to fish through the wall partitions.

Fig. 20–6   Multi-outlet assembly starter is mounted on an outlet box and is used to supply power to the assembly track.

Fig. 20–7   Components of a jet pump.

UTILITY ROOM

In the utility room, the water pump and water heater require special-purpose outlets.

WATER PUMP CIRCUIT B

All dwellings need a good water supply. City dwellings generally are connected to the city water system. In rural areas, where there is no public water supply, each dwelling has its own system in the form of a well. The residence plans show that a deep-well jet pump is used to pump water from the well to the various plumbing outlets. The outlet for this jet pump is shown by the symbol ⒜B. An electric motor drives the jet pump. A circuit must be installed for this system.

JET PUMP OPERATION

Fig. 20–7 shows the major parts of a typical jet pump. The pump impeller wheel ① forces water down a drive pipe ② at a high velocity and pressure to a point just above the water level in the well casing. (Some well casings may be driven to a depth of more than 30.5 m before striking water.) Just above the

water level, the drive pipe curves sharply upward and enters a larger vertical suction pipe ③. The drive pipe terminates in a small nozzle or jet ④ in the suction pipe. The water emerges from the jet with great force and flows upward through the suction pipe.

Water rises in the suction pipe, drawn up through the tailpipe ⑤ by the action of the jet. The water rises to the pump inlet and passes through the impeller wheel of the pump. Part of the water is forced down through the drive pipe again. The remaining water passes through a check valve and enters the storage tank ⑥.

The tailpipe is submerged in the well water to a depth of about 3.0 m. A foot valve ⑦ and strainer ⑧ are provided at the end of the tailpipe. The foot valve prevents water in the pumping equipment from draining back into the well when the pump is not operating.

When the pump is operating, the lower part of the storage tank fills with water, and air is trapped in the upper part. As the water rises in the tank, the air is compressed. When the air pressure is 276 kilopascals (kPa), a pressure switch ⑨ disconnects the pump motor. The pressure switch is adjusted so that as the water is used and the air pressure falls to 138 kPa, the pump restarts and fills the tank again.

The Pump Motor

A jet pump for residential use may be driven by a one-horsepower (hp) motor at 3400 rpm. A dual-voltage motor is used so that it can be connected to 120 or 240 volts. For a 1-hp motor, the higher voltage is preferred since the current at this voltage is half that at the lower voltage. It is recommended that a single-phase, capacitor-start motor be used for a residential jet pump. Such a motor is designed to produce a high starting torque, which helps to overcome the back pressure within the tank and the weight of the water being lifted from the well.

The Pump Motor Circuit

Fig. 20–8 is a diagram of the electrical circuit that operates the jet pump motor and pressure switch. This circuit is taken from Panel A, circuits A5 and

(A) 240-VOLT BUS IN PANEL

(B) 20-AMPERE DOUBLE-POLE CIRCUIT BREAKER.

(C) NO. 12 AWG CONDUCTORS

(D) DOUBLE-POLE MOTOR CIRCUIT SWITCH WITH OVERLOAD DEVICE (SEE *TABLE 25*)

(E) DOUBLE-POLE PRESSURE SWITCH

(F) 1-HP, 240-VOLT, SINGLE-PHASE MOTOR

Fig. 20–8   Pump circuit.

A7. *CEC Table 45* indicates that a 1-hp single-phase motor has a rating of 8 amperes at 240 volts. This motor is provided with internal thermal protection. With this information, the conductor size, conduit size, branch-circuit overcurrent protection, and overload protection can be determined for the motor.

Conductor Size (*Rule 28–106(1)*)

Rule 28–106(1) requires that conductors supplying motors have an ampacity of not less than 125 percent of the motor full-load current rating. Therefore, the 8-ampere pump motor would require a minimum of a 10-ampere conductor, which would be a minimum No. 14 AWG. However, the schedule of special-purpose outlets in the specifications states that No. 12 AWG conductors are to be used for the pump circuit.

Conduit Size

If rigid PVC conduit is run from the house to the well head, two No. 12 and one #14 AWG T90/TWN75 conductors require 16-mm conduit, *Rules 12–1014, 12–1120.*

Branch-Circuit Overcurrent Protection (*Rule 28–200, Tables 29 & D16*)

Rule 28–200 requires the jet pump motor to be protected with an overcurrent device sized according to *Table 29*. Therefore, *Table 29*, line 1, can be used.

Motor circuit overcurrent sizing may be determined using *Table D16*, which is based on *Table 29*. However, since not all motor full-load current ratings are listed in this table, the overcurrent devices should be sized according to *Table 29* and *Rule 28–200* as follows:

- Non-time-delay fuses
  Maximum of 300% of the motor FLA (full-load amperes)
  $8 \times 300\% = 24$ amperes
  therefore, use 20-ampere fuse

- Time-delay fuses
  Maximum of 175% of motor FLA
  $8 \times 175\% = 14$ amperes therefore, use 15-ampere fuse (*Rule 28–200(b)*)

- Circuit breakers—inverse time
Maximum of 250% of motor FLA
$8 \times 250\% = 20$ amperes
therefore, use 20-ampere breaker

When overcurrent devices, as selected above, will not allow the motor to start, it is permissible to increase the overcurrent device according to *Rule 28–200(d)* as follows:

- Non-time-delay fuses
Maximum of 400% of motor FLA

- Time-delay fuses
Maximum of 225% of motor FLA

- Circuit breakers—inverse time
Maximum of 400% of motor FLA for motors rated 100 hp or less

The previously mentioned percentage ratings, as indicated in *Table 29* and *Rule 28–200(d)*, are maximum values. If the value calculated by this method does not correspond to a standard size or rating of overcurrent device, the next smaller size must be used. However, overcurrent devices smaller than 15 amperes are not required, *Rule 28–200(b)*. Overcurrent devices may be sized smaller than the maximum allowable size if desired.

Running Overload Protection for the Motor (*Rule 28–300*)

All motors are required to have overload protection (*Rule 28–300*) unless they meet the requirements of *Rule 28–308*.

Overload devices can be time-delay fuses or thermal devices of the fixed or adjustable type. When sizing overload devices, it is important to note the motor service factor, which is indicated on the motor nameplate. If the motor has a marked service factor of 1.15 (115 percent) or greater, the overload devices are sized at 125 percent of the motor full-load current, *Rule 28–306(1)*. If the service factor is less than 1.15 (115 percent) or is unknown, the overload devices will be sized at a maximum of 115 percent of the motor FLA.

Assuming that the pump has, as most modern motors do, a service factor of 1.15 or greater, the overloads will be sized as:

$$8 \text{ amps} \times 125\% = 10 \text{ amperes}$$

Wiring for the Pump Motor Circuit

The water pump circuit is run from Panel A circuits A5 and A7, to a disconnect switch and controller mounted on the wall next to the pump. Thermal overload devices (also called *heaters*) are installed in the controller to provide running overload protection. Two No. 12 conductors are connected to the double-pole, 20-ampere branch-circuit overcurrent protective device in Panel A to form circuit A5/A7. Dual-element, time-delay fuses may be installed in the disconnect switch. These fuses will serve the dual function of providing both branch-circuit overcurrent protection for the pump motor and backup protection for the thermal overload protecting the motor. These fuses may be rated at not more than 125 percent of the full-load running current of the motor.

A short length of flexible metal conduit is connected between the load side of the thermal overload switch and the pressure switch on the pump. This pressure switch is usually connected to the motor at the factory. Fig. 20–8 shows a double-pole pressure switch at (E). When the contacts of the pressure switch are open, all conductors feeding the motor are disconnected.

SUBMERSIBLE PUMP

A submersible pump, Fig. 20–9, consists of a centrifugal pump driven by an electric motor. The pump and the motor are contained in one housing, submersed below the permanent water level within the well casing. When running, the pump raises the water upward through the piping to the water tank. Proper pressure is maintained in the system by a pressure switch. The disconnect switch, the pressure and limit switches, and the controller are installed in a logical and convenient location near the water tank.

Some water tanks have a precharged air chamber containing a vinyl bag that separates the air from the water. This assures that air is not absorbed by the water. The absorption of air causes a slow reduction of the water pressure, and, ultimately, the tank will fill completely with water with no room for the air. The use of a vinyl bag to hold the air ensures that the initial air charge is always maintained.

Fig. 20–9 Submersible pump.

Power to the motor is supplied by a cable especially designed for use with submersible pumps. Marked "submersible pump cable," this cable is generally supplied with the pump and its other components. The cable is cut to the proper length to reach between the pump and its controller. When needed, the cable may be spliced according to the manufacturer's specifications.

Submersible water pump cable is "tag-marked" for use within the well casing for wiring deep-well water pumps where the cable is not subject to repetitive handling caused by frequent servicing of the pump unit.

The wiring from the well head to the main distribution panel must be in accordance with *CEC Section 12*. Should it be required to run the pump's circuit underground for any distance, it is necessary to install underground feeder cable type NMWU, or a raceway with suitable conductors, and then make up the necessary splices in an approved weatherproof junction box.

Fig. 20–10 illustrates how to provide proper bonding of a submersible water pump and the well casing. Proper grounding of the well casing and the submersible pump motor will minimize or eliminate stray voltage problems that could occur if the pump is not grounded.

Key *CEC* rules that relate to grounding water pumping equipment are:

- *26–954(d)*: pumps shall be bonded to ground according to *Section 10*
- *10–402(1a)*: motors operating at more than 30 volts must be grounded
- *10–808*: bonding conductor size
- *10–406(2)*: bond water-piping system to ground with # 6 copper

See Unit 25 for information about underground wiring.

The controller contains the motor's starting relay, overload protection, starting and running capacitors, lightning arrester, and terminals for making the necessary electrical connections. Thus, there are no moving electrical parts with the pump itself, such as would be found in a typical single-phase, split-phase induction motor that would require a centrifugal starting switch.

The calculations for sizing the conductors, and the requirements for the disconnect switch, the motor's branch circuit fuses, and the grounding connections are the same as those for the jet pump motor. The nameplate data and instructions furnished with the pump must be followed.

WATER HEATER CIRCUIT ▲c

All homes require a continuous supply of hot water. To meet this need, one or more automatic water heaters are installed close to the areas in the home with greatest need for hot water. The proper size and type of water piping is installed to carry the heated water from the water heater to the plumbing fixtures and to the appliances that require hot water, such as clothes washers and dishwashers.

Modern water heaters have adjustable temperature-regulating controls that maintain the water temperature at the desired setting, generally within a range of 45°–75°C.

SPLICES MADE IN SPACE
PROVIDED IN WELL CAP
OR IN JUNCTION BOX.

WELL
CAP

BONDING CONDUCTOR
CONNECTED TO CASING
AND TO PUMP, *RULE 10–406(2).*

CASE BONDING, MIN. NO. 6 COPPER FROM
CASE OF MAIN PANEL TO CASE OF WELL PUMP.

WELL
CASING

TYPE NMWU OR CABLE OR
RACEWAY WITH PROPERLY SIZED
CONDUCTORS. MUST CARRY
BONDING CONDUCTOR SIZED PER
TABLE 16.

PUMP SHALL BE GROUNDED,
RULE 26–954(d).

INTAKE

SUBMERSIBLE PUMP

PUMP CONTROLLER AND
DISCONNECT SWITCH

• DISCONNECT MUST BE IN
  SIGHT OF CONTROLLER—
  RULE 28–604(3)

• DISCONNECT MUST BE IN
  SIGHT OF MOTOR OR BE
  CAPABLE OF BEING LOCKED
  IN THE "OFF" POSITION—
  RULE 28–604(4)

• DISCONNECT MUST OPEN
  ALL UNGROUNDED ("HOT")
  CONDUCTORS—
  RULE 14–010(b)

• DISCONNECT MAY BE IN
  SAME ENCLOSURE AS
  CONTROLLER—
  RULE 28–604 1(b)

• DISCONNECT MUST BE
  HORSEPOWER RATED—
  RULE 28–602 1(a)

Fig. 20–10   Bonding requirements for submersible pumps. *Rules 26–954(d)* and *10–406(2)* require that submersible pumps be bonded to ground.

For safety reasons, the *CEC* and the CSA require that electric water heaters be equipped with a high-temperature cutoff. This high-temperature control limits the maximum water temperature to about 96°C. This control is set at the factory and cannot be changed in the field. According to *Rule 26–750(1)*, electric water heaters shall be equipped with a temperature limiting means in addition to the control thermostat. This temperature limiting means must disconnect all ungrounded conductors.

In most residential water heaters, the upper thermostat is set so that it will not exceed 90°C, and the high-temperature cutoff is set so that the water temperature does not exceed 96°C. They are usually combined in one device, *Rule 26–750(2)*.

Most customers will, for safety reasons, ask for a setting less than the maximum. See the section on scalding from hot water later in this unit.

The plumber usually installs pressure/temperature relief valves into an opening provided and marked for the purpose on the water heater. The plumber also installs tubing from the pressure/temperature relief valve downward to within 150 mm of the floor so that any discharge will exit close to the floor. This reduces the possibility of scalding someone who might be standing nearby. The water heater manufacturer can also provide a factory-installed pressure/temperature relief valve.

Pressure/temperature relief valves are installed to prevent the water heater tank from rupturing under excessive pressures and temperatures should the adjustable thermostat and the high-temperature cutoff fail to open. This failure would allow the water to boil (100°C) and create steam, and could lead to the tank rupturing.

Residential water heaters typically are available in 60-, 101-, 170-, 225-, and 270-litre sizes.

To reduce corrosion, most water heaters have glass-lined steel tanks and contain one or two magnesium anodes (rods) permanently submerged in the water. Their purpose is to provide "cathodic protection." Simply stated, corrosion problems are minimized as long as the magnesium anodes are in an active state. Information about replacing the anode is in the manufacturer's literature included with every electric water heater.

HEATING ELEMENTS

The wattage ratings of electric water heaters vary greatly, depending on the size of the heater in litres, the speed of recovery desired, local electric utility regulations, and local plumbing and building codes.

Wattage ratings are usually 1650, 2000, 2500, 3000, 3800, 4500, or 5500 watts; other wattage ratings are available.

The heating elements are generally rated 240 volts. Elements are also available with 120-volt ratings and 208-volt ratings. These elements will produce rated wattage output at rated voltage. When operated at less than rated voltage, the wattage output will be reduced. The effect of voltage variation is discussed later in this unit. Most heating elements will burn out if operated at voltages five percent higher than that for which they are rated.

To reduce the element burnout situation, some manufacturers will supply water heaters with heating elements that are actually rated for 250 volts, but will mark the water heater 240 volts with its corresponding wattage at 240 volts. This allows a cushion should higher than normal voltages be experienced. *CSA Standard CAN/CSA-C22.2 No. 110-M90* titled *Construction and Test of Electric Storage-Tank Water Heaters* details the specifica-

tions and test procedures for water heater storage tanks and elements.

Heating elements consist of a high-resistance nickel-chrome (nichrome) alloy wire that is coiled and embedded (compacted) in magnesium oxide in a copper-clad, or stainless steel, tubular housing, Fig. 20–11. The magnesium oxide is an excellent insulator of electricity, yet it effectively conducts the heat from the nichrome wire to the housing. The construction of heating elements is the same as that of the surface units of an electric range. Terminals are provided on the heating element for the connection of the conductors. A threaded hub allows the element to be screwed securely into the side of the water tank. Water heater resistance heating elements are in direct contact with the water for efficient transfer of heat from the element to the water.

Electric water heaters are available with one or two heating elements, Fig. 20–12. Single-element water heaters have the heating element located near the bottom of the tank. Two-element water heaters have one heating element near the bottom of the tank and the other about halfway up.

Thermostats, as illustrated in Fig. 20–13, turn the heating elements on and off, thereby maintaining the desired water temperature. In two-element water heaters, the top thermostat is an interlocking type, sometimes referred to as a snap-over type. The bottom thermostat is a single-pole, on/off type. Both thermostats have silver contacts and have quick make and break characteristics to prevent arcing. The various wiring diagrams shown in this unit show that both heating elements cannot be energized at the same time.

Fig. 20–11 Heating element. Courtesy of Emerson Heating Products, Emerson Electric Company.

Fig. 20–12 Typical electric water heater showing location of heating elements and thermostats.

SPEED OF RECOVERY

Speed of recovery is the time required to bring the water temperature to satisfactory levels in a given time. The current accepted industry standard is based on a 32.22°C rise. For example, Table 20–3 shows that a 3500-watt element can raise the water temperature 90°F (32.22°C) of a little more than 16 gallons an hour. A 5500-watt element can raise water temperature 90°F (50°C) of about 25 gallons an hour. Speed of recovery is affected by the type and amount of insulation surrounding the tank, the supply voltage, and the temperature of the incoming cold water.

SCALDING FROM HOT WATER

The following are plumbing code issues but are discussed briefly because of their seriousness.

Temperature settings for water heaters pose a dilemma. Set the water heater thermostat high enough to kill bacteria on the dishes in the dishwasher and possibly get scalded in the shower. Or set the thermostat to a lower setting and not get scalded in the shower, but live with the possibility that bacteria will still be present on washed dishes.

Fig. 20–13 Electric water heater controls. This seven-terminal control combines water temperature control plus high-temperature limit control. This is the type commonly used to control the upper heating element and provide a high temperature limiting feature. The two-terminal control is the type used to control water temperature for the lower heating element. Courtesy of THERM-O-DISC, Inc.

LITRES PER HOUR RECOVERY FOR INDICATED TEMPERATURE RISE

Heating element wattage	30°C	35°C	40°C	45°C	50°C	55°C
750	21	18	16	14	13	12
1000	29	25	21	19	17	16
1250	36	31	26	24	22	20
1500	43	37	32	29	26	24
2000	57	49	43	38	34	31
2250	65	55	48	43	39	35
2500	72	61	54	48	43	39
3000	86	74	65	57	52	47
3500	100	86	75	67	60	55
4000	115	98	86	77	69	63
4500	129	111	97	86	77	70
5000	143	123	108	96	86	78
5500	158	135	118	105	95	86
6000	172	147	129	115	103	94

NOTES: If the incoming water temperature is 5°C, use the 45° column to raise the water temperature to 50°C.

For example, about how long would it initially take to raise the water temperature in a 160-litre electric water heater to 50°C where the incoming water temperature is 5°C? The water heater has a 2500-watt heating element.

Answer: 160 ÷ 48 = 3.33 hours (about 3 hours and 20 minutes)

Table 20–3 Electric water heater recovery rate per litre of water for various wattages.

In the United States, the Consumer Product Safety Commission (CPSC) reports that 3800 injuries and 34 deaths occur each year due to scalding from excessively hot tap water. Unfortunately, the victims are mainly the elderly and children under the age of 5. They recommend that the setting should not be higher than 49°C. Some state laws require that the thermostat be preset at no higher than 49°C. Residential water heaters are all preset by the manufacturer at 49°C.

Manufacturers post caution labels on their water heaters that read something like this:

SCALD HAZARD: WATER TEMPERATURE OVER 49°C CAN CAUSE SEVERE BURNS INSTANTLY OR DEATH FROM SCALDS. SEE INSTRUCTION MANUAL BEFORE CHANGING TEMPERATURE SETTINGS.

To reduce the possibility of scalding, water heaters can be equipped with a thermostatic non-scald mixing valve or pressure-balancing valve that holds a selected water temperature to within one degree regardless of incoming water pressure changes. This prevents sudden unanticipated changes in water temperatures.

Table 20–4 shows time/temperature relationships relating to scalding.

WHAT ABOUT WASHING DISHES?

Older automatic dishwashers required incoming water temperature of 60°C or greater to get the dishes clean. A water temperature of 49°C is not hot enough to dissolve grease, activate power detergents, and kill bacteria. That is why commercial

TEMPERATURE	TIME TO PRODUCE A SERIOUS BURN
37°C	Safe temperature for bathing
48°C	5 minutes
51°C	3 minutes
52°C	1 minute
56°C	15 seconds
60°C	5 seconds
64°C	2 seconds
68°C	1 second
71°C	0.5 seconds

Table 20–4   Time/temperature relationships related to scalding an adult subjected to moving water. For children, the time to produce a serious burn is less than for adults.

and restaurant dishwashers boost temperatures to as high as 82°C.

Newer residential dishwashers have their own built-in water heaters to boost the incoming water temperature in the dishwasher from 49°C to 60°C – 63°C. This heating element also provides the heat for the drying cycle. Today, a thermostat setting of 49°C on the water heater should be satisfactory.

SEQUENCE OF OPERATION

Water heaters can be connected in many ways, usually depending on the local power company's regulations. This book illustrates only three types of connections. Consult your power utility and/or your supplier of electric water heaters to find out what connections are available in your area.

Figs. 20–14, 20–15, and 20–16 illustrate wiring for water heaters that are two-element, limited-demand operation type.

Figs. 20–14 and 20–15 show the wiring for flat rate connection, where electricity to the water heater is "off" during peak load periods of the day. This is controlled by the hydro utility. Generally, the flat rate charged depends on the size of the hot water tank in the residence. Fig. 20–16 shows the wiring for connection to the main panel, so that the kWh power consumption of the water heater is metered, but has the advantage of having 24-hour operation.

When all of the water in the tank is cold, the upper element heats only the water in the upper half of the tank. The upper element is energized through the lower contacts of the upper thermostat. The lower thermostat also indicates that heat is needed, but heat cannot be supplied because of the open condition of the upper contacts of the upper thermostat.

When the water in the upper half of the tank reaches a temperature of about 65°C, the upper thermostat snaps over; the upper contacts are closed and the lower contacts of the upper thermostat are opened. The lower element now begins to heat the cold water in the bottom of the tank. When the entire tank is heated to 65°C, the lower thermostat shuts off. Electrical energy is not being used by either element at this point. The heat-insulating jacket around the water heater is designed to keep the heat loss to a minimum. Thus, it takes many hours for the water to cool only a few degrees.

(A) 120/240-VOLT SERVICE WITH FLAT RATE CONTROL

(B) MAIN METER SOCKET (METER REMOVED)

(C) NO. 10 AWG RED WIRES IN 21-mm RIGID CONDUIT

(D) TO MAIN SWITCH

(E) DOUBLE-POLE FUSED DISCONNECT SWITCH

(F) HIGH-TEMPERATURE CUTOFF

(G) INTERLOCKING THERMOSTAT

(H) UPPER HEATING ELEMENT

(I) LOWER THERMOSTAT

(J) LOWER HEATING ELEMENT

(K) HOT WATER TANK

Fig. 20–14 Wiring for typical off-peak water heating installation.

(L) RELAY CONTACT CONTROLLED BY UTILITY DURING PEAK DEMAND PERIODS TO SHUT OFF WATER HEATER.

NOTE: CIRCLED LETTERS REFER TO THE COMPONENTS LISTED IN FIG. 20–14.

Fig. 20–15 Feedthrough wiring for off-peak control.

Fig. 20–16 **Water heater connected to two-pole circuit in the main panel.**

Most of the time, the bottom heating element will keep the water at the desired temperature. Should a large amount of hot water be used, the top element comes on, resulting in a fast recovery of the water in the upper portion of the tank, which is where the hot water is drawn from.

METERING AND SEQUENCE OF OPERATION

Electric water heaters are available in a number of styles:

- single element
- double element, simultaneous operation
- double element, nonsimultaneous operation

There are several ways that a utility company can provide metering and for the electrician to make the connections.

1. Connect the water heating to one of the two-pole, 240-volt circuits in the main distribution panel, Fig. 20–16. This is probably the most common method because the wiring is simple, and additional switches, breakers, time clocks, or meters are not required. This is the connection for the residence in this text. This method provides electricity to the water heater 24 hours of every day.

2. Connect the water heater in a flat rate arrangement, Fig. 20–14. The utility supplies power to the consumer via a four-wire cable. This contains the three feeder lines and a fourth wire to control the water heater during peak demand periods. This wire is connected to a separate terminal block located at the 3 o'clock or 9 o'clock position in the meter base. The electrician then connects two red No. 10 AWG solid wires to the terminal block at the 3 o'clock position and to the water heater lug on the line side of the meter base. These wires pass through the service conduit, and through the main disconnect into a separate 30-ampere, fusible disconnect. The No. 10/2 AC90 or NMD90 then connects to the water heater.

3. Connect the water heater circuit, as in Fig. 20–16, but install a time clock in the circuit leading to the water heater. The time clock would be set to be "off" during certain times of the day deemed peak hours by the utility. The utility gives special incentives to the homeowner for agreeing to this arrangement. This is a good conservation method that many supply authorities are promoting.

4. Some utilities use radio-controlled switching to control water heater loads. They send a carrier signal through the power line, which is received by the electronic controlling device, thereby turning the power to the water heater "on" and "off" as required by the utility, Fig. 20–15. This is commonly called a flat rate water heater. The utility will charge a fixed rate regardless of the energy used. This is still very common in many centres.

WATER HEATER LOAD DEMAND

These *CEC* rules apply to load demands placed on an electric water heater branch circuit.

Rule 8–104(2) requires that a branch circuit shall have a rating not less than the calculated load of the heater.

Rule 8–104(5)(a) requires that a continuously loaded appliance must have a branch-circuit rating not less than 125 percent of the marked rating of the appliance.

Rule 26–750(4) requires that all electric storage-tank water heaters shall be supplied from a branch circuit used specifically for the heater.

Disconnecting Means

Rule 26–746 requires that the appliance have a means of disconnect. This would normally be located in the panel (load centre), or it could be provided by a separate disconnect switch. *Rule 14–416* allows the unit to have a switch that is an integral part of the appliance having an "off" position and capable of disconnecting all ungrounded conductors from the source, to be considered the required disconnecting means. In many local codes, however, this is not acceptable.

Rules 14–300 and *14–408* require that switches and circuit breakers used as the disconnecting means be of the indicating type, meaning they must clearly show that they are in the "on" or "off" position.

Branch-Circuit Overcurrent Protection

The rating or setting of overcurrent devices shall not exceed the allowable ampacity of the conductors they protect, *Rule 14–104*.

EXAMPLE: A water heater nameplate indicates 4500 watts, 240 volts. What is the maximum size overcurrent protection permitted by the *CEC*?

ANSWER:

$$\frac{4500}{240} \text{ amperes} = 18.75 \text{ amperes}$$

$$18.75 \times 1.25 \text{ amperes} = 23.44 \text{ amperes}$$

Therefore, the *CEC* permits the use of a mimimum 25-ampere fuse or circuit breaker, *Table 13*. If the calculated ampacity of an overcurrent device is not a standard value, the next larger standard size must be used.

Rule 8–104(2) states that the branch-circuit conductors shall have an ampacity not less then the maximum load to be served.

Rule 8–104(1) states that the branch-circuit rating is the lesser of the wire ampacity or that of the overcurrent device.

The water heater for the residence is connected to circuit A6/A8, a 20-ampere, double-pole overcurrent device located in the main panel in the workshop. This circuit is straight 240 volts and does not require a neutral conductor.

Checking the Schedule of Special-Purpose Outlets in Appendix A on page 428, we find that the electric water heater has two heating elements: a 2000-watt element and a 3000-watt element. Because of the thermostatic devices on the water heater, both elements cannot be energized at the same time. Therefore, the maximum load demand is:

$$\text{Amperes} = \frac{\text{watts}}{\text{volts}} = \frac{3000 \text{ watts}}{240 \text{ volts}} = 12.5 \text{ amperes}$$

For water heaters having two heating elements connected for non-simultaneous (limited demand) operation, the nameplate on the water heater will be marked with the largest element's wattage at rated voltage. For water heaters having two heating elements connected for simultaneous operation, the nameplate on the water heater will be marked with the total wattage of both elements at rated voltage.

Conductor Size

Table 2 indicates that the conductors will be No. 12 AWG. The specifications for the residence call for conductors to be type T90/TWN75 or NMD90.

Bonding

If flexible conduit is run from a junction box to the water heater, the water heater is bonded through a separate bonding conductor sized according to *Table 16*, which is run in the flex to the box.

EFFECT OF VOLTAGE VARIATION

The heating elements in the heater in this residence are rated 240 volts. They will operate at a lower voltage with a reduction in wattage. If connected to voltages above their rating, they will have a very short life.

Ohm's law and the wattage formula show how the wattage and current depend upon the applied voltage.

$$R = \frac{E^2}{W} = \frac{240 \times 240}{3000} \text{ ohms} = 19.2 \text{ ohms}$$

$$I = \frac{E}{R} = \frac{240}{19.2} \text{ amperes} = 12.5 \text{ amperes}$$

If a different voltage is substituted, the wattage and current values change accordingly.

At 220 volts:

$$W = \frac{E^2}{R} = \frac{220 \times 220}{19.2} \text{ watts} = 2521 \text{ watts}$$

$$I = \frac{W}{E} = \frac{2521}{220} \text{ amperes} = 11.46 \text{ amperes}$$

At 230 volts:

$$W = \frac{E^2}{R} = \frac{230 \times 230}{19.2} \text{ watts} = 2755 \text{ watts}$$

$$I = \frac{W}{E} = \frac{2755}{230} \text{ amperes} = 11.98 \text{ amperes}$$

In resistive circuits, current is directly proportional to voltage and can be simply calculated using a ratio and proportion formula. For example, if a 240-volt heating element draws 12.7 amperes at rated voltage, the current draw can be determined at any other applied voltage, say 208 volts.

$$\frac{208}{240} = \frac{x}{12.7}$$

$$\frac{208 \times 12.7}{240} \text{ amperes} = 11.01 \text{ amperes}$$

To calculate this example if the applied voltage is 120 volts:

$$\frac{120}{240} = \frac{x}{12.7}$$

$$\frac{120 \times 12.7}{240} \text{ amperes} = 6.35 \text{ amperes}$$

Also, in a resistive circuit, wattage varies as the square of the current. Therefore, when the voltage on a heating element is doubled, the current also doubles and the wattage increases four times. When the voltage is reduced to one-half, the current is halved, and the wattage is reduced to one-fourth.

According to these formulas, a 240-volt heating element connected to 208 volts will have about three-quarters of the wattage rating it has at the rated voltage.

REVIEW

Note: Refer to the *CEC* or the plans where necessary.

WORKSHOP

1. a. What circuit supplies the workshop lighting? _____

    b. What circuit supplies the plug-in strip over the workbench? _____

    c. What circuit supplies the freezer receptacle? _____

2. What type of lighting fixtures are installed on the workshop ceiling? _____
_____
_____

3. Does a receptacle for a freezer have to be GFCI protected? Explain. _____
_____
_____
_____
_____
_____

4. Does a receptacle for a sump pump have to be GFCI protected? Explain. _____
_____
_____
_____
_____
_____

5. When a 120-volt receptacle is provided for the laundry equipment in a basement, what is the *minimum* number of additional receptacle outlets required? What is the *CEC* reference? _____
_____
_____
_____

6. When a single receptacle is installed on an individual branch circuit, the receptacle must have a rating _____ the rating of the branch circuit.

7. Calculate the total current draw of circuit A17 if all lighting fixtures and the exhaust fan were turned on.

8. If a No. 10 NMD90 cable is located in an ambient temperature of 42°C, what would be the temperature correction factor? _____
_____
_____

9. What size box would you use where circuit A17 enters the junction box on the ceiling?
_____
_____
_____

10. The following is a layout of the lighting circuits for the workshop. Using the conduit layout in Fig. 20–1, make a complete wiring diagram. Use coloured pencils or pens to indicate conductors.

WATER PUMP CIRCUIT ▲B

1. How is the motor disconnected if pumping is no longer required? _____

   _____

   _____

   _____

2. Why is a 240-volt motor preferable to a 120-volt motor for use in this residence?

   _____

   _____

   _____

3. How many amperes does a 1-hp, 240-volt, single-phase motor draw? (See *Table 45.*)

   _____

4. What size are the conductors used for this circuit? _____

5. a. What is the branch-circuit protective device? _____

   b. What furnishes the running protection for the pump motor? _____

   _____

   c. What is the maximum ampere setting required for running protection of the 1-hp,
      240-volt pump motor? _____

6. Submersible water pumps operate with the electrical motor and actual pump located
   (circle correct answer)

   a. above permanent water level.

   b. below permanent water level.

   c. half above and half below permanent water level.

7. Because the controller contains the motor starting relay and the running and starting
   capacitors, the motor itself contains _____.

8. What type of pump moves the water inside the deep-well pipe? _____

   _____

   _____

9. Proper pressure of the submersible pump system is maintained by a _____

   _____.

10. Fill in the data for a 16-ampere electric motor, single-phase, 1.15 service factor.

    a. Branch-circuit protection non-time-delay fuses.

       Normal size   _____ A   Switch size _____ A

       Maximum size _____ A   Switch size _____ A

    b. Branch-circuit protection dual-element, time-delay fuses.

       Normal size   _____ A   Switch size _____ A

       Maximum size _____ A   Switch size _____ A

    c. Branch-circuit protection instant-trip breaker.
       Maximum setting _____ A

    d. Branch-circuit protection—inverse time breaker.
       Maximum setting _____ B

    e. Branch-circuit conductor size. _____
       Ampacity _____ A

    f. Motor running protection using dual-element time-delay fuses.
       Maximum size _____ A

11. The *CEC* is very specific in its requirement that submersible electric water pump motors
    be grounded. Where is this specific requirement found? _____

    _____

12. Does the *CEC* allow submersible pump cable to be buried directly in the ground?

_____

_____

WATER HEATER CIRCUIT ▲c

1. At what temperature is the water in a residential heater usually maintained? _____

2. Magnesium rods are installed inside the water tank to reduce _____.

3. Is a separate meter required to record the amount of energy used to heat the water?

_____

4. What is meant by "flat rate" installation? _____

_____

_____

_____

_____

_____

_____

5. What is the most common method of metering water heater loads? _____

_____

_____

_____

_____

_____

6. Two thermostats are generally used in an electric water heater.

   a. What is the location of each thermostat? _____

   _____

   b. What type of thermostat is used at each location? _____

   _____

7. a. How many heating elements are provided in the heater in the residence discussed in this text? _____

   b. Are these heating elements allowed to operate at the same time? _____

8. When does the lower heater operate? _____

   _____

9. The *CEC* states that water heaters must have a temperature regulating device set at _____ °C maximum. Secondary protection is by a temperature cutoff set at the factory at _____ °C.

10. Why does the storage tank hold the heat so long? _____

_____

_____

11. The electric water heater is connected for "limited demand" so that only one heating element can be on at one time.

   a. What size wire is used to connect the water heater? _____

   b. What size overcurrent device is used? _____

12. a. If both elements of the water heater are energized at the same time, how much current will they draw? (Assume the elements are rated at 240 volts.)

   Element No. 1 = 3000 W

   Element No. 2 = 2000 W

   b. What size wire is required for the load of both elements? Show calculations.

13. a. How much current do the elements in Question 12 draw if connected to 220 volts? Show calculations.

   b. What is the wattage value at 220 volts? Show calculations.

14. A 240-volt heater is rated at 1500 watts.

    a. What is its current rating at rated voltage? _____

    b. What current draw would occur if connected to a 120-volt supply?

    c. What is the wattage output at 120 volts?

15. A customer asks the electrician to set the water heater thermostat 30°C below the maximum permitted by *CEC*. The thermostat is set at

    a. 70°C

    b. 66°C

    c. 60°C

    d. 49°C

UNIT 21

Heat and Smoke Detectors, Carbon Monoxide Detectors, and Security Systems

OBJECTIVES

After studying this unit, the student will be able to

- identify the Underwriters Laboratories of Canada standard that refers to household fire warning equipment
- name the two basic types of smoke detectors
- discuss the location requirements for the installation of heat and smoke detectors
- list the locations where heat and smoke detectors may not be installed
- describe the wiring requirements for the installation of heat and smoke detectors
- list the major components of a typical residential security system
- identify the *Rule 32* requirements for the installation of residential fire alarm systems
- identify the requirement for the installation of carbon monoxide detectors

THE IMPORTANCE OF HEAT AND SMOKE DETECTORS

Fire is the third leading cause of accidental death. Home fires account for the biggest share of these fatalities, most of which occur at night during sleeping hours.

In recent years, the standards of the ULC, combined with the *Canadian Building Code* and associated provincial and municipal codes, have made mandatory the installation of smoke alarms, smoke detectors, and/or heat detectors in critical areas of a home. A sufficient number of detectors must be installed to provide full coverage.

Smoke alarms are installed in a residence to give the occupants an early warning of the presence of a fire. Fires produce smoke and toxic gases, which can overcome the occupants while they sleep. Most fatalities result from the inhalation of smoke and toxic gases rather than from burns. Heavy smoke reduces visibility.

Fire-sensing devices commonly used in a residence are heat detectors, Fig. 21–1, and smoke detectors, Fig. 21–2. These must be connected to a central fire alarm control panel that will process the information and send an output signal to alarm bells or buzzers.

The Underwriters Laboratories of Canada (ULC)

The *Underwriters Laboratories of Canada Standard CAN/ULC-S524-M91* discusses the proper

Fig. 21–1   Heat detector.

Fig. 21–2   Smoke detector.

selection, installation, operation, and maintenance of fire warning equipment commonly used in residential occupancies. Minimum requirements are stated in the standard.

In some provincial jurisdictions, it is mandatory that qualified technicians install, verify, and maintain fire alarm systems. The certification to become a fire alarm technician is available through a variety of government-approved courses, through colleges, local electrical unions (IBEW), and the Canadian Fire Alarm Association (CFAA).

Smoke alarms must be installed outside each separate sleeping area in the immediate vicinity of the bedrooms, and on each additional storey of the dwelling, including basements, but excluding crawl spaces and unfinished attics.

Residential-type smoke alarms incorporate both the initiating circuit (detector) and the alarm circuit (buzzer) in the same device. If more than one of these are installed, they must be interconnected so that all will go into alarm at the same time.

It is recommended that smoke and heat detectors be placed strategically to provide protection for living rooms, dining rooms, bedrooms, hallways, attics, furnace and utility rooms, basements, and heated attached garages. Detectors in kitchens must not be directly above the stove.

TYPES OF SMOKE DETECTORS

Two types of smoke detectors commonly used are the photoelectronic type and the ionization type.

Photoelectric Type

In the photoelectric type of smoke detector, a light sensor measures the amount of light in a chamber. When smoke is present, an alarm sounds, indicating that light is being reflected off the smoke in the chamber. This type of sensor detects smoke from burning materials that produce great quantities of white or grey smoke, such as furniture, mattresses, and rags. It is less effective for gasoline and alcohol fires, which do not produce heavy smoke. It is also not as effective with black smoke, which absorbs light rather than reflecting it as grey or white smoke does.

Ionization Type

The ionization type of smoke detector contains a low-level radioactive source that supplies particles that ionize the air in the smoke chamber of the detector. Plates in this chamber are oppositely charged. Because the air is ionized, an extremely small amount of current (millionths of an ampere) flows between the plates. Smoke entering the chamber impedes the movement of the ions, reducing the current flow, which causes an alarm to go off. This type of detector is able to sense products of combustion (POC) in the earliest stages of a fire. It can detect airborne particles (aerosols) as small as 5 microns in diameter.

The ionization type of detector is effective for detecting small amounts of smoke, as in gasoline and alcohol fires, which are fast-flaming. It is also more effective for detecting dark or black smoke than the photoelectric type of detector.

INSTALLATION REQUIREMENTS

The following information concerns specific recommendations for installing smoke or heat detectors, Figs. 21–3, 21–4, and 21–5. Complete data are found in the ULC standards.

It is **mandatory to install smoke detectors and alarms** in all new homes (*Canadian Building Code 3.2.4.21.1/9.10.18.1*).

- **Do** install smoke alarms on each floor of a residence for *minimum* protection. Make sure that one of these alarms is located between the sleeping area and the rest of the house. These alarms must be interconnected so that if one alarm detects smoke they all sound.

- **Do** install smoke alarms in all rooms, hallways, attics, basements, and storage areas of a residence to provide protection that *exceeds* the minimum recommendation above.

- **Do** install the detector on the ceiling as close as possible to the centre of the room or hallway.

- **Do** install a smoke detector on a basement ceiling close to the stairway to the first floor.

- **Do** place a detector within 900 mm of the peak on a sloped ceiling that has greater than 300 mm rise per 2400 mm horizontally (1:8).

- **Do** install smoke detectors and heat detectors at the *top* of an *open* stairway, because heat and smoke travel up.

- **Do** install hard-wired detectors in a *new* construction. These detectors are directly connected to a 120-volt AC source. For existing homes, either battery-powered units or hard-wired 120-volt AC units may be used.

- **Do not** install detectors in the *dead* airspace at the top of a stairway that can be closed off by a door.

- **Do not** place the edge of a detector mounted on the ceiling closer than 150 mm from the wall.

- **Do not** place a wall-mounted detector closer than 150 mm from the ceiling and not farther

Fig. 21–3   Recommended location of heat or smoke detector between the sleeping areas and the rest of the house. Locate outside the bedroom, but near the bedrooms.

Fig. 21–4   Recommendations for the installation of heat and smoke detectors.

than 300 mm from the ceiling. The area where the ceiling meets the wall is considered dead airspace where smoke and heat may not reach the detector.

- **Do not** connect detectors to wiring that is controlled by a wall switch.

- **Do not** connect detectors to a circuit that is protected by a ground-fault circuit interrupter (GFCI) or an arc fault circuit interrupter (AFCI), *Rule 32–110(a)*.

- **Do not** install detectors where relative humidity exceeds 85 percent, e.g., adjacent to showers, laundry areas, or other areas where large amounts of visible water vapour exist.

- **Do not** install detectors in front of air ducts, air conditioners, or any high-draft areas where the moving air will keep the smoke or heat from entering the detector.

- **Do not** install detectors in kitchens where the accumulation of household smoke can result in false alarms with certain types of detectors. The photoelectric type *may* be installed in kitchens, but must *not* be installed directly over the range or cooking appliance.

- **Do not** install where temperature can fall below 0°C or rise above 50°C.

- **Do not** install in garages (vehicle exhaust can set off the detector).

- **Do not** install detectors in airstreams that will pass air originating at the kitchen cooking appliances across the detector. False alarms will result.

MANUFACTURERS' REQUIREMENTS

Manufacturers of smoke and heat detectors have certain responsibilities:

- The power supply must be capable of operating the signal for at least four minutes continuously.

- Battery-powered units must be capable of a low battery warning "beep" of at least one beep per minute for seven consecutive days.

- Direct-connected 120-volt AC detectors must have a visible indicator that shows "power-on."

- Detectors must not signal when a power loss occurs or when power is restored.

Features of Smoke Alarms and Heat Detectors

Smoke alarms may contain an indicating light to show that the unit is functioning properly. They may also have a test button to test the circuitry and alarm capability: When the button is pushed, it is the circuitry and alarm that are tested, not the smoke-detecting ability.

Heat detectors are available that sense a specific *fixed* temperature, such as 57°C or 90°C. Available also are *rate-of-rise* heat detectors that sense rapid changes in temperature (6°C per minute), such as those caused by flash fires.

Fixed and rate-of-rise temperature detectors are available as a combination unit. Combination smoke, fixed, and rate-of-rise temperature detectors are also available in one unit.

The spacing of heat detectors shall be as recommended by the manufacturer, because detectors are capable of sensing heat only within a limited space in a given time.

150 mm MIN.

EDGE OF DETECTOR MUST BE NO CLOSER THAN 150 mm TO WALL.

150 mm MIN.

TOP OF DETECTOR MUST BE NO CLOSER THAN 150 mm FROM CEILING, OR MORE THAN 300 mm FROM CEILING.

300 mm MAX.

THE SHADED AREA IS CONSIDERED TO BE *DEAD* AIRSPACE. DO NOT INSTALL DETECTORS IN THIS SPACE.

Fig. 21–5   Do not mount detectors in the *dead* airspace where the ceiling meets the wall.

Wiring Requirements

Direct-Connected Units: All new construction requires smoke alarms to be connected to a 120-volt circuit that feeds a lighting circuit, *Rule 32–110(a)*. Smoke alarms must never be connected to a dedicated circuit. This type of smoke alarm will usually be of the interconnected type and may include other features such as battery backup or a built-in carbon monoxide detector. A direct connect unit with battery backup is shown in Fig. 21–6.

When using interconnected smoke alarms, it is important to refer to the manufacturer's literature for the maximum number of devices allowed. Each manufacturer and model may allow different maximum numbers. When devices are interconnected, all alarms will sound at the same time, regardless of which device actually detected smoke. All new residential construction requires an interconnected smoke alarm on each level of the home and near the bedrooms. If bedrooms are located at opposite ends of a home, it may be necessary to install more than one alarm on a level.

Interconnected smoke alarms have one white wire, one black wire, and one red or yellow wire, which is used to interconnect them. Refer to Fig. 21–7 for wiring these devices.

Battery-Operated Units: These units require no wiring; they are simply mounted in the desired

Fig. 21–6   An AC–DC smoke alarm that operates on 120 volts AC as the primary source of power and a 9-volt battery as the secondary source of power. In a power outage, the alarm continues to operate on the battery. Courtesy of BRK Electronics.

locations. Batteries last for about a year. Battery-operated units should be tested periodically, as recommended by the manufacturer.

Always follow the installation requirements and recommendations of the unit manufacturer. Refer to the ULC standards for additional technical data relating to the installation of smoke and heat detectors.

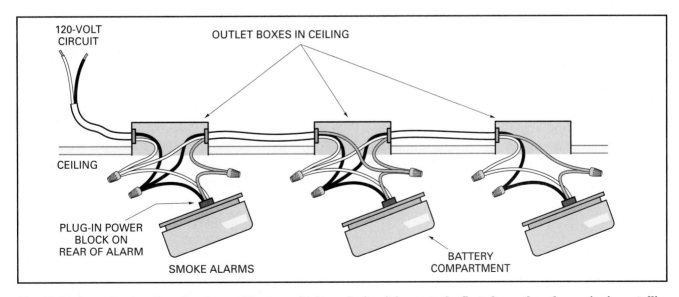

Fig. 21–7   Interconnected smoke alarms. The two-wire branch circuit is run to the first alarm, then three-wire in-metallic-sheathed cable interconnects the other alarms. If one alarm in the series is triggered, all other alarms will sound off.

COMBINATION DIRECT/BATTERY/FEED-THROUGH DETECTORS

These units are connected directly to a 120-volt AC circuit (not GFCI or AFCI protected) and are equipped with a 9-volt DC battery, Fig. 21–6. In a power outage, these detectors will continue to operate on the 9-volt DC battery. The ability to interconnect a number of detectors offers "whole house" protection, because when one unit senses smoke, it places all the other interconnected units into alarm mode as well.

Because of the vital importance of smoke and fire detectors to prevent loss of life, you should very carefully read the application and installation manuals that are prepared by the manufacturer of the smoke and heat detectors.

The residence in this text has four smoke detectors, hard-wired directly to a 120-volt AC branch circuit. They are interconnected feed-through units that will set off signalling in all units should one unit trigger. These interconnected smoke detectors are also powered by a 9-volt DC battery that allows the unit to operate in the event of a failure in the 120-volt AC power supply.

CARBON MONOXIDE DETECTORS

The *National Building Code of Canada* may require the installation of carbon monoxide (CO) detectors in dwelling units. These may be free-standing units, permanently wired to a source of supply, or part of a smoke detector. If the building code requires a CO detector, a battery-operated or plug-in device is not acceptable. It must be permanently wired.

A CO detector is required when:

- The dwelling has a solid-fuel-burning appliance such as a wood stove, wood-burning fireplace, or wood or corn furnaces.

- The dwelling contains a fuel-burning appliance such as a furnace or fireplace operating from a non-solid fuel such as natural gas, propane, or oil.

- The dwelling shares a wall with a garage.

Location and Connection of Carbon Monoxide Detectors

If a solid-fuel appliance is used, a detector must be located on or near the ceiling in the room or area where the appliance is located. When the dwelling uses another source of fuel, or shares a wall with a garage, a detector must be located in each bedroom or outside each bedroom within five metres of the bedroom doors, this distance being measured along the corridor. If the house is large and bedrooms are located on multiple floors or at opposite ends of the building, it may require multiple CO detectors. If the dwelling has a solid-fuel-burning appliance, a detector in the room with the appliance and detectors near the bedrooms are required.

CO detectors may be connected to most 15-ampere lighting and receptacle branch circuits except those used to feed kitchen counter, refrigerator, or dining area receptacles or outlets in garages or located outdoors. They must not be connected to GFCI- or AFCI-protected circuits or circuits that supply only receptacles. A switching or disconnecting means between the device and the circuit breaker is not allowed.

SECURITY SYSTEMS

It is beyond the scope of this text to cover every type of residential security system; we focus on some of the features available from the manufacturers of these systems.

Installed security systems range from simple to very complex. Features and options include intruder detectors for doors and windows, motion detectors, infrared detectors, and under-the-carpet floor mat detectors for "space" protection. Inside and outdoor horns, bells, electronic buzzers, and strobe lights can provide audio as well as visual alarm. Telephone interconnection to preselected telephone numbers, such as the police or fire departments, is also available. (See Figs. 21–8 to 21–15.)

The homeowner and electrician should discuss specific features and benefits of security systems. Most electricians are familiar with a particular manufacturer's selection of systems. Fig. 21–16 shows the range of devices available for such a system.

These systems are generally on display at lighting fixture stores that offer the homeowner an opportunity to see a security system "in action."

Wiring a security system consists of small, easy-to-install, extra-low-voltage, multiconductor cables made up of No. 18 AWG conductors. The actual installation of these conductors should be done after

Fig. 21–8   The wireless key allows the end user to disarm and arm the system with the press of a button. There is no need to remember a security code, and it can also be used to operate lights and appliances.

Fig. 21–9   Monitored smoke alarms are on 24 hours a day, even if the burglary protection is turned off.

Fig. 21–10   Magnetic contacts are often located at vulnerable entry points such as doors.

Fig. 21–11   Sirens, such as these placed inside the home, alert families to emergencies.

Fig. 21–12   Motion detectors are interior security devices that sense motion inside the home.

Fig. 21–13   A wireless panic button allows the user to press a button to signal for help.

Fig. 21–14   The wireless keypad is portable so you can arm or disarm the system from the yard or driveway.

Fig. 21–15   Glass break detectors, available in a variety of sizes, will sound the alarm when they detect the sound of breaking glass.

Fig. 21–16   Typical residential security system showing some of the devices available. Complete wiring and installation instructions are included with these systems. Check local code requirements and follow the detailed instructions furnished with the system. Most of the interconnecting conductors are No. 18 AWG.

the regular house wiring is completed to prevent damage to these smaller cables. Usually the wiring can be done at the same time as the chime wiring is being installed. Security system wiring comes under the scope of *Section 16*.

When such detectors as door-entry, glass break, floor mat, and window foil are used, circuits are electrically connected in series so that if any part of the circuit is opened, the security system will detect the open circuit. These circuits are generally referred to as "closed" or "closed loop." Alarms, horns, and other signalling devices are connected in parallel, since they will all signal at the same time when the security system is set off. Heat detectors and smoke detectors are generally con-

nected in parallel because all of these devices will "close" the circuit to the security master control unit, setting off the alarms. Fire detectors are available for series connection for use in security systems that require normally closed contacts for circuit operation.

The instructions furnished with all security systems cover the installation requirements in detail, alerting the installer to *CEC* regulations, clearances, suggested locations, and mounting heights of the systems components.

Always check with the local electrical inspection authority to determine if there are any special requirements in your locality for the installation of security systems.

REVIEW

Note: Refer to the *CEC* or the plans where necessary.

1. What is the name and number of the ULC standard that gives information about smoke and heat detectors? _____

   _____

2. Name the two basic types of smoke detectors. _____

   _____

   _____

3. Why is it important to mount a smoke alarm on the ceiling not closer than 150 mm from a wall? _____

   _____

   _____

   _____

   _____

4. A basic rule is to install smoke alarms (circle one)

   a. between the sleeping area and the rest of the house.

   b. at the top of a basement stairway that has a closed door at the top of the stairs.

   c. in a garage that is subject to subzero temperatures.

5. List two types of thermal detectors that are used in residential fire alarm systems.

   _____

   _____

6. Although the ULC gives many rules for the installation of smoke and heat detectors, always follow the installation recommendations of _____.

7. Security systems are usually installed with wire that is much smaller than normal house wires. These wires are generally No. _____ AWG, which is similar in size to the wiring for chimes.

8. Because the wire used in wiring security systems is rather small and cannot stand rough service and abuse, security wiring should be installed during the rough-in stages of a new house (before) (after) the other power-circuit wiring is completed. Circle the correct answer.

9. When several smoke alarms are installed in a new home, these units must be (circle the correct answer)

   a. connected to separate 120-volt circuits.

   b. connected to one 120-volt circuit.

   c. interconnected so that if one smoke alarm is set off, the others will also sound a warning.

10. It is advisable to connect smoke alarms to circuits that are protected by GFCI devices. (Circle the correct answer.) TRUE   FALSE

11. Because a smoke alarm might need servicing or cleaning, be sure to connect it so that it is controlled by a wall switch. (Circle the correct answer.) TRUE   FALSE

12. Smoke alarms of the type installed in homes must be able to sound a continuous alarm when set off for at least (circle one)

   a. 4 minutes.

   b. 30 minutes.

   c. 60 minutes.

13. If a carbon monoxide detector is located in the hall outside the bedrooms, what is the maximum distance that the detector may be from their doors?

14. If a home has an attached garage, is a CO detector required?

15. In a home with a wood-burning fireplace, where must the CO detector(s) be located?

UNIT 22

Extra-Low-Voltage, Remote-Control and Signal Systems

OBJECTIVES

After studying this unit, the student will be able to

- explain the operation of an extra-low-voltage, remote-control system for lighting circuits
- interpret the wiring diagrams of various types of extra-low-voltage, remote-control systems
- install an extra-low-voltage, remote-control system to comply with the requirements of the *CEC*
- define what is meant by a signal circuit
- describe a chime circuit with one main chime and one or more extention chimes

The general lighting circuits described in this text are wired using standard methods. Another way of installing lighting circuits is to use a remote-control, or low-voltage, switching system. This system provides very flexible switching.

Section 0 defines any voltage from 0 to 30 volts as extra-low voltage. However, many electricians incorrectly refer to voltages of up to 30 volts as low voltage.

This unit discusses the remote-control wiring of one manufacturer's system. In general, it is difficult to mix components of one manufacturer with those of another.

This is how typical remote-control, extra-low-voltage systems operate. There are differences in the form, fit, and function of systems from the many manufacturers of these systems. Electrical wholesalers and distributors usually carry one or two brands of these systems. Once you have decided to install a remote-control, extra-low-voltage system, it is best to stay with one manufacturer's components.

Another type of remote-control system plus many added features is covered in Unit 24.

A remote-control wiring system uses controlling devices, such as relays, with the equipment to be controlled. An auxiliary means is provided to operate the controlling device, such as the switches shown in Fig. 22–1 or the master control in Fig. 22–2.

An extra-low-voltage, remote-control system can be installed to provide multiswitch control of lighting convenience outlets or small appliances in dwellings. This system is only slightly more expensive than a conventional wiring system. Extra-low-voltage switches are easy to install and can be used for light outlets controlled from one or more locations.

The remote-control system for lighting circuits consists of 24-volt relays. These devices have built-in 120-volt contacts, which can be controlled by extra-low-voltage controlling switches. The relays are rated at 25 volts AC for use at a nominal voltage of 24 volts. Since an extra-low voltage is used, the

Fig. 22–1   Low-voltage switches.

Fig. 22–2   Master selector switch.

shock hazard is less than for a 120-volt system. Extra-low-voltage, remote-control circuits may be installed with two-wire or three-wire No. 18 AWG cable, depending on the manufacturer's requirements.

This type of installation lends itself to large residences as well as commercial and industrial buildings.

REMOTE-CONTROL SWITCHES

The extra-low-voltage, remote-control system uses a normally open, single-pole, double-throw momentary contact switch. This switch is roughly one-third the size of a standard single-pole switch. It may have three or four terminals or three colour-coded lead wires. Fig. 22–3 is the wiring diagram of an extra-low-voltage switch. Fig. 22–1 shows typical extra-low-voltage switches.

Remote-control switches for lighting systems can be mounted in switch boxes at the switch locations, or without boxes on standard raised plaster rings selected for the thickness of the finished walls. The switch boxes or plaster rings must be mounted horizontally for some types of extra-low-voltage switches. The master selector switch requires a horizontally mounted, two-gang raised plaster ring. (The master selector switch is covered later in this unit.)

Some local building codes require the use of boxes at the switch locations. The appearance of the installation generally is improved when switch boxes are used. Many devices have pigtail leads that must be spliced to the extra-low-voltage cables. Switch boxes provide the room necessary for splicing. They are especially desirable for outside walls so that the splices do not have to be pushed back into the insulation or vapour barrier.

The white lead of the extra-low-voltage switch is connected to the 24-volt transformer. The red lead of the switch is connected to the red lead of the relay (on), and the black lead is connected to the black lead of the relay (off), Fig. 22–3.

The master selector switch, Fig. 22–2, is another type of remote-control switch. This switch can operate from one to 12 circuits. The switch knob is pressed and rotated through the 12 numbers to turn all of the circuits on or off. To turn specific circuits on or off, the knob is turned to the proper number on the dial and then depressed.

Some extra-low-voltage switches have pilot lights or locator lights to show when the circuit is on. The master selector switch is available with a pilot-lighted directory, which allows the user to see immediately which circuits are on or off.

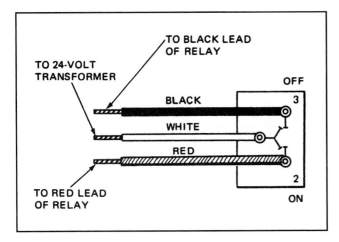

Fig. 22-3   Wiring diagram of an extra-low-voltage switch.

Fig. 22-4   Motorized master control.

The master selector switch is a group of individual switches mounted on a single plate, making it easy to select one or all of the individual circuits. When all of the circuits are to be controlled, the master selector switch can sweep all of them. This method is faster and easier than operating each circuit alone. Thus, the master selector switch is convenient for controlling many circuits from a single point. In this residence, a master selector switch can be located in one of the bedrooms. This permits the owner to turn on all of the outdoor lighting to provide some protection against prowlers.

A *motorized master control*, Fig. 22-4, can provide control for a maximum of 25 circuits. For each touch of the on/off position of a standard extra-low-voltage switch, the motorized master control makes a complete sweep of the 25 positions. Thus, the switch controls the 25 relays connected to these positions.

The extra-low-voltage relay is a double solenoid. When the coil connected to the red and blue leads is energized, the iron core tries to centre itself within the coil. The core moves to the left until it closes the switch contacts. When the power supply to the coil is removed, the switch contacts remain closed. They are held in the on position by a mechanical latching device in the relay. When the coil connected to the black and blue leads is energized, the iron core tries to centre itself in this coil. The core now moves to the right to open the switch contacts. The relay is mechanically latched in the off position.

EXTRA-LOW-VOLTAGE RELAY

Fig. 22-5 shows the wiring of a relay used in extra-low-voltage, remote-control systems. This relay operates contacts in the 120-volt lighting circuit. The relay has a split coil design: One coil is used to close the 120-volt circuit; the other opens the contactors in the 120-volt circuit. This relay operates on 24 volts. There are two No. 12 AWG black leads for the 120-volt circuits and three No. 20 AWG coloured leads for the 24-volt connections to the relay. The relay may have screw terminals for the 120-volt connections in place of the wire tails.

Relay Mounting

The extra-low-voltage relay is small and can be mounted in a 16-mm knockout opening from the inside of a standard outlet box. For quiet operation, the relay may be mounted in a 21-mm knockout opening with a rubber grommet. When mounted from the inside of the box, the two low-voltage (120 V) leads of the relay remain inside the outlet box. The extra-low-voltage wires are kept outside the box. The wall of the box serves as a partition between the low and extra-low voltages. The two low-voltage leads inside the box are connected like

Fig. 22–5   Wiring for an extra-low-voltage relay.

a standard single-pole switch. The hot or black wire from the 120-volt source goes to one of these leads. The other lead wire from the relay is connected to the lamp terminal. The blue lead of the three extra-low-voltage wires is common to both the off coil and the on coil. The red lead connects internally to the on coil, and the black lead connects internally to the off coil.

Plug-in Relay

Plug-in relays (Fig. 22–6) may be used when a number of relays are to be mounted in one enclosure designed for this purpose. It is not necessary to splice the line voltage leads for this type of relay.

The relays in Fig. 22–6 are rated for one horsepower, 20 amperes at 125 volts AC and for 20 amperes at 277 volts AC. The 20-ampere AC rating for both 125 volts and 277 volts means that the relay can be used to control AC loads, including tungsten filament loads and fluorescent loads, up to the full rating of the relays.

TRANSFORMERS

An extra-low-voltage, remote-control system requires just one transformer, Fig. 22–7, which steps down the lighting circuit voltage from 120 volts to 24 volts for the extra-low-voltage control circuit. The transformer rating is no larger than 100 volt-amperes, and must be marked as Class 2. When the transformer is overloaded, its output voltage decreases, and there is less current output. In other words, the transformer has energy-limiting characteristics that counteract any overload. A specially designed core limits the amount of electrical energy that can be delivered at the secondary or output terminals. Some transformers have a thermal breaker which opens the primary circuit to protect the transformer from overheating. As soon as it cools down, the thermal breaker resets itself and the transformer is automatically reconnected to the line.

Rarely is more than one switch pressed at the same time in an extra-low-voltage, remote-control system. Although 25 or more relays may be con-

Fig. 22–6   Relays.

Fig. 22–7   Transformer for remote-control systems.

nected to the transformer, the effective load is only one. The transformer might seem to be underrated for the connected load, but it can actually supply many relays. The transformer manufacturer usually recommends the maximum number of relays that may be operated at one time.

Some remote-control system manufacturers suggest that a rectifier be added to the secondary circuit. This device changes the AC supply to DC or pulsating DC, depending on whether a half-wave or full-wave rectifier is used. If a relay is energized for a long time, the AC supply will produce eddy currents inside the laminations, resulting in a temperature buildup that may damage the relay. (Recall that this type of relay is meant to provide momentary contact operation only.) Direct current or pulsating DC does not produce as much heat in this type of relay as does AC.

CONDUCTORS

No. 18 AWG conductors are generally used for extra-low-voltage, remote-control systems, Fig. 22–8. Larger conductors may be required for long runs. A two- or three-conductor cable is used for the installation. Insulation on the individual conductors is generally a thermoplastic material, colour-coded for ease in identification.

The installation of a remote-control system is simplified by using colour-coded extra-low-voltage cables. Cables are available with such colour combinations as blue–white, red–black, or black–white–red. To install these cables correctly in a dwelling, connect the wires like colour to like colour. A cable suitable for either overhead or underground installations is available for use in an extra-low-voltage, remote-control system installed outdoors.

INSTALLATION PROCEDURE

The installation of a remote-control system in new construction begins with roughing in the 120-volt system. All of the switch legs or switch loops are omitted for the system: The 120-volt supply conductors are brought directly to each outlet or switch box that will have a switched lighting fixture or switched outlet attached to it. Although the various devices used may be numbered or colour-coded in different ways, the electrician must remember that only three extra-low-voltage conductors are required between the relay and the switch. In addition, two extra-low-voltage conductors must be run from the transformer to either the relay or the switch to carry the extra-low-voltage supply. These wires are spliced like colour to like colour. Another method is to assign certain colours to correspond to the numbers on the relays and switches. The wiring is then completed by slipping the relay through the 6-mm knockout in the outlet or switch box. The 120-volt connections are made in the standard way.

A relay with a built-in transformer can be used. No. 18/3 cable is run between the relay and the

THREE-CONDUCTOR CABLE   TWO-CONDUCTOR CABLE

Fig. 22–8   Cable for remote-control systems.

switch. The connections are made like colour to like colour: The black and white line voltage relay leads are connected to the supply circuit, and the red line voltage relay lead is connected to the lamp terminal. (This red lead is actually the switch leg.)

The extra-low-voltage cables are stapled in place in new construction. When existing buildings are rewired, the extra-low-voltage cables may be fished through the walls and ceiling or run behind mouldings and baseboards.

Many or all of the relays are sometimes placed at one location, for example, when a motorized master control is used. Special relay ganging boxes can be used. In this type of box, all of the relays are inserted through a metal barrier. The line voltage connections are made on one side of this barrier, and the extra-low-voltage connections are made on the other side.

TYPICAL WIRING INSTALLATION

Fig. 22–9 shows one method of installing remote-control, extra-low-voltage switches. Multipoint control of the light results when the extra-low-voltage, remote-control switches are connected in parallel.

Fig. 22–10 shows one method of connecting a master selector control with individual extra-low-voltage switches. The 120-volt lighting outlets can be placed in different parts of the building. These outlets are then controlled by individual switches or the master selector control. Three lighting circuits are shown, but the master selector switch can control up to 12 different relays.

If a master selector switch is installed in the residence, it can control the shrub lights, front entrance light, post light (front), garage lights (front), garage light and post light (rear), garage lights (inside), cornice lights, terrace entrance light, and rear porch light. Thus, the entire outside of the residence can be flooded with light instantly, if necessary.

CEC RULES

The extra-low-voltage, remote-control system is not subject to the same *CEC* restrictions as the standard 120/240-volt system. The extra-low-voltage (24-volt) portion of the remote-control system is a Class 2 system regulated by *Rules 16–200* through *16–226*.

Fig. 22–9   Typical cable layout and wiring diagram for low-voltage switches.

Fig. 22–10 Cable layout and wiring diagrams for a remote-control system.

For a remote-control system, all wiring on the supply side of the transformer (120 volts) must conform to the wiring methods given in *Section 12*. All wiring on the load side of the transformer (24 volts) must conform to *Rule 16–202* through *16–226*.

The transformer for a remote-control, extra-low-voltage system is designed so that the power output is limited to 100 volt-amperes, as required by *Rule 16–200*. The maximum nameplate rating of the transformer is 100 volt-amperes and must be marked as Class 2.

Conductors with thermoplastic insulation are normally used for an extra-low-voltage system. However, the *CEC* does require that the wire be listed in *Table 19*. Conductors for Class 2 systems may not be run with regular light and power wiring, *Rule 16–212*.

Since the insulation on remote-control conductors is rather thin, the electrician must be careful to not damage it. Any damage can result in a short circuit or a relay malfunction. The electrician usually bores separate holes through the studs and joists of the residence for the small remote-control conductors. They are not pulled through the holes used for the cables for the regular wiring because of the possibility of mechanical damage to the insulation. The electrician must use care when stapling these extra-low-voltage conductors. If the insulation is pinched, a short may occur between the conductors.

The extra-low-voltage system conductors and installation methods are similar to those used to install the chimes in this residence.

SIGNAL SYSTEM (CHIMES)

The *CEC* describes a signal circuit as any electrical circuit, other than a communications circuit, that supplies energy to a device that gives a recognizable signal—a chime, doorbell, buzzer, code-calling system, or signal light.

Door Chimes (Symbol CH)

Modern dwellings more often use chimes than doorbells or buzzers, sounding a musical note rather than a harsh ringing or buzzing. Chimes are available in single-note, two-note, eight-note (four-tube),

and repeater tone styles. In a repeater tone chime, both notes sound as long as the push button is depressed. In an eight-note chime, contacts on a motor-driven cam are arranged in sequence to sound the notes of a simple melody when the chime button is pushed. This type of chime is usually installed in dwellings having three entrances. The chime can be connected so that the eight-note melody sounds for the front door, two notes sound for the side door, and a single note sounds for the rear door. Chimes are also available with clocks and lights.

Electronic chimes may relay their chime tones through the various speakers of an intercom system. The manufacturer's instructions must be followed in any chime installation.

The plans show that two-note chimes are installed in the residence. Each chime has two solenoids and two iron plungers. When one solenoid is energized, the iron plunger is drawn into the opening of the solenoid. A plastic peg in the end of the plunger strikes one chime tone bar. When the solenoid is de-energized, spring action returns the plunger where it comes to rest against a soft felt pad so that it does not strike the other chime tone bar. Thus, a single chime tone sounds. As the second solenoid is energized, one chime tone bar is struck. When the second solenoid is de-energized, the plunger returns and strikes the second tone bar. A two-tone signal is produced. The plunger then comes to rest between the two tone bars. Generally, two notes indicate front door signalling, and one note indicates rear or side door signalling.

Fig. 22–11 shows some of the many push-button styles used for chimes. Fig. 22–12 shows several wall-mounted cover plates for chime units. The symbols used to indicate push buttons and audible signals on the plans are shown in Fig. 22–13.

Figs. 22–14 and 22–15 show one way in which to provide proper backing for chimes, during the rough-in stages of the electrical installation before the walls are closed up.

Transformers

Because of the extra-low-voltage ratings and power limitations of chime transformers, the secondary wiring and the transformer itself are classified as Class 2 circuits.

Fig. 22–11   Push buttons for door chimes. Courtesy of Broan-NuTone.

Fig. 22–12   Residential door chimes. Courtesy of Broan-NuTone.

The transformers used for door chimes have a greater capacity than those used with bell and buzzer circuits, Fig. 22–16. The voltage output of chime transformers ranges between 10 and 24 volts. These transformers are rated from 5 to 30 volt-amperes. Bell transformers have a voltage output range of 6 to 10 volts and a rating of 5 to 20 volt-amperes.

Fig. 22–13   Symbols for chimes and push buttons.

Chime transformers used in dwellings are available with a 16-volt rating. Transformers that give a combination of voltages, such as 4, 8, 12, and 24 volts, also are used.

CSA lists the types of chime transformers normally used for chime wiring. The Class 2 *transformer* is rated at 100 volt-amperes or less. This transformer may have an overcurrent protective device, *Rule 16–206(2)*, which limits the voltage and current to the values specified in *Rule 16–200(1)(b)*. The open circuit voltage of both Class 2 transformers must not exceed 30 volts. Most Class 2 transformers are constructed so that they cannot be overloaded. Should a fault occur on the secondary

BE SURE TO PROVIDE ADEQUATE
BACKING TO SUPPORT THE CHIME

RECOMMENDED HEIGHT EYE LEVEL TO
CENTRE OF CHIME, BUT NO HIGHER
THAN 2 m TO CENTRE OF CHIME.

CHIMES WITH TUBES MIGHT HAVE A
TOTAL LENGTH OF 125 cm OR
MORE, IN WHICH CASE EYE LEVEL
MIGHT BE TOO LOW. MANUFACTURERS
OF CHIMES OFFER MOUNTING HEIGHT
SUGGESTIONS IN THEIR CATALOGUES.

Fig. 22–14 One method of providing support for a surface-mounted chime, in the rough-in stage, before the walls are closed.

BE SURE TO PROVIDE
ADEQUATE BACKING FOR
SUPPORTING THE CHIME.

RECOMMENDED HEIGHT EYE LEVEL TO
CENTRE OF CHIME, BUT NO HIGHER THAN
2 m TO CENTRE OF CHIME. MANUFACTURERS
OF CHIMES PROVIDE MOUNTING HEIGHT
SUGGESTIONS IN THEIR CATALOGUES.

ROUGH-IN BOX FOR
FLUSH-MOUNTED CHIME.

Fig. 22–15 Roughing-in for a flush-mounted chime.

Fig. 22–16 Chime transformer.

Fig. 22–17 Circuit for chime installation.

of the transformer, the increased current causes an increase in voltage drop in the transformer. This reduces the transformer terminal voltage and thereby reduces current to a level that does not exceed the transformer rating.

Additional Chimes

To extend a chime system to cover a larger area, a second or third chime may be added. In the residence used in this text, two chimes are used: one is mounted in the front hall, and an extension is mounted in the recreation room. The extension chime is wired in parallel to the first chime. The wires are run from one chime terminal board to the terminal board of the other. The terminals are connected transformer to transformer, front to front, and rear to rear, Fig. 22–17.

When chimes are added to a chime circuit, a transformer with a higher volt-ampere may be required to energize the greater number of solenoids used at one time. Also, more wire is used in the circuit, and the voltage drop and power loss are greater. This increase is the result of resistance in the wire between the chimes and the transformer. *Rule 16–200(3)* does not allow transformers supplying Class 2 systems to be connected in parallel.

If a buzzer (or bell) and a chime are connected to a single transformer and are used at the same time, the transformer will put out a fluctuating voltage, which allows neither the buzzer nor the chime to operate properly. The use of a transformer with a larger rating may solve this problem.

The power consumption of chimes varies with the manufacturer. Typically standard two-note, repeating, and internally lighted two-lamp chimes all consume 10 volt-amperes. Internally lighted, four-lamp, combination chime and clock, motor-driven, and electronic chimes all typically consume 15 volt-amperes.

Transformers with ratings of 5, 10, 15, 20, and 30 volt-amperes are available. For a multiple chime installation, volt-ampere ratings for the individual chimes are added: The total value is the minimum transformer rating needed to do the job properly. If you still have technical questions or problems, check the literature supplied by the chime and transformer manufacturers.

Wiring for Chime Installation

Bell Wire and Cable: The wire used for extra-low-voltage bell and chime circuits and for connecting extra-low-voltage thermostats is called *bell wire, annunciator wire,* or *thermostat wire.*

These wires are generally made of copper and have a thermoplastic insulation suitable for use at 30 volts or less.

Since the current required for bell and chime circuits is rather small, No. 18 AWG conductors are often used. If the circuit is very long, use a larger conductor.

Multiconductor cables consist of two, three, or more single wires and may be covered with a single protective insulation. This type of cable is often used for electrical installations because there is less danger of damage to individual wires, and it gives a neat appearance to the wiring. The conductors within the cables are colour coded to make circuit identification easy.

Bell wire and cable may be fastened directly to surfaces with insulated staples or cleats, or may be installed in the raceways. Be careful to not pierce or crush the wires. The job requirements will determine how the conductors are to be attached. In the residence in the plans, the bell wire is run along the sides of the floor joists in the basement and on the sides of the studs in the walls.

Wiring Circuit: The circuit shown in Fig. 22–17 is recommended for this chime installation because it provides a "hot" extra-low-voltage circuit at the front hall location. However, it is not the only way in which these chimes may be connected.

Fig. 22–17 shows that a two-wire cable runs from the transformer in the workshop to the front hall chime. A two-wire cable then runs from the chime to both the front and rear door push buttons.

A three-wire cable also runs between the front hall chime and the recreation room chime. Because of the "hot" extra-low-voltage circuit at the front hall location, a chime with a built-in clock can be used. A four-conductor cable may be run to the extension chime so that the owner may install a clock-chime at this location also.

The plans show that the chime transformer is mounted on the main panel on the branch-circuit part of the panel. The electrician decides where to mount the transformer after considering the factors of convenience, economy, and good wiring practice.

Rules for Signal Systems (*Rules 16–210 and 16–212*)

Bell wire with extra-low-voltage insulation must not be installed in the same enclosure or raceway with light or power conductors. Bell wires must be not closer than 50 mm to open light or power conductors, *Rule 16–212(1)*, unless the bell wires are permanently separated from the other conductors by some approved insulation used in addition to the insulation on the wire. Such insulation is provided by the outer jacket on non-metallic-sheathed cable and armoured cable. Furthermore, bell wire with extra-low-voltage insulation may not enter an outlet box or switch box containing light or power conductors unless a metal barrier is used to separate the two types of wiring.

CEC requirements for Class 2, low-voltage bell circuits are contained in *Rules 16–210* and *16–212*.

Remember, extra-low-voltage Class 2 circuits are not permitted to be run in the same raceway as power-circuit conductors.

REVIEW

Note: Refer to the *CEC* or the plans where necessary.

REMOTE-CONTROL SYSTEMS

1. How does the *CEC* define this voltage? _____

   _____

   _____

   _____

2. What are the advantages of an extra-low-voltage system? _____
_____
_____
_____
_____
_____
_____

3. What type of switch is used to control the relays of an extra-low-voltage system?
_____

4. a. Do the relays connected to a motorized master control operate at the same time?
   Explain your answer. _____
   _____
   _____

   b. Do the relays connected to a master selector switch operate at the same time? Explain
   your answer. _____
   _____
   _____

5. Briefly explain the operating principle of an extra-low-voltage relay. _____
_____
_____
_____
_____
_____

6. Is it permitted to run extra-low-voltage conductors and line-voltage (120-volt) conductors into the same box? Explain your answer. _____
_____
_____
_____
_____
_____

7. The maximum nameplate rating of a Class 2 transformer is _____ volt-amperes.

8. Since the insulation on extra-low-voltage wire is much thinner than the insulation on regular building wire, the electrician should

   a. install the wires at the same time as the rough-in

   b. carefully install the conductors to avoid damaging the insulation

   c. install the wiring after the power and lighting are roughed in

   d. both b and c

9. a. What is the purpose of a rectifier? _____

_____

_____

   b. Must rectifiers be used on extra-low-voltage systems? _____

_____

_____

10. What type of conductors are generally used for an extra-low-voltage, remote-control system?

_____

_____

_____

11. The type of extra-low-voltage wiring discussed in this unit is classified as (Class 1) (Class 2). (Circle one.)

12. When overcurrent protection is not provided in the secondary circuits of the transformers described in this unit, the transformers must limit their power output to

    _____ VA for Class 2 circuits.   *Rule* _____.

13. Conductors supplying transformers are governed by what *CEC* rule?

_____

14. Complete all line voltage and extra-low-voltage connections in the following diagram. Use coloured pencils to indicate conductors. Show extra-low-voltage connections with a dot. Switch No. 1 controls lamps A and C, switch No. 2 controls lamp B, and switch No. 3 controls the bottom of each receptacle. The top of each receptacle is to be hot at all times. The line-voltage wiring is in armoured cable.

SIGNAL SYSTEMS

1. What is a signal circuit? _____

   _____

2. What style of chime is used in the residence used in this text? _____

3. a. How many solenoids are contained in a two-tone chime? _____

   b. What closes the circuit to the solenoid of a chime? _____

4. Explain briefly how two notes are sounded by depressing one push button (when two solenoids are provided). _____

   _____

   _____

   _____

   _____

5. a. Sketch the symbol for a push button. _____

   b. Sketch the symbol for a chime. _____

6. a. At what voltage do residence chimes generally operate? _____

   b. How is this voltage obtained? _____

7. What is the maximum volt-ampere rating of transformers supplying Class 2 systems?

   _____

8. What type of transformer is used for Class 2 circuits on a door chime? _____

   _____

9. Is the extension chime connected with the front hall chime in series or in parallel?

   _____

10. How many bell wires terminate at

    a. the transformer? _____

    b. the front hall chime? _____

    c. the extension chime? _____

    d. each push button? _____

11. a. What change in equipment may be necessary when more than one chime is con-
    nected to sound at the same time on one circuit? _____

    _____

    b. Why? _____

    _____

    _____

12. What type of insulation is usually found on extra-low-voltage wires? _____

_____

13. What size wire is installed for signal systems of the type in this residence? _____

_____

14. a. How many wires are run between the front hall chime and the extension chime in the recreation room? _____

    b. How many wires are required to provide a "hot" extra-low-voltage circuit at the extension chime? _____

15. Why is it recommended that the extra-low-voltage secondary of the transformer be run to the front hall chime location and separate two-wire cables be installed to each push button? _____

_____

16. a. Is it permissible to install extra-low-voltage Class 2 systems in the same raceway or enclosure with lighting and power wiring? _____

    b. What *CEC* section covers this? _____

17. a. Where is the transformer in the residence mounted? _____

    b. To what circuit is the transformer connected? _____

18. a. How many metres of two-conductor bell wire cable are required? _____

    _____

    b. How many metres of three-conductor bell wire cable are required? _____

    _____

19. How many insulated staples are needed for the bell wire if it is stapled every 600 mm?

_____

UNIT 23

Swimming Pools, Spas, Hot Tubs

OBJECTIVES

After studying this unit, the student will be able to

- recognize the importance of proper swimming pool wiring for human safety
- discuss the hazards of electrical shock associated with faulty wiring in, on, or near pools
- describe the differences between permanently installed pools and pools that are portable or storable
- understand and apply the basic *CEC* requirements for the wiring of swimming pools, spas, hot tubs, and hydromassage bathtubs

POOL WIRING (*SECTION 68*)

For easy reference, a detailed drawing of the *CEC* requirements for swimming pool wiring (Plan 6) is included with this text. Refer to this drawing as you study this unit.

Swimming pools, wading pools, hydromassage bathtubs, therapeutic pools, decorative pools, hot tubs, and spas must be wired according to *Section 68*. To protect people using such pools, rules specific to pool wiring have been developed over the years. Extreme care is required when wiring all equipment associated with pools.

ELECTRICAL HAZARDS

A person can suffer an immobilizing or lethal shock in a residential-type pool in two ways.

1. An electrical shock can be transmitted to someone in a pool who touches a live wire or the live casing or enclosure of an appliance, such as a hair dryer, radio, or extension cord, Fig. 23–1.

2. If an appliance falls into the pool, an electrical shock can be transmitted to a person in the water, by means of voltage gradients in the water. Refer to Fig. 23–2 for an illustration of this life-threatening hazard.

As the figure shows, "rings" of voltage radiate outwardly from the radio to the pool walls. These rings can be likened to the rings that form when a rock is thrown into the water. The voltage gradients range from 120 volts at the radio to zero volts at the pool walls. The pool walls are assumed to be at ground or zero potential. The gradients, in varying degrees, are found in the entire body of water.

Fig. 23–2 shows voltage gradients in the pool of 90 volts and 60 volts. (This figure is a simplification of the actual situation, in which there are many voltage gradients.) In this case, the voltage differential, 30 volts, is an extremely hazardous value. **The person in the pool, who is surrounded by these voltage gradients, is subject to severe shock, immobilization (which can result in drowning),**

Fig. 23–1   Touching a "live" faulty appliance can cause a lethal shock.

Fig. 23–2   Voltage gradients surrounding a person in the pool can cause severe shock, drowning, or electrocution.

or actual electrocution. Tests conducted over the years have shown that a voltage gradient of 1½ volts per 300 mm can cause paralysis.

Study Plan 6, supplied with this text. Note that, in general, underwater swimming-pool lighting fixtures must be positioned not more than 600 mm below the normal water level, so that when a person is in the water close to an underwater lighting fixture, the relative position of the fixture is well below the person's heart. Fixtures that are permitted to be installed no less than 10 mm below the normal water level have undergone tougher, abnormal impact tests, such as those tests given to cleaning tools or other mechanical objects. The lenses on these fixtures can withstand these abnormal impact tests,

whereas the standard underwater fixtures that are required to be 600 mm below the water level are subjected to normal impact tests that duplicate the impact of a person kicking the lens.

The shock hazard to a person in a pool is quite different from that of the normal "touch" shock hazard to a person not submersed in water. The water makes contact with the entire skin surface of the body rather than just at one "touch" point. Also, skin wounds, such as cuts and scratches, reduce the body's resistance to shock to a much lower value than that of the skin alone. Body openings such as ears, nose, and mouth further reduce the body resistance. As Ohm's Law states, for a given voltage, the lower the resistance, the higher the current.

Another hazard associated with spas and hot tubs is prolonged immersion in hot water. If the water is too hot and/or the immersion too long, lethargy (drowsiness, hyperthermia) can set in. This can increase the risk of drowning. The CSA standard establishes maximum water temperature at 40°C. The suggested maximum time of immersion is generally 15 minutes.

Instructions furnished with spas and hot tubs specify the maximum temperature and time permitted for using the spa or hot tub. Hot water, in combination with drugs and alcohol, presents a hazard to life. Caution must be observed at all times.

The figures that follow show the *CEC* rules that relate to safety procedures for wiring pools.

Wiring Methods

Throughout *Section 68,* you will find that the wiring must be installed in rigid metal conduit, flexible metal conduit, rigid non-metallic conduit, flexible cord for wet locations, or, in some cases, EMT. For wet-niche fixtures, brass, copper, or other corrosion-resistant metal must be used, *Rule 68–066(5).*

Rule 68–100(3) allows any type of wiring method recognized in *Section 12* for that portion of the interior wiring of one-family dwelling installations that supplies pool-associated pump motors, provided the wiring method contains an equipment bonding conductor no smaller than that required by *Table 16.*

Be sure to read the requirements of *Rule 68–100* closely to determine the proper wiring method to be used for a particular situation.

CODE-DEFINED POOLS

The *CEC* describes a *permanently installed swimming pool* as one that cannot be disassembled readily for storage, *Rule 68–050.*

The *CEC* describes a *storable swimming pool* as one that can be disassembled for storage and reassembled to its original form, *Rule 68–050.*

Pools that have non-metallic inflatable walls are considered to be storable pools regardless of their dimensions.

GROUNDING AND BONDING OF SWIMMING POOLS

Bonding to Ground (*Rule 68–058*)

Rule 68–058 requires that all of the following items *must* be bonded to ground, as illustrated in Fig. 23–3.

- wet- and dry-niche lighting fixtures
- any electrical equipment within 1.5 m of the inside wall of the pool
- pool-reinforcing steel
- any metal parts of the pool located within 1.5 m of the inside wall of the pool
- all electrical equipment associated with the recirculating system of the pool
- junction boxes
- transformer enclosures
- ground-fault circuit interrupters
- panelboards that are not part of the service equipment and supply the electrical equipment associated with the pool

Proper grounding and bonding ensures that all metal parts in and around the pool area are at the same ground potential, thus reducing the shock hazard. Proper grounding and bonding practices also facilitate the opening of the overcurrent protective device (fuse or circuit breaker) in the event of a fault in the circuit. Grounding is covered in other units of this text.

Bonding Conductors: Bonding conductors *must*

- when smaller than No. 6 AWG, be run in the same conduit with the circuit conductors, or be part of an approved flexible cord assembly, as used for the connection of wet-niche underwater lighting fixtures;
- terminate on equipment bonding terminals provided in the junction box, transformer, ground-fault circuit interrupter, subpanel, or other specific equipment.

Metal conduit by itself is *not* considered to be an adequate grounding means for bonding in and around pools.

Fig. 23–3   Grounding and bonding of important metal parts of a swimming pool, as covered in *Rule 68–058*. Refer to Plan 6 of the drawings supplied with this text.

Bonding (*Rule 68–058*)

Rule 68–058 requires that all metal parts of a pool installation be bonded together. For in-ground pools, a No. 6 AWG copper conductor is required. For above-ground pools a copper conductor sized in accordance with *Table 16* is acceptable, Fig. 23–4. The No. 6 or larger copper bonding wire must be run to the main distribution panel board and "tie everything together." This helps keep all metal parts in and around the pool at the same voltage potential, reducing the shock hazard brought about by stray voltages and voltage gradients.

Bonding Conductors: The bonding conductor must be connected to

- the steel reinforcing bars in the concrete using a minimum of four equally spaced connections

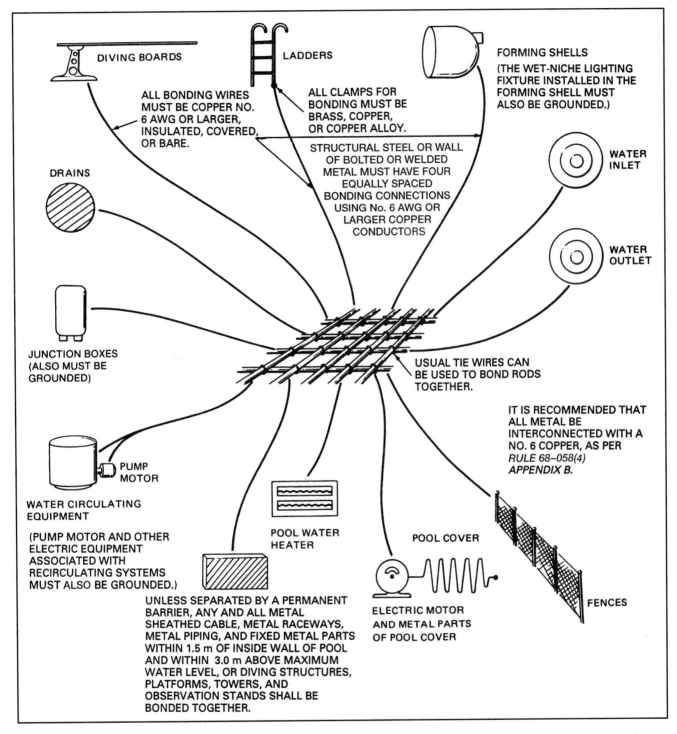

Fig. 23–4 **Bonding the metal parts of swimming pool installations, *Rule 68–058*. In addition to bonding, some of the above items must also be grounded. Refer to *Rules 68–058* and *68–060* for specifics. See Plan 6 of the drawings supplied with this text.**

- when epoxy-coated reinforcing steel is used in the construction of a swimming pool, placing a loop of #6 AWG copper around the pool below the normal water level could be an alternative to bonding the reinforcing steel at four locations, *Rule 68–058(3)*

- the wall of a bolted or welded metal pool
- any deck boxes
- ladders
- other metal parts of the pool or non-electric equipment

Bonding conductors

- need *not* be installed in conduit
- may be connected directly to the equipment that requires bonding, by means of brass, copper, or copper-alloy clamps.

LIGHTING FIXTURES UNDER WATER

Rule 68–066 covers underwater lighting fixtures for permanently installed pools. There are three types of underwater lighting fixtures:

1. *Dry-niche* lighting fixtures are mounted in the side walls of the pool and are designed to be relamped from the rear. These fixtures are waterproof.

2. *Wet-niche* lighting fixtures are mounted in the side wall of the pool and are designed to be relamped from the front. The supply cord is stored inside the fixture. The cord should be long enough to reach the top of the pool deck for lamp replacement. The forming shell to which the fixture attaches is intended to fill up with pool water.

3. *No-niche* lighting fixtures are mounted on a bracket on the inside wall of the pool. The supply conduit terminates on this bracket. The supply cord runs through this conduit to a deck box for connection to the circuit wiring. The extra supply cord is stored in the space behind the no-niche fixture. The cord should be long enough to reach the top of the pool deck for lamp replacement. This type of fixture is mainly used in retrofit installations and for above-ground pools.

Extra-low-voltage lighting has recently become more popular for pool installations. This type of lighting receives its power from an extra-low-voltage transformer fed from a GFCI-protected circuit. The transformer and the GFCI are generally located over three metres from the pool.

ELECTRIC HEATING OF SWIMMING POOL DECKS

Recommendations for deck-area electric heating units are illustrated in Fig. 23–5.

UNIT HEATER MUST BE:
- RIGIDLY MOUNTED ON STRUCTURE
- AT LEAST 1.5 m BACK FROM EDGE OF POOL
- OF THE TOTALLY ENCLOSED OR GUARDED TYPES
- APPROVED FOR WET LOCATIONS, *RULE 62–104*
- PROTECTED BY A GFCI IF WITHIN 3 m OF POOL

PERMANENTLY WIRED RADIANT HEATER MUST BE:
- AT LEAST 3 m ABOVE DECK
- SUITABLY GUARDED
- SECURELY FASTENED
- AT LEAST 1.5 m BACK FROM EDGE OF POOL
- PERMANENTLY WIRED
- APPROVED FOR WET LOCATIONS, *RULE 62–104*
- GFCI PROTECTED, *RULE 62–300(4).*

MIN 1.5 m

MIN 1.5 m

MIN 3.0 m

UNIT HEATERS OR RADIANT HEATERS *NOT* PERMITTED OVER THE POOL

MAXIMUM WATER LEVEL

POOL

RADIANT HEATING CABLES EMBEDDED IN OR BELOW THE DECK MUST BE APPROVED FOR THIS APPLICATION AND G.F.C.I PROTECTED, *RULE 62–300(4)*

Fig. 23–5   Recommendations for deck-area electric heating 3 m to 6 m from the inside edge of swimming pools.

Circuit sizing for electric heaters: The branch circuit overcurrent protective devices shall be rated not less than 125 percent of the heater's nameplate rating according to *Rule 62–114(8)*. The branch circuit conductors must have an ampacity not less than the connected load supplied, and not less than 80 percent of the rating of the overcurrent devices protecting the circuit, *Rules 62–108(1)(b)* and *62–114(6)(b)*.

All heating cable sets used for surface heating sidewalks, driveways, and pool decks must be fed from a GFCI-protected circuit, *Rule 62–300(4)*.

SPAS AND HOT TUBS (*RULES 68–400 TO 68–408*)

The basic difference between a spa and a hot tub is that a hot tub is constructed of wood, such as redwood, teak, cypress, or oak, whereas a spa is made of plastic, fibreglass, concrete, tile, or other manmade products.

Spas and hot tubs contain electrical equipment for heating and recirculating water. They have no provisions for direct connection to the building plumbing system.

Spas and hot tubs are intended to be filled and used. They are not drained after each use, as in the case of a regular bathtub.

Some spas and hot tubs are furnished with a cord-and-plug set when shipped by the manufacturer. If these can be converted from cord-and-plug connection to "hard wiring," they are listed and identified as "convertible" spas and hot tubs. The manufacturer must furnish clear instructions on how to convert from cord-and-plug set to hard wiring.

Outdoor Installation

Spas and hot tubs installed outdoors *must* conform to the installation requirements for regular swimming pools, *Rule 68–404*.

The metal parts of spas and hot tubs must be bonded together and to ground in accordance with *Rule 68–058*.

Rule 68–402(2) states that the metal bands and hoops that secure the wood staves of a spa or hot tub do not have to be bonded.

Indoor Installation

Spas and hot tubs installed indoors *must* conform to the following requirements:

- must be connected using the wiring methods covered in *Section 12*.
- may be cord-and-plug connected, provided the placement of the receptacle complies with the provisions of *Rules 68–064* and *68–404(2)*.

Receptacles (*Rule 68–064*)

- Any receptacles located between 1.5 and 3 metres of the inside walls of a pool, hot tub, or spa shall be protected by a GFCI of the Class-A type, *Rule 68–064(2)*.
- No receptacle shall be located less than 1.5 metre from the edge of the pool, *Rule 68–064(1)*.
- Any receptacles that supply power to a spa or hot tub must be GFCI protected no matter how far they are from the spa or hot tub.

Lighting Fixtures (*Rule 68–066*)

1. All lighting outlets and fixtures located within 3 metres of the inside wall or the water surface of the spa or hot tub must be GFCI protected. GFCI protection is not required for fixtures more than 3 metres above maximum water level, *Rule 68–066(6)*.

2. Recessed lighting fixtures may be installed less than 3 metres above the hot tub or spa if the fixtures are GFCI protected, and are suitable for use in damp locations.

3. Surface-mounted fixtures may be installed less than 3 metres above the hot tub or spa if the fixtures are GFCI protected, and are suitable for use in damp locations.

4. If any underwater lighting is to be installed, the rules for regular swimming pools apply.

Bonding to Ground, *Rule 68–402*

The requirements for grounding spas and hot tubs are:

- Ground all electrical equipment within 1.5 metres of the inside wall of the spa or hot tub.

- Ground all electrical equipment associated with the circulating system, including the pump motor.

- Grounding *must* conform to all of the applicable *CEC* rules.

- Grounding conductor connections *must* be made according to the applicable requirements of the *CEC*.

- If equipment is connected by a flexible cord, the bonding conductor must be part of the flexible cord, and must be fastened to a fixed metal part of the equipment.

Bonding (*Rule 68–402*)

The bonding requirements for spas and hot tubs are similar to those for regular swimming pools. Bond together:

- all metal fittings within or attached to the spa or hot tub;

- all metal parts of electrical equipment associated with the circulating system, including pump motors;

- all metal pipes, conduits, and metal surfaces within 1.5 m of the inside edge of the spa or hot tub (this bonding is not required if the materials are separated from the spa or hot tub by a permanent barrier, such as a wall or building);

- all electrical devices and controls *not* associated with the spa or hot tub located less than 1.5 m from the inside edge of the spa or hot tub. If located 1.5 m or more from the hot tub or spa, bonding is not required.

Bonding is to be accomplished by means of threaded metal piping and fittings, metal-to-metal mounting on a common base or frame, or by means of a solid, insulated, covered, or bare bonding conductor sized by *Table 16*.

Electric Water Heaters (*Rule 68–408(2)*)

Electric water heaters must be approved for the purpose.

They must have overcurrent protection rated not less than 125 percent of the total load, as indicated on the nameplate of the heater, due to the panel being marked for 80 percent continuous operation.

Wiring and Circuit Protection

Most spas and hot tubs require a 120/240-volt circuit with an overcurrent device ranging from 40 to 60 amperes. All spas and hot tubs must be fed from a GFCI-protected circuit. As with pools, this GFCI device must be located more than 3 metres from the pool water. Since the spa or hot tub has a motor, a motor-disconnecting means must also be supplied. This is typically accomplished by installing the GFCI circuit breaker in a weatherproof enclosure, located in a convenient location for testing purposes, and within sight of the hot tub or spa.

HYDROMASSAGE BATHTUBS (*SUBSECTION 68–300*)

Hydromassage bathtubs, also called whirlpool tubs or bathtubs, together with their associated electrical components, must be GFCI protected.

Because a hydromassage bathtub does not present any more of a shock hazard than a regular bathtub, the *CEC* permits all other wiring (fixtures, switches, receptacles, and other equipment) in the same room but not directly associated with the hydromassage bathtub to be installed according to all the normal *CEC* requirements covering installation of that equipment in bathrooms.

This residence has a hydromassage tub in the master bathroom. It is powered by a $\frac{1}{2}$-hp, 115-volt, 3450-rpm, single-phase motor rated at 10 amperes.

A hydromassage tub and all of its associated equipment (generally provided by the manufacturer) must be connected to a circuit that has GFCI protection. Some manufacturers provide a GFCI device built into the hydromassage tub control system. Otherwise, the GFCI device must be supplied and installed by the electrician. This can be a feed-through device installed in a device box in the bathroom, a GFCI receptacle located under the tub, or a GFCI circuit breaker. The device *must* be located where it is accessible for testing.

If the hydromassage tub is cordcap connected, the receptacle located under the tub must be located at least 300 mm above the floor, *68–306(1)*.

Fig. 13–16 illustrates one manufacturer's hydromassage bathtub.

By definition in *Rule 68–050*, a hydromassage bathtub is intended to be filled (used) and then drained after each use, whereas the spa or hot tub is filled with water and not drained after each use.

FOUNTAINS

The home discussed in this text does not have a fountain. When fountains share water with a regular swimming pool, the fountain wiring must conform to pool wiring requirements. If the house has a water pump for a waterfall or decorative pool in the garden, the pump must be protected by a GFCI, *Rule 26–952*.

A decorative pool is one in which any dimension is greater than 1.5 metres, *Rule 68–050*. Decorative pools must be wired in the same manner as swimming pools.

Leakage Current Collectors (*Rule 68–406*)

Leakage current collectors must be installed in the water inlets or water outlets of spas or hot tubs. They must be electrically insulated from the tub and bonded to the control panel of the unit or to the main service ground with a minimum No. 6 copper bonding conductor, or the bonding conductor may be sized from *Table 16*, provided it is mechanically protected in the same manner as the circuit conductors.

SUMMARY

The figures in this unit and the plan-sized diagram included with the text present a detailed overview of the *CEC* requirements for pool wiring.

REVIEW

Note: Refer to the *CEC* or the plans where necessary.

1. The *CEC* rule that covers most of the requirements for wiring swimming pools is *Rule* _____.

2. Name two ways in which a person can sustain an electrical shock when in a pool.

   _____

   _____

   _____

   _____

3. *Rule 68–050* discusses types of pools. Name and describe two types. _____

   _____

   _____

   _____

4. Use *Section 68* to determine if the following items must be bonded.

	YES	NO
a. Wet- and dry-niche lighting fixtures	____	____
b. Electrical equipment located beyond 1.5 m of inside edge of the pool	____	____
c. Electrical equipment located within 3 m of inside edge of the pool	____	____
d. Recirculating equipment and pumps	____	____
e. Lighting fixtures installed more than 7.5 m from inside edge of pool	____	____
f. Junction boxes, transformers, and GFCI enclosures	____	____
g. Panelboards that supply the electrical equipment for the pool	____	____
h. Panelboards 6 m from the pool that do not supply the electrical equipment for the pool	____	____

5. Bonding conductors (must) (may) be run in the same conduit as the circuit conductors. (Circle the correct answer.)

6. Bonding conductors (may) (may not) be spliced with the wire-nut types of wire connectors. (Circle the correct answer.)

7. The purpose of grounding and bonding is to _____
   _____
   _____
   _____

8. What parts of a pool must be bonded together? _____
   _____
   _____
   _____
   _____
   _____
   _____
   _____

9. May electrical wires be run above the pool? Explain. _____
   _____
   _____
   _____
   _____

10. What is the closest distance that a receptacle may be installed to the inside edge of a pool?

_____

_____

_____

_____

_____

_____

11. Receptacles located between 1.5 m and 3 m from the inside edge of a pool must be protected by a _____.

12. Lighting fixtures installed over a pool must be mounted at least (3 m) (4 m) (4.5 m) above the maximum water level. (Circle the correct answer.) (_Rule 68–054_)

13. Grounding conductor terminations in wet-niche metal-forming shells, as well as the conduits entering junction boxes or transformer enclosures where the conduit runs directly to the wet-niche lighting fixture, must be _____ with a (an) _____ to prevent corrosion to the terminal and to prevent the passage of water through the conduit, which could result in corrosion, _Rule 68–060(7)._

14. Wet-niche lighting fixtures are accessible (from the inside of the pool) (from a tunnel) (on top of a pole). (Circle the correct answer.)

15. Wet-niche lighting fixtures shall not be submersed to a depth of more than (_Rule 68–066_). (Circle the correct answer.)

    a. 600 mm   b. 300 mm   c. 450 mm   d. 1 m

16. Dry-niche lighting fixtures are accessible (from a tunnel, or passageway, or deck) (from the inside of the pool) (on top of the pole). Circle the correct answer.

17. In general, it is _not_ permitted to install conduits under the pool or within 1.5 m measured horizontally from the inside edge of the pool. True or false? _Rule 68–056, Table 61._ Explain. _____

_____

_____

_____

_____

_____

_____

18. Junction boxes, transformers, and GFCI enclosures have one thing in common. They all (are made of bronze) (have threaded hubs) (must have a bonding wire or terminal screw). Circle the correct answer.

19. Lighting fixtures are permitted to be mounted less than 3 m measured horizontally from the inside edge of the pool only if they are (approved for wet locations) (rigidly fastened to an existing structure) (GFCI protected). Circle the correct answer.

20. The following statements about indoor spas and hot tubs are either true or false. Check one.

	TRUE	FALSE
a. Receptacles may be installed within 1.5 m of the edge of the spa or hot tub.	_____	_____
b. All receptacles within 3 m of the spa or hot tub must be GFCI protected.	_____	_____
c. Any receptacles that supply power to pool equipment must be GFCI protected if within 3 m of the pool.	_____	_____
d. Wall switches must be located at least 1 m from the spa or hot tub.	_____	_____
e. Lighting fixtures above the pool or within 1.5 m from the inside edge of the pool must be GFCI protected.	_____	_____

21. Bonding and grounding of electrical equipment in and around spas and hot tubs (are required by the *CEC*) (are not required by the *CEC*) (are decided by the electrician). Circle the correct answer.

22. When installing a hot tub, a control shall be located behind a barrier not less than _____m from the tub.

23. For the circuit supplying a hydromassage or whirlpool bathtub, be sure that the circuit (does not have GFCI protection that could result in nuisance tripping) (is GFCI protected). Circle one.

24. Leakage current collectors shall be bonded to the control panel with a minimum No. 6 AWG copper bonding conductor. (TRUE) (FALSE) Circle the correct answer.

25. The maximum voltage that wet-niche fixtures are intended to be connected to is
   a. 120 volts.
   b. 15 volts.
   c. 240 volts.

26. Swimming pool pump motors shall be specifically approved, and, if within 3 m of the pool wall, shall be _____

   _____

   _____

   Quote rule numbers.

UNIT 24

Television, Telephone, Data, and Home Automation Systems

OBJECTIVES

After studying this unit, the student will be able to

- install television outlets, antennas, cables, and lead-in wires
- list CATV installation requirements
- describe the basic installation of satellite antennas
- install telephone conductors, outlet boxes, and outlets
- install and terminate computer network cable
- understand the basics of "wireless"
- understand the basic terminology of home automation
- identify types of devices that can be controlled by a home automation system

TELEVISION (SYMBOL $\boxed{\text{TV}}$)

Television signals can be received by means of antenna, cable, or satellite dish. Because some systems may not be available in the region of the country or the area in which you will be working, the text discusses all three installations.

General Wiring

Television outlets may be connected in several ways, depending on their proposed locations. In general, for installation in dwellings, the electrician uses standard sectional switch boxes or 102-mm square, 38-mm deep outlet boxes with single-gang raised plaster covers. A box is installed at each point on the system where an outlet is located. Use non-metallic boxes if the system is to be wired with unshielded cable, Fig. 24–1.

Television Installation Methods

Fig. 24–2 shows a master amplifier distribution system. The lead-in wire from the antenna is connected to an amplifier placed in an accessible area, such as the basement. Twin-lead, 300-ohm cable runs from the amplifier to each of the outlet boxes. The cable is connected to tap-off units, which have a 300-ohm input and a 300-ohm output. The tap-off units have terminals or plug-in arrangements so that the 300-ohm, twin-lead cable can be connected between the television outlet and the television.

Amplified systems are generally not necessary in typical residential installations, where cable television is available.

With a *multiset coupler*, Fig. 24–3, the lead-in wire from the antenna is connected to a pair of terminals on the coupler. The cables to each television are connected to other pairs of terminals.

Fig. 24–1   Non-metallic boxes and a non-metallic raised cover.

Fig. 24–2   Television master amplifier distribution system.

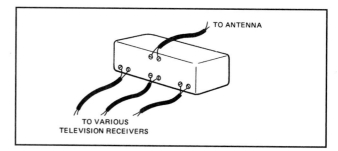

Fig. 24–3   Multiset coupler.

Two to four televisions can be connected to a coupler, depending on the number of terminal pairs provided. Because an amplifier is not required, a coupler system is less expensive than an amplifier distribution system.

A third method of installing a television system uses shielded 75-ohm coaxial cable, which is connected to impedance-matching transformers with 75- or 300-ohm outputs. Additional cable is then connected between the matching transformer and the television. Many older televisions have 300-ohm inputs; newer models generally have 75-ohm inputs. Some manufacturers recommend shielded coaxial cable to prevent interference and keep the colour signal strong. Both shielded and non-shielded lead-in wire deliver good television reception when properly installed.

Fig. 24–4 shows common locations for the television antenna on a residence.

Faceplates are provided for tap-off units to match conventional switch and convenience outlet faceplates. Combination two-gang faceplates also are available for tap-off units, Figs. 24–5 and 24–5A. Both a 120-volt convenience receptacle and a television outlet can be mounted in one outlet box using this combination faceplate.

However, a combination installation requires that a metal barrier be installed between the 120-volt section and the television section of the outlet box, *Rule 54–400(3)(a)*. This barrier prevents line voltage interference with the television signal and keeps the 120-volt wiring away from the television wiring. In addition, line voltage cables and lead-in wire must be separated to prevent interference when they are run in the same space within the walls and ceilings. The line voltage cables should be fastened to one side of the space and the lead-in cable to the other side.

Fig. 24–4   Typical mounting of television antennas.

Fig. 24–5 Mounting of power/communication box with coaxial outlet and duplex receptacle. Courtesy of Thomas & Betts Incorporated.

Fig. 24–5A Coaxial (F connector) and telephone combination wall jacks and cover plate. Courtesy of Leviton.

Code Rules for Cable Television (CATV) (*Section 54*)

Cable, or community antenna television (CATV), systems are installed both overhead and underground in a community. To supply an individual customer, the CATV company runs coaxial cables through the wall of a residence at a convenient point. Up to this point of entry, inside the building, and underground the cable company must conform to the requirements of *Section 54*, plus local codes if applicable. These are some of the key rules:

1. The outer conductive shield of the coaxial cables must be grounded as close to the point of entry as possible, *Rule 54–200(1)*.

2. Coaxial cables shall not be run in the same conduits or box as electric light and power conductors, *Rule 54–400(3)*.

3. The grounding conductor (*Rule 54–300*):

   a. must be insulated

   b. must not be smaller than No. 14 AWG

   c. may be solid or stranded but must be of copper

   d. must be run in a line as straight as practicable to the grounding electrode

   e. shall be guarded from mechanical damage

   f. shall be connected to the nearest accessible location on: the building or structure grounding electrode (*Rules 10–700* and *54–304*); the grounded interior water pipes (*Rule 10–700*); or the grounding electrode conductor

   g. If none of *f* is available, then ground to any of the electrodes per *Rule 10–702*, such as the metal frame of the building, a concrete-encased electrode, or a plate electrode.

   h. If none of *f* or *g* is available, then ground to any of the electrodes per *Rule 54–302*. The electrodes shall be driven to a depth of at least 2 metres. The connection may be by ground clamp or by a wire lead with a pressure connector. Refer to *CSA Standard C22.1–09*.

   i. *CAUTION:* If any of the preceding results in one grounding electrode for the CATV shield and another grounding electrode for the electrical system, bond the two electrodes together with a bonding jumper no smaller than No. 6 copper, *Rule 10–702*. Grounding and bonding the coaxial cable shield to the same grounding electrode as the main service ground prevents voltage differences between the shield and other grounded objects, such as a water pipe.

SATELLITE ANTENNAS

A satellite antenna system consists of a parabolic dish, a LNBF (low noise block amplifier with integral feed), mounting hardware, and a receiver. At one time, satellite systems were most popular in the rural areas where choices of stations on a conventional antenna were limited. The dish was two or three metres in diameter and was usually mounted on a large steel pole cemented into the ground (Fig. 24–6). Many of these systems had rotors that rotated and pointed the dish to different satellites. These systems were prone to lightning strikes and very expensive repairs.

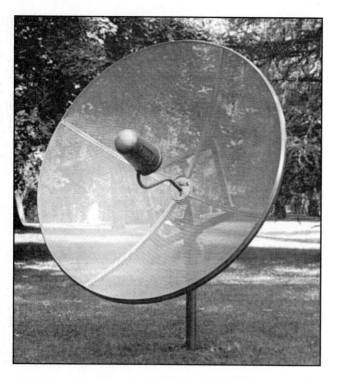

Fig. 24–6 Satellite antenna. Courtesy of Winegard Company.

Modern satellites systems have dishes ranging from 450 to 600 mm. They are light and easy to install, with many users choosing to perform their own installation. These satellite systems are "fixed" on one satellite, which is in a geosynchronous orbit above the earth (Fig. 24–7). Because of their light weight, many dishes are installed on the side of the residence, where they are less likely to be struck by lightning (Fig. 24–8).

Selecting a Mounting Location for the Dish

The dish can be mounted on nearly any surface, including brick, concrete, masonry block, wood siding, roof, or a pole. When selecting the mounting location, check the documentation included with the system. The dish must have an unobstructed "view" of the satellite. Be sure to consider growth of trees and possible construction of adjacent buildings. Be sure to protect the antenna from wind, as strong winds can bend mounting hardware or pull out mounting screws. The mounting surface must also be strong enough to prevent movement of the dish during high winds as even the smallest amount of movement may affect the signal reception. Because the system will be grounded to earth, it will provide a low resistance path to ground for lightning. Locating the dish high on the building or on a pole increases the likelihood of a lightning strike.

Mounting the Satellite Dish

Each manufacturer of satellite dishes provides detailed installation instructions with their systems. The mounting base must be securely fastened to the mounting surface. The mounting pole is attached to the mounting base and adjusted as recommended by the manufacturer. The support arm is attached to the mounting pole with the dish and the LNBF attached to the arm. After installation of the system is complete, aim the dish as recommended by the manufacturer.

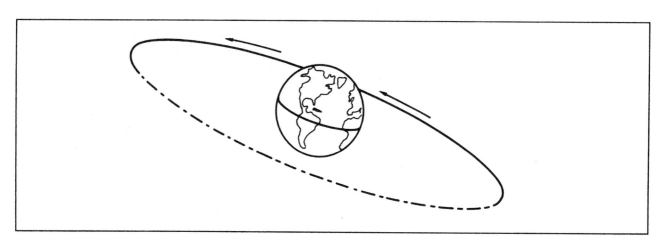

Fig. 24–7 Television satellites appear to be stationary in space because they are rotating at the same speed as the earth. This is called geosynchronous orbit. The satellite receives the uplink signal from earth, amplifies it, and transmits it back to earth. The downlink signal is picked up by the satellite antenna.

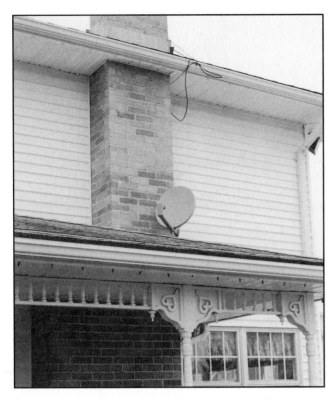

Fig. 24–8　Satellite antenna installed on the side of a home.

Fig. 24–9　Satellite coaxial cable grounding blocks.

Grounding of a Satellite System

The system must be grounded according to the manufacturer's recommendations and all applicable codes. To provide proper grounding, install a grounding block in the coaxial cable that enters the building from the dish. The distance from the copper ground rod and the grounding block should be as short as possible, and the grounding block must be connected to the rod using a copper grounding conductor of not less than No. 14 AWG (Fig. 24–9).

Installation of the Satellite Receiver

The satellite receiver is a remote-controlled device the viewer uses to control the channel. It therefore must be installed close to the television. More than one TV may be connected to this system; however, the same program will be viewed on all. It is possible to view up to two different programs at the same time on different televisions, but this will require that the LNBF be a dual type and a second receiver must be purchased. Also, a second coaxial cable must be supplied from the dish to the second receiver. Satellite television companies offer pay-per-view service on their systems, which require

the receiver to be connected to a telephone line. Manufacturers recommend that the phone line be left connected to the receiver at all times. Refer to Fig. 24–10 for a diagram of the installation. The coaxial cable from the satellite dish to the receiver must be type RG-6. After the receiver, type RG-59 may be used to distribute the signal throughout the residence. However, the higher-quality video provided by digital and high-definition television (HDTV) require RG-6 throughout the home.

Many signals received by a satellite dish provide the audio portion of the program in stereo. Many satellite services even supply specialty audio (radio) stations. The satellite receiver has audio output jacks on the rear panel for connection to a conventional stereo system for improved audio. This connection usually requires two audio patch cables with RCA-type jacks on both ends.

CEC RULES FOR INSTALLING ANTENNAS AND LEAD-IN WIRES (SECTION 54)

Home television sets and AM and FM radios generally come complete with built-in antennas. For those locations in outlying areas, out of reach of strong signals, it is quite common to install a separate antenna system.

Either indoor or outdoor antennas may be used with televisions. The front of an outdoor antenna is aimed at the television transmitting station. When transmitting stations are located in different directions, a rotor on the mast turns the antenna to face

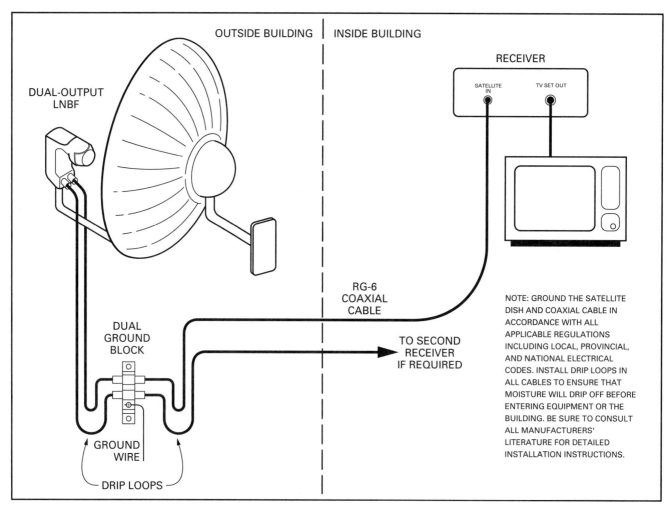

OUTSIDE BUILDING | INSIDE BUILDING

RECEIVER

DUAL-OUTPUT LNBF

SATELLITE IN   TV SET OUT

RG-6 COAXIAL CABLE

DUAL GROUND BLOCK

TO SECOND RECEIVER IF REQUIRED

NOTE: GROUND THE SATELLITE DISH AND COAXIAL CABLE IN ACCORDANCE WITH ALL APPLICABLE REGULATIONS INCLUDING LOCAL, PROVINCIAL, AND NATIONAL ELECTRICAL CODES. INSTALL DRIP LOOPS IN ALL CABLES TO ENSURE THAT MOISTURE WILL DRIP OFF BEFORE ENTERING EQUIPMENT OR THE BUILDING. BE SURE TO CONSULT ALL MANUFACTURERS' LITERATURE FOR DETAILED INSTALLATION INSTRUCTIONS.

GROUND WIRE

DRIP LOOPS

Fig. 24–10   Typical wiring diagram for a dual-receiver satellite system.

the direction of each transmitter. The rotor is controlled from inside the building. The controller's cord is plugged into a regular 120-volt receptacle to obtain power. A four-wire cable is usually installed between the rotor motor and the control unit. The wiring for a rotor may be installed during the roughing-in stage of construction by running the rotor's four-wire cable into a device wall box, allowing 1.5–1.8 m of extra cable, and then installing a regular single-gang switch plate when finishing.

Section 54 covers this subject. Although instructions are supplied with antennas, be sure to follow these *CEC* rules.

1. Antennas and lead-in wires shall be securely supported, *Rule 54–108*.

2. Antennas and lead-in wires shall not be attached to the electric service mast.

3. Antennas and lead-in wires shall be kept away from all light and power conductors to avoid accidental contact with them.

4. Lead-in wires shall be securely attached to the antenna.

5. Outdoor antennas and lead-in conductors shall not cross over light and power wires.

6. Lead-in conductors shall be kept at least 600 mm away from open light and power conductors.

7. Where practicable, antenna conductors shall not be run under open light and power conductors.

8. On the outside of a building:

   a. position and fasten lead-in wires so they cannot swing closer than 300 mm to light and power wires having **not** over 300 volts between conductors.

    b. keep lead-in conductors at least 2 metres away from the lightning rod system, or bond together according to *Rule 10–702*.

9. On the inside of a building:

    a. keep antenna and lead-in wires at least 50 mm from other open wiring (as in old houses) unless the other wiring is in metal raceway or in cable armour.

    b. keep lead-in wires out of electric boxes unless there is an effective, permanently installed barrier to separate the light and power wires from the lead-in wire.

10. Grounding:

    a. The grounding wire must be copper.

    b. The grounding wire need not be insulated. It must be securely fastened in place; may be attached directly to a surface without the need for insulating supports; shall be protected from physical damage or be large enough to compensate for lack of protection; and shall be run in as straight a line as is practicable.

    c. The grounding conductor shall be connected to the nearest accessible location on: the building or structure grounding electrode, *Rule 54–200(1)*; or the grounded interior water pipe, *Rule 54–302*; or the metallic power service raceway; or the service equipment enclosure; or the grounding electrode conductor or its metal enclosure.

    d. If none of the options in *c* is available, ground to any one of the electrodes per *Rule 10–700*, such as a metal underground waterpipe, an independent metal waterpiping system, or a concrete-encased electrode.

    e. The grounding conductor may be run inside or outside the building, but must be protected where exposed to mechanical injury.

    f. The grounding conductor shall not be smaller than No. 14 copper.

    g. *CAUTION:* If any of the preceding rules results in one grounding electrode for the antenna and another grounding electrode for the electrical system, bond the two electrodes together with a bonding jumper not smaller than No. 6 copper or its equivalent.

Grounding and bonding to the same grounding electrode as the main service ground prevents voltage differences between the two systems.

TELEPHONE WIRING (SYMBOL ▶I) (*SECTION 60*)

Since the deregulation of telephone companies, residential do-it-yourself telephone wiring has become quite common.

A telephone company installs the service line to a residence up to a protector device, which protects the system from hazardous voltages. The company decides if the protector is mounted outside or inside the home. Always check with the phone company before starting your installation.

The telephone company's wiring ends and the homeowner's interior wiring begins at the *demarcation point*. The telephone wiring inside the home may be done by the telephone company, the homeowner, or the electrician. Whatever the case, the rules in *Section 60* apply. Refer to Figs. 24–11, 24–12, and 24–13.

Section 60 covers the installation of communication systems. For the typical residence, the electrician or homeowner will rough in the boxes wherever a telephone outlet is wanted. The most common interior telephone wiring for a new home is with four-conductor No. 22 or 24 cable, known as "station wire." The outer jacket is usually of a thermoplastic material, most often in a neutral colour that blends in with decorator colours in the home. This is particularly important in existing homes where the telephone cable may have to be exposed.

Voice communication does not require a high-quality cable within the residence. However, many consumers use their telephone lines for data communication through a modem connected to their computer. The phone lines themselves are sometimes used as a computer data network within the home using special equipment. In these cases, the

Fig. 24–11 Typical telephone installation. The telephone company installs and connects its underground cable to the protector. In this illustration, the protector is mounted to the service raceway. This establishes the ground that offers protection against hazardous voltages such as lightning surge on the incoming line. A cable is then run from the protector to a standard network interface (SNI). The electrician or homeowner plugs the interior wiring into the SNI by means of the modular plug. The term *modular* means that most of the phones are plugged-in rather than hard-wired connections.

performance of station wire may not be satisfactory. In these situations, a 3- or 4-pair UTP (unshielded twisted pair) cable of the Category 3 type (or higher) is recommended. Twisting the conductors within the pairs and twisting the pairs themselves help to reduce electromagnetic interference with data communication.

Cables, mounting boxes, junction boxes, terminal blocks, jacks, adapters, faceplates, cords, hardware, plugs, and so on are all available through electrical distributors, builders' supply outlets, telephone stores, electronic stores, hardware stores, and similar wholesale and retail outlets.

Telephone Conductors

The colour coding of telephone cables is shown in Fig. 24–14. Fig. 24–15 shows some types of telephone cable.

Cables are available in these sizes:

2 pair (4 conductors)
3 pair (6 conductors)
6 pair (12 conductors)
12 pair (24 conductors)
25 pair (50 conductors)
50 pair (100 conductors)

Telephone circuits require a separate pair of conductors from the telephone all the way back to the phone company's central switching centre. To minimize interference that could come from other electrical equipment, such as electric motors and fluorescent fixtures, each pair of telephone wires is twisted together.

The recommended circuit lengths are:

No. 24 gauge: not over 61 m

No. 22 gauge: not over 76 m

(A)

(B) JUNCTION BOX FEATURES A SHORT, PREWIRED CORD THAT PLUGS INTO A STANDARD NETWORK INTERFACE (SNI). IT ALLOWS EASY CONNECTION OF ADDITIONAL TELEPHONE EXTENSIONS.

Fig. 24–12   Individual telephone cables run to each telephone outlet from the telephone company's connection point. KEPtel, Armiger.

Fig. 24–13   A complete "loop system" of the telephone cable. If something happens to one section of the cable, the circuit can be fed from the other direction.

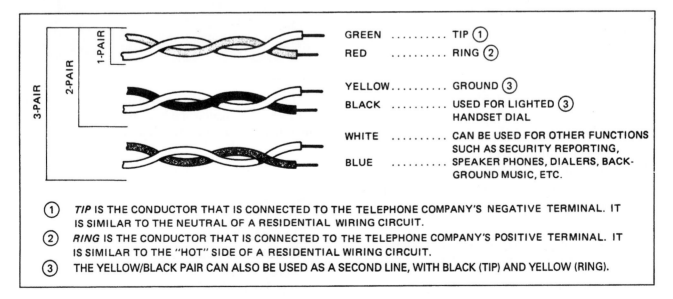

GREEN	TIP ①
RED	RING ②
YELLOW	GROUND ③
BLACK	USED FOR LIGHTED ③ HANDSET DIAL
WHITE	CAN BE USED FOR OTHER FUNCTIONS SUCH AS SECURITY REPORTING,
BLUE	SPEAKER PHONES, DIALERS, BACK-GROUND MUSIC, ETC.

① TIP IS THE CONDUCTOR THAT IS CONNECTED TO THE TELEPHONE COMPANY'S NEGATIVE TERMINAL. IT IS SIMILAR TO THE NEUTRAL OF A RESIDENTIAL WIRING CIRCUIT.

② RING IS THE CONDUCTOR THAT IS CONNECTED TO THE TELEPHONE COMPANY'S POSITIVE TERMINAL. IT IS SIMILAR TO THE "HOT" SIDE OF A RESIDENTIAL WIRING CIRCUIT.

③ THE YELLOW/BLACK PAIR CAN ALSO BE USED AS A SECOND LINE, WITH BLACK (TIP) AND YELLOW (RING).

Fig. 24–14   Colour coding of telephone cables.

Fig. 24–15   Telephone cable. These cords are available in round and flat configurations, depending on the number of conductors in the cable.

Fig. 24–16   Telephone modular jacks. Courtesy of Leviton.

Fig. 24–17   Three styles of wall plates for modular telephone jacks: rectangular stainless steel, weatherproof for outdoor use, and circular. KEPtel, Armiger.

When telephones had electromechanical ringers, one line had enough power to ring up to five phones. Most telephones now have electronic ringers, which use very little power; therefore, more than five phones can be installed on one line. (Some telephones are labelled with a ringer equivalence number (REN), with one REN equal to the power required by one electromechanical ringer.) Check this with the telephone company before installation.

It is recommended that at least one telephone be permanently installed, probably the wall phone in the kitchen. The remaining phones may be portable, able to be plugged into any of the phone jacks, Figs. 24–16 and 24–17. If all of the phones were portable, there might be a time when none of the phones were connected to phone jacks. As a result there would be no audible signal. However, most telephone companies no longer direct connect phones, resulting in all phones being cord and jack connected.

The electrical plans show that nine telephone outlets are provided, one in each of these areas:

Front bedroom
Kitchen
Laundry
Living room
Master bedroom
Rear outdoor patio
Recreation room
Study/bedroom
Workshop

Installation of Telephone Conductors, *Rules 60–300* to *60–334*

Telephone conductors:

1. shall be of a type listed in *Table 19* as suitable for communications and suitable for the installation environment;

2. shall be separated by at least 50 mm from light and power conductors operating at 300 volts or less, unless the light and power conductors are in a raceway, in non-metallic-sheathed cable, or type AC90 cable, *Table 19;*

3. shall not be placed in any conduit or boxes with electric light and power conductors unless the conductors are separated by a partition;

4. may be terminated in either metallic or non-metallic boxes—check with the local telephone company or local electrical inspector for specific requirements;

5. should not share the same stud space as the electrical branch circuit wiring;*

6. should be kept at least 300 mm away from the electrical branch circuit wiring where the telephone cable is run parallel to the branch circuit wiring to guard against induction noise;*

7. should not share the same bored holes as electrical wiring, plumbing, or gas pipes;*

8. should be kept away from hot-water pipes, hot-air ducts, and other heat sources that might harm the insulation;*

9. should be carefully attached to the sides of joists and studs with insulated staples.*

*These are recommendations from residential telephone manufacturers' installation data.

Conduit

In some communities, electricians prefer to install regular device boxes at a telephone outlet location, then "stub" a 21-mm conduit to the basement or attic from this box. At a later date, the telephone cable can be "fished" through the conduit.

Grounding (Rule 60–202)

The telephone company will provide the proper grounding of its incoming cable sheath and protector, generally using wire not smaller than No. 14 AWG copper or equal. Or, the company might clamp the protector to the grounded metal service raceway conduit to establish ground. The grounding requirements are practically the same as for CATV and antennas, which were covered earlier in this unit.

Safety

Open circuit voltage between telephone conductors of an idle pair can range from 50 to 60 volts DC. The ringing voltage can reach 90 volts AC. Therefore, always work carefully with insulated tools and stay clear of bare terminals and grounded surfaces. Disconnect the interior telephone wiring if you must work on the circuit, or take the phone off the hook, in which case the DC voltage level will drop and there should be no AC ringing voltage delivered.

DATA SYSTEMS

As more families enter the world of computers and the Internet, they quickly discover the usefulness of this technology. It provides the opportunity to find information on nearly any subject, to engage in multiplayer computer games, and to buy or sell just about anything. Many families have more than one computer and opt to install a home data network similar to those installed in small businesses, which allows all the computers to share peripheral devices such as drives, printers, and scanners, as well as Internet access and files.

Data Rate

Many forms of networks have been used over the years. The most popular of these is Ethernet, using UTP (unshielded twisted pair) cable (Fig. 24–18). Ethernet began with a data rate of 10 Mbps, which is quite adequate for most residential applications.

Ethernet expanded to a data rate of 100 Mbps and is now up to 1000 Mbps, or 1 Gbps (gigabits per second). Equipment for Gigabit networks is still quite expensive, while 100 Mbps equipment is only slightly more expensive than 10 Mbps equipment.

We recommend 100 Mbps equipment as a good compromise between communication speed and

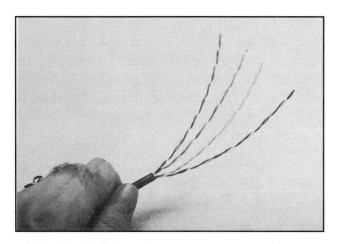

Fig. 24–18   Category 5e UTP data cable.

affordability. The equipment is not permanently installed and can be easily, although not necessarily inexpensively, replaced. The wiring, which is inside the walls, is considerably more difficult to replace when it becomes inadequate.

Data Cable Type

Category 5 cable has been the standard for Ethernet applications for many years, providing data communication rates of 100 Mbps. Category 5e (Cat 5e), or Category 5 enhanced, cable can support data rates up to 1 Gbps. Cat 6, which supports 1 Gbps but at a higher frequency, allows for greater efficiency in the network. Cat 6 is recommended for all new business applications but is considered excessive for most residential networks.

Category 5e cabling is recommended for these reasons:

1. Cat 5e cable and terminations are only slightly more expensive than Category 5.

2. Cat 5e cabling and terminations are relatively simple to install. Cat 6 terminations require a higher level of expertise to install successfully.

3. Affordable network interface cards (NIC) for the computers and network switches (Fig. 24–19) or hubs can be used. These devices can be either 10 Mbps or 100 Mbps devices.

4. If the consumer later needs a 1000 Mbps network, the installed cable and terminations can support the higher speed.

Installation of Data Cable

There are only a few basic rules to remember for a successful installation of a computer data network.

1. All cable and terminations must be of the same type: Cat 5e cable requires Cat 5e terminations. If cable and terminations of different categories are used, the system will perform only to the lowest category standard.

2. The maximum pulling force applied to a data cable is 11 kilograms.

3. When installing UTP cable, maintain a separation of 300 mm from the cable and any electrical equipment whenever possible.

4. A separation of 300 mm must be maintained between the data cable and ballasts of any type of fixtures, including fluorescent and transformers of low-voltage spotlights and track lights.

5. Data cable may be supported in many ways, including straps and plastic ties. Take care to not deform the cable when installing supports. If the supports are too tight, the twisting of the conductors inside the cable will be altered and the performance of the system will decrease.

6. The minimum radius of bend of data cable is not to be less than 25 mm. A smaller bend radius can damage the conductors or insulation and alter the twist rates of the conductors or pairs.

Fig. 24–19   A popular choice for the home network. This router includes a 4-port 10/100 Mbps switch, a print server for printer sharing, and a WAN port for connection to the Internet by cable or ADSL. Courtesy of SMC.

Installation of Termination Jacks

Jacks for data network installations are available in many forms from many manufacturers (Fig. 24–20). Some require the use of a punchdown tool, shown in Fig. 24–21, while others snap together. Jacks are available as both surface and flushmount. All jacks are clearly identified as to the colour coding for termination. Two configurations are used for termination: 568A most common in Canada; 568B most common in the United States. Either configuration is acceptable if you use the same one throughout the network. Some manufacturers make jacks that are specifically 568A or 568B: Be sure to order carefully and check that they are all the same when the order is received. Other manufacturers make jacks that can be used for either configuration and provide both colour coding layouts.

When terminating the conductor to the jack, ensure that no more than 12 mm conduct is exposed from the jacket. Each pair of conductors in UTP is twisted together at different numbers of twists per unit length. The four twisted pairs are then twisted together. No more twisting shall be removed than is necessary to complete the termination. Do not alter

Fig. 24–20   Category 5e data jacks.

Fig. 24–21   Punchdown tools.

the twist rates of the pairs in any way. Fig. 24–22 demonstrates the use of a punchdown tool to terminate a jack, while Figs. 24–23 to Fig. 24–27 show a detailed installation of a modular type of snap-together jack.

Wireless Data Networks

Wireless data networks are now affordable to the average consumer. The equipment used in wireless systems, such as routers and wireless network adapters, are only slightly more expensive than their wired counterparts. However, the "up front" cost of the installation of the network cabling can be eliminated.

With a wired network system, the consumer is "tied" to a location with the computer. Relocating the computer may require installing additional cabling and network jacks or using a long network cable to connect to an existing jack. Most consumers do not want the long cables lying on their floors because of the appearance and trip hazard. Wireless networks allow the consumers to locate their computers anywhere they like.

The advantages of wireless networks are:

- Increased flexibility when locating computers and peripherals such as printers.
- Portability when using laptop or portable computing devices, even into the yard surrounding the building.
- No wiring costs.
- Ease of adding devices to the network.

Its disadvantages are:

- Equipment is slightly more expensive than comparable wired equipment.
- The network can be tapped into by outside computers. It is **essential** that the security features supplied with the equipment be used.
- Wireless systems generally have a slower data transfer rate than wired systems; however, new technologies allow connection speeds as high as 108 Mbps.
- Connection speed is affected by the signal strength of the transmitter: It drops as the distance from the transmitter increases.

Fig. 24–22 Termination of punchdown type data jack.

Fig. 24–25 Lay conductors in modular wiring jig and cut off excess.

Fig. 24–23 Strip back and cut off outer jacket and nylon cord.

Fig. 24–26 Assemble wiring jig and jack body.

Fig. 24–24 Fan out individual conductors.

Fig. 24–27 Click assembly together. Some jacks may require a special tool for assembly.

- Network connectivity is affected by building construction. When heating system ductwork, refrigerators, or other metal objects are located between the user and the wireless transmitter, the metal will cast a "shadow" that makes wireless connection not possible. Wood frame construction is generally better than concrete or steel for the use of wireless networking.

- Connectivity is affected by other systems operating in the same wireless frequency range, such as microwave ovens and wireless telephones.

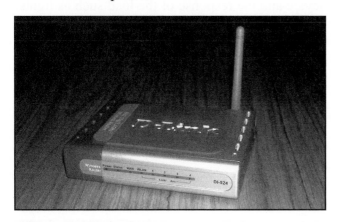

Fig. 24–28   Wireless router

HOME AUTOMATION

To keep up with the technological demands of our time, wiring in our homes has become very complex. Homes have telephone jacks in most rooms as well as wiring for computer networks and Internet access. As consumers, we are fascinated by automatic or remote-controlled devices.

One of the most popular types of home automation controls is based on the X-10 system technology developed in the late 1970s. Many of the popular systems that are developed on this technology are interchangeable between manufacturers, including X-10, IBM, and Leviton Decora Home Control.

Leviton offers a complete and versatile line of devices to control almost any type of load, such as lights (including dimmers), receptacles, heating systems, lawn sprinkler systems, garage door openers, photocontrols, motion sensors, and interfaces with other systems, such as security systems and personal computers. Handheld remote controls

and wall-mounted and tabletop controllers are also available, so that multiple devices can be operated from almost anywhere in the home.

The X-10 technology uses devices that take a small amount of energy from the 120-volt system in the house, convert it to a higher frequency, and send it back into the electrical distribution system. These devices are considered to be transmitters. When a receiver detects the signal on the distribution system, it will complete the intended action, such as turning a light on or off. A direct electrical connection between the transmitter and receiver is not required; only that they be on the same distribution system. In large homes, where the signal may not be strong enough, an amplifier/coupler is installed at the main electrical panel to bridge the phase conductors together for better signal distribution and to amplify the signal.

The X-10 system is sensitive to electrical noise on the distribution system, created by such devices as motors and fluorescent ballasts. The noise created within a home is usually insufficient to create problems on the home automation system. However, disturbances coming into the home from other buildings connected to the same distribution transformers may cause significant difficulty. If these are suspected as the cause of poor system operation, filters can be installed at the electrical panel.

Each transmitter and receiver is addressable using an alphanumeric system. This allows for a maximum of 256 unique addresses and, therefore, 256 unique control devices. Multiple devices may be addressed the same so that a single transmitter can operate several control devices, multiple transmitters can operate a single control device, or multiple transmitters can operate multiple control devices. The automation control devices generally mount in a standard device box. Check the size of the device to meet *Rule 12–3034* requirements for box fill.

You can use a personal computer and software from the home automation control company to run very complex programs. The computer is connected to an interface that is plugged into any 120-volt receptacle in the home. As the program runs, the devices turn on and off at the programmed times. This is very useful for security lighting or for making a home look occupied during vacations. Conventional timers are set so that the connected lights turn on and off at the same times each night.

The programs developed in a computer can be set to turn lights on or off at different times or at several intervals each night.

Leviton Decora Home Controls have developed scene controllers, which have become quite popular. As an example, home automation controls are to be installed in a home theatre room, which has four sources of lighting that will have individual control. The homeowner has identified four uses for the room, each requiring a different illumination level:

1. full illumination of all lights for use when cleaning the room
2. a level for entertaining company
3. a level for use when reading
4. a level for use when watching television

When each light is individually controlled, it is difficult to consistently get the level just right for each of these applications. Scene controllers allow you to program each custom lighting level so that the press of a single button will automatically adjust each light to a preset amount. Each light remains individually controllable for those unexpected lighting requirements.

A disadvantage of this system is the relatively slow operating speed. The data rate is about 60 bits per second, compared to 10 or 100 Mbps for a computer network. This slow data rate results in a noticeable delay between the operation of the control device and the response of the load such as lights. However, the flexibility of the system and the wide range of control options usually make this delay tolerable.

REVIEW

Note: Refer to the *CEC* or the plans where necessary.

TELEVISION CIRCUIT

1. How many television outlets are installed in this residence? _____

2. What types of boxes are recommended when unshielded lead-in wire is used?
   _____

3. What determines the design of the faceplates used? _____
   _____

4. What must be provided when installing a television outlet and receptacle outlet in one wall box? _____
   _____
   _____
   _____
   _____

5. Which system is more economical to install: a master amplifier distribution system or a multiset coupler? Explain the basic differences between these two systems.

_____

_____

_____

_____

_____

_____

6. What precautions should you take when installing a television antenna and its lead-in wires? _____

_____

_____

_____

_____

7. List the requirements for cable television inside the house. _____

_____

_____

_____

_____

_____

8. What section of the *CEC* references the requirements for cable television antenna installation? _____

9. Grounding and bonding together all metal parts of an electrical system and the metal shield of the cable television cable to the same grounding reference point in a residence will keep both systems at the same voltage level should a surge, such as lightning, occur. If the CATV company installer grounds the metal shield of the incoming cable (to a house) to a driven ground rod, does this installation conform to the *CEC*?

_____

_____

10. The basics of cable television are that a transmitter on earth sends a(n) _____ signal to a(n) _____ where the signal is reamplified and sent back to earth via the _____ signal. This signal is picked up by the _____ on the satellite antenna, which is sometimes referred to as a(n) _____. The feedhorn funnels the signal into a(n) _____, then onto a(n) _____, where the high-frequency signal is converted to a(n) _____-frequency signal suitable for the television receiver.

11. All television satellites revolve above the earth in (the same orbit) (different orbits). Circle the correct answer.

12. Television satellites are set in orbit (16 093) (28 970) (36 532) km above the earth, which results in their revolving around the earth at (precisely the same) (different) rotational speed as/from the earth. This is done so that the satellite dish can be focused on a specific satellite (once) (one time each month) (whenever the television set is used). Circle the correct answers.

TELEPHONE SYSTEM

1. How many locations are provided for telephones in the residence? _____
   _____

2. At what height are the telephone outlets mounted? _____

3. Sketch the symbol for a telephone outlet.

4. Is the telephone system regulated by the *CEC*? _____

5. a. Who is to furnish the outlet boxes required at each telephone outlet? _____
      _____

   b. Who is to furnish the faceplates? _____

6. Who is to furnish the telephones? _____

7. Who does the actual installation of the telephone equipment? _____
   _____

8. How are the telephone cables concealed in this residence? _____
   _____
   _____
   _____

9. The point where the telephone company's cable enters the residence and the interior telephone cable wiring meet is called the _____ point.

10. What are the colours contained in a four-conductor telephone cable assembly and what are they used for? _____
    _____
    _____
    _____

11. List the *CEC* rules for the installation of telephone cables in a residence.

_____

_____

12. If finger contact were made between the red conductor and green conductor at the instant a "ring" occurs, what shock voltage would be felt?_____

_____

DATA SYSTEMS

1. What is the data rate for Ethernet networks as discussed in this chapter?

_____

2. What precautions must be taken when installing data cable?

_____

_____

_____

_____

3. Which termination configuration is most commonly used in Canada?

_____

4. What is the maximum data rate of a Category 5e network?

_____

5. What is the maximum pulling force that may be exerted on a data cable?

_____

6. When supporting data cable, why is it important not to strap or tie the cable too tightly?

_____

_____

7. What is the minimum radius of bend of data cable and why?

_____

_____

_____

_____

8. List the factors that affect the connectivity and data transfer rate of wireless network systems.

_____

_____

_____

9. What other equipment can affect a wireless data network?

_____

_____

HOME AUTOMATION

1. What systems can be controlled using an X-10-based home automation system?

_____

_____

_____

2. Is additional wiring between the controller and controlled device required? Explain why.

_____

_____

_____

3. If the home is very large, what additional device may need to be installed?

_____

_____

_____

4. With the alphanumeric addressing system, how many unique devices can be controlled?

_____

_____

_____

Lighting Branch Circuit for the Garage and Outdoor Lighting

OBJECTIVES

After studying this unit, the student will be able to

- understand the fundamentals of providing proper lighting in residential garages
- understand the *CEC* requirements for underground wiring, with either conduit or underground cable
- complete the garage circuit wiring diagram
- discuss typical outdoor lighting, including *CEC* requirements
- describe how conduits and cables are brought through cement foundations to serve loads outside the building structure
- understand the application of GFCI protection for loads fed by underground wiring
- make a proper installation for a residential overhead garage door opener
- make a proper installation for a residential central vacuum system

LIGHTING BRANCH CIRCUIT

Circuit B22 originates at Panel B in the recreation room and is brought into the wall receptacle box on the rear wall of the garage. Fig. 25–1 shows the cable layout for circuit B22. From this point, the circuit is carried to the switch box adjacent to the side door of the garage. From there, the circuit conductors plus the switch leg are carried upward to the ceiling fixtures and then to the wall receptacle outlets on the right-hand wall of the garage.

The garage ceiling's porcelain lampholders are controlled from switches located at all three entrances to the garage.

The post light feed from circuit A15 is controlled by a switch in the front entrance while the post light turns on and off by a photocell, an integral part of the light.

The fixtures on the outside next to the overhead garage door are controlled by two three-way switches, one just inside the overhead door and the other at the door entering the service entrance. These fixtures are not connected to the garage lighting circuit.

All of the wiring in the garage will be concealed because the walls and ceiling are to be covered with $3/4$-hour fire-rated drywall.

Table 25–1 summarizes the outlets and the estimated load for the garage circuit.

LIGHTING A RESIDENTIAL GARAGE

To provide adequate lighting in a residential garage, you should install a light fixture on the ceiling above

Fig. 25–1   Cable layout for the garage, circuit B22.

DESCRIPTION	QUANTITY	WATTS	VOLT-AMPERES
Receptacles @ 120 W each	3	360	360
Ceiling fixtures @ 60 W each	3	180	180
Outdoor garage bracket fixture (rear)	1	100	100
Overhead door opener Two 40-W lamps Motor (5.8 A @ 120 V)	1 1	80 373	80 696
TOTALS	9	1093	1416

Table 25–1   Garage: outlet count and estimated load, circuit B22.

each side of an automobile, Fig. 25–2. Fig. 25–3 shows typical lampholders commonly mounted in garages.

- For a one-car garage, a minimum of two fixtures is recommended.

- For a two-car garage, a minimum of three fixtures is recommended.

- For a three-car garage, a minimum of four fixtures is recommended.

Lighting fixtures arranged this way eliminate shadows between automobiles, which can hide objects that may cause a person to trip or fall.

These ceiling fixtures should be mounted toward the front end of the automobile as it is usually parked in the garage. This provides better lighting where it is most needed. Do not place lights where they will be covered by the open overhead door.

RECEPTACLE OUTLETS

At least one receptacle must be installed for each car space in an attached residential garage or carport, *Rule 26–714(b)*. The garage in this residence has three receptacles, and if standard receptacles are installed, they must be of a tamper-resistant nature to comply with *Rule 26–712(g)*. These must be on a separate circuit, but the garage door opener and the lighting fixtures for the garage may be on the same circuit, *Rule 26–724(b)*. When electricity is provided in a detached garage, all pertinent *CEC* rules become effective and must be followed. These include rules relating to grounding, lighting outlets, receptacle outlets, and GFCI protection, and so on.

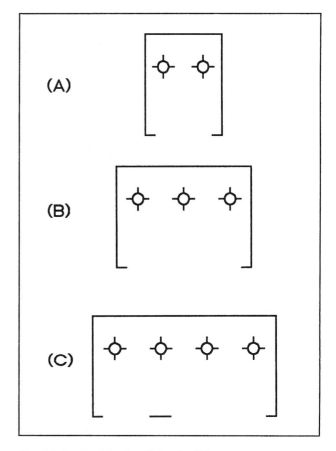

Fig. 25–2   Positioning lights in (A) a one-car garage, (B) a two-car garage, and (C) a three-car garage.

A. KEYLESS   B. PULL CHAIN

Fig. 25–3   Typical plastic lampholders of the type installed in garages, attics, basements, crawl spaces, and similar locations.

OUTDOOR WIRING

If the homeowner requests that 120-volt receptacles be installed away from the building structure, the electrician can provide weatherproof receptacle outlets, Fig. 25–4. These must be GFCI protected if within 2.5 m of ground level, *Rule 26–710(o)*.

LOW-VOLTAGE
DECORATIVE
LAWN LIGHTING

120- TO 12-VOLT
TRANSFORMER

WEATHERPROOF OUTLETS
STUBBED OUT OF GROUND

(COVERS
SHOWN
CLOSED)

Fig. 25–4 Weatherproof receptacle outlets stubbed out of the ground: Either low-voltage decorative lighting or 120-volt PAR lighting fixtures may be plugged into these outlets.

These types of boxes have 16-mm threaded openings in which conduit fittings secure the conduit "stub-ups" to the box. Any unused openings are closed with plugs that are screwed tightly into the threads, Fig. 25–5.

Wiring with Type NMWU Cable

The plans show that type NMWU non-metallic sheathed cable, Fig. 25–6, is used to connect the post light. The current-carrying capacity (ampacity) of NMWU cable is found in *Table 2,* using the 60°C column. According to *Table 19* and *Table D1,* NMWU cable:

- is available in sizes from No. 14 AWG through No. 2 (for copper conductors), (See *Table 2* for the ampacity ratings);

- may be used with non-metallic-sheathed cable fittings;

- is suitable for category 1 and 2 locations, *Table 19, Section 22;*

- is moisture and fungus resistant;

- may be buried directly in the earth;

- may be used for interior wiring in wet, dry, or corrosive areas;

- is installed using the methods for non-metallic-sheathed cable, *Subsection 12–500*;

- must not be used as service entrance cable;

- must not be embedded in concrete, cement, or aggregate.

The grounding of equipment fed by type NMWU cable is accomplished by properly connecting the bare bonding conductor in the cable to the equipment to be grounded, Fig. 25–7.

UNDERGROUND WIRING

Underground wiring is common in such residential applications as decorative landscape lighting and wiring to post lamps and detached buildings such as garages or tool sheds.

THIS BODY IS CONSIDERED TO
BE ADEQUATELY SUPPORTED.

(A) WHEN TWO OR MORE CONDUITS ARE TIGHTLY THREADED INTO THE HUBS OF A CONDUIT BODY, THE CONDUIT BODY
IS CONSIDERED TO BE ADEQUATELY SUPPORTED. *RULE 12–3012(2).*

THIS BODY IS NOT CONSIDERED TO BE
ADEQUATELY SUPPORTED. IF VOLUME
IS GREATER THAN 1640 mL IT MUST BE
SUPPORTED.

(B) THIS CONDUIT BODY IS NOT ADEQUATELY SUPPORTED BY THE ONE CONDUIT THREADED INTO THE HUB. THIS
CONDUIT BODY COULD TWIST VERY EASILY, RESULTING IN DAMAGED INSULATION ON THE CONDUCTORS AND A POOR
GROUND CONNECTION BETWEEN THE CONDUIT AND THE CONDUIT BODY.

SPLICES MAY BE MADE IN SUCH
CONDUIT BODIES WHEN MARKED WITH
THEIR MILLILITRE CAPACITY.

(C) CONDUCTORS MAY BE SPLICED IN THESE CONDUIT BODIES. IF THE CONDUIT BODY IS NOT MARKED WITH ITS
CUBIC-INCH CAPACITY, THE BOX VOLUME MAY BE CALCULATED AND THE PERMISSIBLE CONDUCTOR FILL
DETERMINED USING THE CONDUCTOR VOLUME FOUND IN *TABLE 22.*

Fig. 25–5 Supporting threaded conduit bodies.

Fig. 25–6 Type NMWU underground cable.

Table 53 shows the minimum depths required for the various types of wiring methods, Figs. 25–8A and 25–8B.

The table specifies the minimum cover in millimetres for the three approved wiring methods, with further division into non-vehicular and vehicular areas. Vehicular areas are understood to refer to alleys, residential driveways, and parking lots, as well as streets and highways.

Pay careful attention to the note that specifies that *minimum cover* means the distance between the top surface of the conductor and the finished grade. For example, the basic rule for direct buried cable is that the cover must be a minimum of

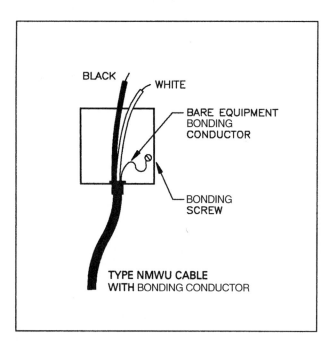

Fig. 25–7   How the bare equipment bonding conductor of a type NMWU cable is used to bond the metal box.

600 mm in depth. Thus the trench must have a depth of 600 mm (cover), plus the diameter of the cable, plus 75 mm for screened sand or earth, *Table 53.*

$$600 \text{ mm} + \text{cable diameter} + 75 \text{ mm} = \text{required depth}$$

Do not backfill a trench with rocks, debris, or similar coarse material, *Rule 12–012(10).*

Rule 12–012(3) describes methods for the mechanical protection of buried conductor, cable, or raceway. *Appendix B* adds two important points:

- Planks should be pressure-treated with pentachlorophenol if used to provide protection.

- Black poly water pipe is acceptable for mechanical protection, *Rule 12–012(3)(e).*

Rule 12–012(2) indicates that approved mechanical protection can reduce the cover requirements of *Table 53* by 150 mm.

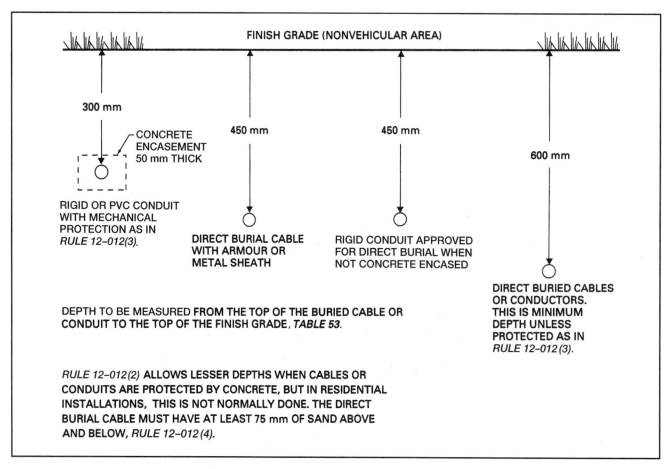

Fig. 25–8A   Minimum depths for cables and conduits installed underground. See *Table 53* for depths for other conditions.

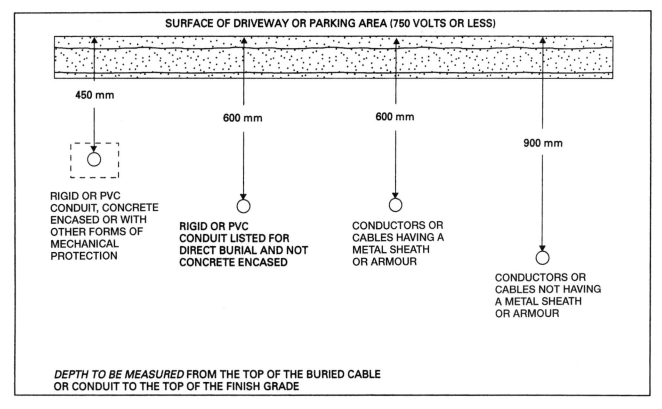

Fig. 25–8B   Depth requirements under residential driveways and parking areas, *Rule 12–012.*

Installation of Conduit Underground

Some specifications require the installation of conduit for all underground wiring, using conductors approved for wet locations, such as type TW, TW75, or RW90.

All metal conduit installed underground must be protected against corrosion. The manufacturer of the conduit and *Section 12* of the *CEC* furnish the information as to whether the conduit is suitable for direct burial.

When metal conduit is used as the wiring method for an underground installation, the metal boxes, post lights, and so on that are connected to the metal raceways are not considered to be grounded because the conduit cannot serve as the equipment bonding conductor, *Rule 10–804(c)*, Fig. 25–9. See *Rule 10–510* for details on grounding requirements for fixtures.

Thus a separate bonding conductor is required when either metallic or non-metallic raceways are installed. The bonding conductor may be bare if the length of the run does not exceed 15 m and has no more than two quarter bends, or is insulated, *Rule 10–808(3)*. When an insulated conductor is

used, it must be green or green with a yellow stripe. The size of a bonding conductor is determined from the ampacity of the circuit conductors and *Table 16*, Fig. 25–10, and *Rule 10–814*.

Fig. 25–9   For direct earth burial the *CEC* **requires that a separate bonding conductor be installed in all rigid metal conduits,** *Rule 10–804(c).*

Fig. 25–10   Bonding a metal box with a separate equipment bonding conductor when PVC conduit is used, *Rules 10–404(1)* **and** *10–814(1).*

Fig. 25–11 shows two methods of bringing the conduit into the basement: (1) It can be run below ground level and then brought through the basement wall; or (2) it can be run up the side of the building and through the basement wall at the ceiling joist level. When the conduit is run through the basement wall, the opening must be sealed to prevent moisture from seeping into the basement. The electrician must decide which of the two methods is more suitable for the installation.

Section 0 defines a wet location as an area where electrical equipment will be exposed to liquids that may drip, splash, or flow against the equipment.

Outdoor fixtures must be constructed so that water cannot enter or accumulate in lampholders, wiring compartments, or other electrical parts. These fixtures must be marked "Suitable for Wet Locations" or approved for outdoor use.

Fig. 25–11   Methods of bringing cable and/or conduit through a concrete wall and/or upward from a concrete wall into the hollow space within a framed wall.

The *CEC* defines partially protected areas, such as roofed open porches or areas under canopies, as damp locations. Fixtures to be used in these locations must be marked "Suitable for damp locations." Many types of fixtures are available, so be sure to check the approval label on the fixture or check with the supplier to determine the suitability of the fixture for a wet or damp location. Fig. 25–12 shows an outdoor fixture with lamps suitable for damp locations.

The post in Fig. 25–11 may be embedded in concrete or not, depending on the consistency of the soil, the height of the post, and the size of the lighting fixture. Most electricians prefer to embed the base of the post in concrete to prevent rotting of wood posts and rusting of metal posts.

Fig. 25–13 illustrates some typical post lights.

OVERHEAD GARAGE DOOR OPENER ▲E

The overhead garage door opener in this residence is plugged into the receptacle provided for this purpose on the garage ceiling. The opener is rated at 5.8 amperes at 120 volts. The installation of the overhead door will be done by the carpenter or a specially trained overhead door installer.

Principles of Operation

The overhead door opener contains a motor, gear reduction unit, motor reversing switch, and electrical clutch. This unit is preassembled and wired by the manufacturer.

Almost all residential overhead door openers use split-phase, capacitor-start motors in sizes from $\frac{1}{4}$ to $\frac{1}{2}$ hp. The motor size depends on the size (weight) of the door to be raised and lowered. If springs are used to counterbalance the weight of the door, a small motor with a gear reduction unit can lift a fairly heavy door.

There are two ways to change the direction of a split-phase motor: (1) Reverse the starting winding with respect to the running winding; or (2) reverse the running winding with respect to the starting winding. Although method (2) is correct, it is not commonly used. The motor can be run in either direction by properly connecting the starting and running winding leads of the motor to the reversing switch. When the motor shaft is connected to a chain drive or screw drive, the door can be raised or lowered according to the direction in which the motor is rotating.

Fig. 25–12   Outdoor fixture with lamps.

Fig. 25–13   Post lights.

The electrically operated reversing switch can be controlled by several means, including push-button stations (marked "open-close-stop," "up-down-stop," or "open-close"), indoor and outdoor weatherproof push buttons, and key-operated stations. Radio-operated controllers are popular because they permit overhead doors to be controlled from an automobile.

Wiring of Door Openers

Wiring overhead door openers for residential use is quite simple, since these units are completely prewired by the manufacturer. The electrician must provide a 120-volt circuit close to where the overhead door operators are to be installed. This 120-volt circuit can be wired directly into the overhead door unit, or a receptacle can be provided into which the cord on the overhead door operator is plugged. The outlet for the door opener is normally placed about 3.2 m in from the centre of the door opening.

The push buttons used to operate a residential overhead door opener can be placed in any convenient location. They are wired with low-voltage wiring, usually 24 volts, Fig. 25–14. This is the same type of wire used for wiring bells and chimes. From each button, the electrician runs a two-wire, low-voltage cable back to the control unit. Since the residential openers use a simple control system, a two-wire cable is all that is necessary between the operator and the push buttons. Commercial overhead door openers can require three or more conductors.

It is desirable to install push buttons at each door leading into the garage. The actual connection of the low-voltage wiring between the controller and the push buttons is a parallel circuit connection, Fig. 25–14.

CENTRAL VACUUM SYSTEM ▲Q

It is common practice to make provision for a central vacuum system in new homes, and to install it when renovating or upgrading existing homes. The ductwork is installed soon after the house is framed and provides a series of centrally located outlets for vacuuming with the 10-m hose provided with the system.

The system turns on when the hose-end is inserted into the outlet and turns off when the hose is removed. This is achieved by low-voltage termi-

Fig. 25–14   Push-button wiring for overhead door operator.

nals that are built into the outlet. All the ducts are connected to the main unit, which is located in the garage of this residence so that any noise and dust that can escape through the exhaust will be outside the house. *Rules 26–710(l)* and *26–712(g)* require that a tamper-resistant receptacle be provided for each cord-connected central vacuum system.

A separate branch circuit, A24, feeds the receptacle in the garage for the power unit. This circuit is required by *Rule 26–720(e)*, and provides power for the motor rated at 12 amperes 120 volts, Fig. 25–15.

Fig. 25–15   The central vac system requires a separate circuit for the unit in the garage, which is a 2-hp motor and draws 12A at 120V.

REVIEW

Note: Refer to the *CEC* or the plans where necessary.

1. Which circuit supplies the garage? _____

2. What is the circuit rating? _____

3. How many receptacle outlets are connected to the garage circuit? _____

   _____

4. a. How many cables enter the wall switch box located next to the side garage door?

   _____

   b. How many circuit conductors enter this box? _____

   c. How many bonding conductors enter this box? _____

5. Show calculations on how to select a proper box for Question 4. What kind of box
   would you use? _____

   _____

   _____

   _____

   _____

   _____

6. a. How many lights are recommended for a one-car garage? _____
      For a two-car garage? _____ For a three-car garage? _____

   b. Where are these lights to be located? _____

   _____

7. From how many points in the garage of this residence are the ceiling lights controlled?

   _____

8. List three methods by which the cover requirements of the *CEC, Table 53* can be reduced.

   _____

   _____

   _____

   _____

   _____

9. The total estimated load of the garage circuit draws how many amperes? Show your
   calculations. _____

   _____

   _____

   _____

10. How high from the floor are the switches and receptacles to be mounted? _____

_____

11. What type of cable feeds the post light? _____

12. Describe the appropriate method of backfilling trenches containing cables, conduits, and/or direct burial conductors. Quote the *CEC* rule:_____

_____

_____

_____

13. In the spaces provided, fill in the cover (depth) for the following residential underground installations (120/240 V)

    a. Type NMWU with no other protection.   _____ mm

    b. Type NMWU below driveway.   _____ mm

    c. Rigid conduit under lawn.   _____ mm

    d. Rigid 240-volt circuit conduit under driveway.   _____ mm

    e. Cable in poly pipe between house and detached garage.   _____ mm

    f. TECK cable under lawn; circuit is 120 volts, 20 amperes, GFCI protected.   _____ mm

    g. TECK cable with an approved plank covering under driveway; circuit is 240 volts AC.   _____ mm

    h. Rigid non-metallic conduit, concrete encased (50 mm); circuit is 120 volts, 20 amperes, GFCI protected.   _____ mm

    i. Rigid conduit embedded in rock and grouted with concrete to the level of the rock surface.   _____ mm

14. What rule of the *CEC* prohibits embedding type NMWU cable in concrete? _____

_____

_____

15. The following is a layout of the garage circuit. Using the suggested cable layout, make a complete wiring diagram of this circuit, using coloured pens or pencils to indicate conductor insulation colours.

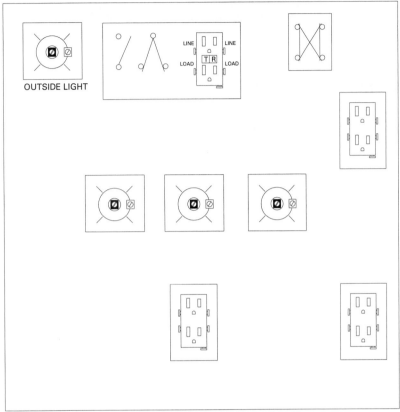

GARAGE CCT B22

16. What type of motor is generally used for garage door openers? _____

    _____

17. How is the direction of a split-phase motor reversed? _____

    _____

    _____

    _____

APPENDIX A

Specifications for Electrical Work — Single-Family Dwelling

1. **GENERAL:** The "General Clause and Conditions" shall be and are hereby made a part of this division.

2. **SCOPE:** The electrical contractor shall furnish and install a complete electrical system as shown on the drawings and/or in the specifications. Where there is no mention of the responsible party to furnish, install, or wire for a specific item on the electrical drawings, the electrical contractor will be responsible completely for all purchases and labour for a complete operating system for this item.

3. **WORKMANSHIP:** All work shall be executed in a neat and workmanlike manner. All exposed conduits shall be routed parallel or perpendicular to walls and structural members. Junction boxes shall be securely fastened, set true and plumb, and set flush with the finished surface when the wiring method is concealed.

4. **LOCATION OF OUTLETS:** The electrical contractor shall verify location, heights, outlet and switch arrangements, and equipment prior to rough-in. No additions to the contract sum will be permitted for outlets in wrong locations, in conflict with other work, and so on. The owner reserves the right to relocate any device up to three metres prior to rough-in, without any charge by the electrical contractor.

5. **CODES:** The electrical installation is to be in accordance with the latest edition of the *Canadian Electrical Code, Part I (CEC)*, all local and provincial electrical codes, and the utility company's requirements.

6. **MATERIALS:** All materials shall be new and shall be listed and bear the appropriate label of the Canadian Standards Association or another nationally recognized testing laboratory for the specific purpose. The material shall be of the size and type specified on the drawings and/or in the specifications.

7. **WIRING METHOD:** Wiring, unless otherwise specified, shall be non-metallic-sheathed cable, armoured cable, or EMT adequately sized and installed according to the *CEC* and local ordinances.

8. **PERMITS AND INSPECTION FEES:** The electrical contractor shall pay for all permit fees, plan review fees, licence fees, inspection fees, and taxes applicable to the electrical installation, and these amounts shall be included in the base bid as part of this contract.

9. **TEMPORARY WIRING:** The electrical contractor shall furnish and install all temporary wiring for hand-held tools and construction lighting according to the *CEC* and Canadian Building Code standards and include all costs in the base bid.

10. **NUMBER OF OUTLETS PER CIRCUIT:** In general, not more than 12 lighting and/or receptacle outlets shall be connected to any one branch circuit. Exceptions may be made for low-current-consuming outlets.

11. **CONDUCTORS:** General lighting and power branch circuits shall be No. 14 AWG copper protected by 15-ampere overcurrent devices. All other circuits: wire and overcurrent device as required by the *CEC*. All conductors shall be part of an approved cable assembly or, if installed in raceway, type T90 or RW90 XLPE, unless specified otherwise.

12. **LOAD BALANCING:** The electrical contractor shall connect all loads, branch circuits, and feeders per Panel Schedule, but shall verify and modify these connections as required to balance connected and computed loads to within 10% variation.

13. **SPARE CONDUITS:** The electrical contractor shall furnish and install two empty 16-mm EMT conduits between the workshop and the attic for future use.

14. **GUARANTEE OF INSTALLATION:** The electrical contractor shall guarantee all work and materials for a period of one full year after final acceptance by the architect/engineer and owner.

15. **APPLIANCE CONNECTIONS:** The electrical contractor shall furnish all wiring materials and make all final electrical connections for all permanently installed appliances such as, but not limited to, furnace, water heater, water pump, built-in ovens and ranges, food waste disposer, dishwasher, central vacuum, food processor power unit, and clothes dryer.
    These appliances are to be furnished by the owner.

16. **CHIMES:** Furnish and install two two-tone door chimes where indicated on the plans, complete with two push buttons and suitable chime transformer. Allow $150 for above items. Chimes and buttons are to be selected by the owner.

17. **DIMMERS:** Furnish and install dimmer switches where indicated.

18. **EXHAUST FANS:** Furnish, install, and provide connections for all exhaust fans indicated on the plans including, but not limited to, ducts, louvres, trims, speed controls, and lamps. Included are recreation room, laundry, rear-entry powder room, range hood, and bedroom hall ceiling fan. Allow a sum of $2000 in the base bid for this. This allowance does not include the two bathroom heat/vent/light/night-light fixtures.

19. **FIXTURES:** A fixture allowance of $2750 shall be included in the electrical contractor's bid. This allowance shall include the furnishing and installation of all surface, recessed, track, strip, pendant, and/or hanging fixtures, complete with lamps. This allowance includes the three bathroom medicine cabinets with lights.
    This allowance does not include the two bathroom ceiling heat/vent/light/night-light fixtures.
    Labour for installation of the fixtures shall be included in the base bid.

20. **HEAT/VENT/LIGHT/NIGHT-LIGHT CEILING FIXTURES:** Furnish and install two heat/vent/light/ nightlight units where indicated on the plans complete with switch assembly, ducts, and louvres required to perform the heating, venting, and lighting operations as recommended by the manufacturer.

21. **PLUG-IN STRIP:** Where noted in the workshop, furnish and install a multi-outlet assembly with outlets 450 mm on centre. Total outlets: 6.

22. **SWITCHES, RECEPTACLES, AND FACEPLATES:** All flush switches shall be of the quiet AC-rated toggle type. They shall be mounted 1300 mm to centre above the finished floor unless otherwise noted.
    Receptacle outlets shall be mounted 300 mm to centre above the finished floor unless otherwise noted. All convenience receptacles shall be of the grounding type. Furnish and install GFCI receptacles where indicated, to provide ground-fault circuit protection as required by the *CEC*. All wiring devices are to be provided with ivory handles or faces and shall be trimmed with ivory faceplates except in the kitchen, where chrome-plated steel faceplates shall be used.
    Receptacle outlets, where indicated, shall be split-switched. All receptacles within the dwelling shall be tamper-resistant (TR), with the exception of microwave, refrigerator, freezer, and kitchen counter receptacles, as per *Rule 26–712(g)*.

23. **TELEVISION OUTLETS:** Furnish and install a 4 × 4 single-gang raised plaster cover at each television outlet where noted on the plans. Provide a vapour boot in all insulated walls. Mount at the same height as receptacle outlets. Furnish and install 75-ohm coaxial cable to each television outlet from a point in the workshop near the main service entrance switch. Allow 2 m of cable in workshop. Furnish and install television plug-in jacks at each location. Faceplates are to match other faceplates in the home. All remaining work is to be done by others.

24. **TELEPHONES:** Furnish and install a 3-in deep device box or 4-inch square box, $1\frac{1}{2}$-in deep with suitable single-gang raised plaster cover, at each telephone location, as indicated on the plans.
    Furnish and install four-conductor No. 22 AWG copper telephone cable to each designated telephone location, terminate in proper modular jack, complete with faceplates. Installation shall be according to any and all applicable *CEC* and local code regulations.

25. **SERVICE ENTRANCE:** Furnish and install one (1) 30-circuit, 200-ampere, 120/240-volt single-phase, three-wire combination panel (rated for continuous operation at 80%) complete with a 200-ampere main breaker in the workshop where indicated on the plans. Branch circuit protection in the panel is to incorporate circuit breakers. The panel is to have a 10 000-ampere interrupting rating.

26. Service entrance underground consumer's service conductors are to be furnished and installed by utility. The meter equipment model number is to be furnished by the utility and installed by the electrical contractor where indicated on plans. The electrical contractor is to furnish and install all panels, conduits, fittings, conductors, and other materials required to complete the service entrance installation from the demarcation point of the utility's equipment to and including the main panel.

27. Service entrance conductors supplied by the electrical contractor shall be three No. 3/0 RW90 XLPE (600 V) or equivalent for the phase conductors and neutral conductor. Install 41-mm trade-size PVC conduit from main Panel A to the meter base.

28. Bond and ground service-entrance equipment in accordance with the *CEC* and local and utility code requirements. Install No. 6 AWG copper system grounding conductor.

29. **SUBPANEL:** Furnish and install one (1) 24-circuit, 120/240-volt, single-phase, three-wire load centre in the recreation room. The load centre is to have 100-ampere mains. Feed the load centre with three No.3 T90 conductors or equivalent protected by a 100-ampere, two-pole overcurrent device in the main panel. Install conductors in 27-mm PVC conduit buried under the basement floor slab. Include a correctly sized bonding conductor with the panel feeds. Branch circuit protection in this panel is to incorporate circuit breakers.

30. **CIRCUIT IDENTIFICATION:** All panel boards shall be furnished with typed-card directories with proper designation of the branch circuit feeder loads and equipment served. The directories shall be located in the panel in a holder for clear viewing. Test charts for GFCI and AFCI breakers will be located on the panel enclosure.

31. The electrical contractor shall seal and weatherproof all penetrations through foundations, exterior walls, and roofs.

32. Upon completion of the installation, the electrical contractor shall review and check the entire installation, clean equipment and devices, and remove surplus materials and rubbish from the owner's property, leaving the workplace neat and clean and in complete working condition. The electrical contractor shall be responsible for the removal of any cartons, debris, and rubbish for equipment installed by the electrical contractor, including equipment furnished by the owner or others and removed from the carton by the electrical contractor.

SYMBOL	DESCRIPTION	VOLTS	HORSE-POWER	APPLIANCE AMPERE RATING	TOTAL APPLIANCE WATTAGE RATING (OR VA)	CIRCUIT AMPERE RATING	POLES	WIRE SIZE NMD90	CIRCUIT NUMBER	COMMENTS
A	Hydromassage tub, master bedroom	120	1/2	10	1200	15	1	14	A9	Connect to Class A GFCI. Separate circuit.
B	Water pump	240	1	8	1920	20	2	12	A5/A7	Run circuit to disconnect switch on wall adjacent to pump; protect with Fusetron dual-element time-delay fuses sized at 125% of motor's FLA.
C	Water heater: top element 2000 W. bottom element 3000 W.	240	–0–	8.33 12.50	2000 3000	20	2	12	A6/A8	Connected for limited demand.
D	Dryer	120/240	120 V 1/6 Motor Only	23.75	5700 Total	30	2	10	B1/B3	Provide flush-mounted 30-A dryer receptacle. 14-30R.
E	Overhead garage door opener	120	1/4	5.8	696	15	1	14	B22	Unit comes with 3 W cord. Unit has integral protection. Connect to garage lighting circuit.
F	Wall-mounted oven	120/240	–0–	27.5	6600	40	2	8	B6/B8	
G	Countertop range	120/240	–0–	31	7450	40	2	8	B2/B4	
H	Food waste disposer	120	1/3	7.2	864	15	1	14	B19	Controlled by single-pole switch on wall.
I	Dishwasher	120	1/4 Motor Only	Motor 5.80 Htr. 6.25 Total 12.05	696 750 1446	15	1	14	B5	Motor and heater are not on simultaneously.
J	Heat/vent/night-light/light master B.R. bath	120		14.9	1788	20	1	12	A13	Minimum circuit ampacity 18.6 A
K	Heat/vent/night-light/light front B.R. bath	120		14.9	1788	20	1	12	A11	Minimum circuit ampacity 18.6 A
L	Attic exhaust fan	120	1/4	5.8	696	15	1	14	A10	Run circuit to 4" square box. Locate near fan in attic. Unit has integral protection.
M	Electric furnace	240	1/3 Motor	Motor 3.5 Htr. 50.7 Total 54.2	13000	70	2	4	A1/A3	The overcurrent device shall not be less than 125% of the total load of the heaters and motor. 54.2 × 1.25 = 67.75 (Section 62-114(7)(a)(b); 62-114(8)).
N	Air conditioner	240	–0–	30	7200	40	2	8	A2/A4	Compressor rated load amperes 27.8 Compressor locked rotor amperes 135.0 Condenser fan full load amperes 2.2 Condenser locked rotor amperes 4.5 Branch-circuit O.C. device 40-A 40.0 Minimum circuit ampacity 37.5
O	Freezer	120	1/4	5.8	696	15	1	14	A 12	Install single receptacle outlet. Do not provide GFCI protection.
P	Microwave oven outlet	120	–0–	12	1440	15	1	14	B17	
Q	Central vac. power unit	125	2	12	1492	15	1	14	B24	Receptacle in garage.

Table S–1   Schedule of Special-Purpose Outlets.

33. **SPECIAL-PURPOSE OUTLETS:** The electrical contractor shall install, provide, and connect all wiring for all special-purpose outlets (see Table S–1). Upon completion of the job, all fixtures and appliances shall be operating properly. See plans and other sections of the specifications for information as to who is to furnish the fixtures and appliances.

34. **ALTERNATIVE BID LOW-VOLTAGE, REMOTE-CONTROL SYSTEM:** The electrical contractor shall submit an alternative bid on the following:
Furnish and install a complete extra-low-voltage, remote-control system to accomplish the same results as would be obtained with the conventional switching arrangement as indicated on the electrical plans.

Furnish and install one 12-position master selector switch in the master bedroom or as directed by the architect or owner. Outlets to be controlled by this switch are to be selected by the owner.

Furnish and install two motorized 25-circuit masters. These motor-operated controls shall be run from the front hall and master bedroom, or as indicated by the architect or owner. Connect the motor-operated master control so that each and every switch-controlled lighting outlet and switch-controlled receptacle outlet can be turned off or on from the two control stations.

All extra-low-voltage wiring is to conform to the *CEC*.

APPENDIX B

JACK PIN DESIGNATIONS AND COLOUR CODES

Standard 4-Pair Wiring Colour Codes

PAIR 1	T	White/Blue
	R	Blue/White
PAIR 2	T	White/Orange
	R	Orange/White
PAIR 3	T	White/Green
	R	Green/White
PAIR 4	T	White/Brown
	R	Brown/White

Note: For 6-wire jacks use pair 1, 2 and 3 colour codes.
For 4-wire jacks use pair 1 and 2 colour codes.

Pin #	Jack Type		
	8P8C/ 8P8C Keyed	6P6C	MMJ
1	Blue (L)	White (W)	Orange (O)
2	Orange (O)	Black (B)	Green (G)
3	Black (B)	Red (R)	Red (R)
4	Red (R)	Green (G)	Yellow (Y)
5	Green (G)	Yellow (Y)	Black (B)
6	Yellow (Y)	Blue (L)	Brown (N)
7	Brown (N)	—	—
8	White (W)	—	—

6P4C

6P6C

MMJ

* Some equipment standards may vary from the standards shown here.

Fig. A–1   Jack pin designations and colour codes for communication jacks and cables.   NEL

APPENDIX C

FORMULAS AND EQUATIONS

USEFUL FORMULAS

TO FIND	SINGLE PHASE	THREE PHASE	DIRECT CURRENT
AMPERES when kVA is known	$\dfrac{kVA \times 1000}{E}$	$\dfrac{kVA \times 1000}{E \times 1.73}$	not applicable
AMPERES when horsepower is known	$\dfrac{hp \times 746}{E \times \% \, eff. \times pf}$	$\dfrac{hp \times 746}{E \times 1.73 \times \% \, eff. \times pf}$	$\dfrac{hp \times 746}{E \times \% \, eff.}$
AMPERES when kilowatts are known	$\dfrac{kW \times 1000}{E \times pf}$	$\dfrac{kW \times 1000}{E \times 1.73 \times pf}$	$\dfrac{kW \times 1000}{E}$
KILOWATTS	$\dfrac{I \times E \times pf}{1000}$	$\dfrac{I \times E \times 1.73 \times pf}{1000}$	$\dfrac{I \times E}{1000}$
KILOVOLT-AMPERES	$\dfrac{I \times E}{1000}$	$\dfrac{I \times E \times 1.73}{1000}$	not applicable
HORSEPOWER	$\dfrac{I \times E \times \% \, eff. \times pf}{746}$	$\dfrac{I \times E \times 1.73 \times \% \, eff. \times pf}{746}$	$\dfrac{I \times E \times \% \, eff.}{746}$
WATTS	$E \times I \times pf$	$E \times I \times 1.73 \times pf$	$E \times I$

I = amperes   E = volts   kW = kilowatts   kVA = kilovolt-amperes

hp = horsepower   % eff. = percent efficiency   pf = power factor

EQUATIONS BASED ON OHM'S LAW:

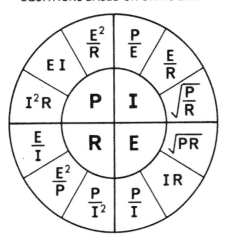

P = POWER, IN WATTS
I = CURRENT, IN AMPERES
R = RESISTANCE, IN OHMS
E = ELECTROMOTIVE FORCE, IN VOLTS

Fig. A–2   Useful formulas and equations based on Ohm's Law.

INDEX

A

AC. *See* Alternating-current (AC) circuits
AC90 cable, 125, 320, 322, 344
AC/DC general-use switch, 150
Accent lighting, 196, 220, 222
ACWU90 cable, 126
AFCIs. *See* Arc-fault circuit interrupters (AFCIs)
Air circulation, 292
Air cleaning, 292
Air conditioning, 243, 299
   exra-low-voltage circuits for, 315
   furnace and, 292
   room units for, 298–299, 300
   terminology for, 299–301
Air conditioning load, 26, 27, 118
   attic heat and, 283
   heating and, 301
Air King products, 243
Alternating-current (AC) circuits, 150
   induction heating and, 159–161
   plug-in relays and, 366
   rectifier to change to DC, 367
   switches for, 149
Aluminum conductors, 27, 114
   connectors with, 114–116, 116
   resistance of, 121
Aluminum plate, induction heating and, 161
Ambient temperatures, conductors and, 116–117
American National Standards Institute (ANSI), 12
American standard unit (ASU), 5–7
American wire gauge (AWG), 6, 7, 112
Ampacity
   colour coding and, 149
   defined, 113
   temperature and, 116, 117, 120–121
Ampacity derating, 101, 117
Ampacity rating, 113
   of branch circuits, 100, 113, 114
   consumer service, 27, 30, 42, 54
   correction factors for, 101, 104
   of multiple meters, 53–54
Amplifier distribution system, 391–392
Annunciator wire, 374
ANSI. *See* American National Standards Institute (ANSI)
Antennas, 392, 393–395
Antishort bushing, 129
Apartment dwelling
   interrupting ratings and, 168
   receptacles in, 167, 168
   service entrance calculations for, 29–30
Appliance loads, 102. *See also* Small appliance circuits
Appliances, 262–271
   cord-connected fixed, 271
   double-insulated, 158, 242–243, 271
   hand-held (*See* Hand-held tools/appliances)
   "hard-wired," 271, 385
   multioutlet assembly for, 334
   nongrounding-type receptacles and, 158–159
   open neutral conductor and, 326–327
   receptacles for cord-and-plug connected, 166

split-switched receptacles and, 204
   TVSS for electronic, 178–179
Aquariums, 243
Arc-fault circuit interrupters (AFCIs), 176–178
   in bathroom, 211
   in bedroom, 201–202
   insulation test and, 178
Architects, specifications by, 72. *See also*
   Specifications
Architect's scale, 6, 18
Architectural drafting symbols, 13–14
Architectural drawings, 18, 19
"Architectural practice," 12, 18
Arcing, 47, 176–178
   bonding and, 53
   series *vs.* parallel, 177
Armoured cable (BX), 116
   bonding and, 155, 158, 203
   clamps for, 133, 136
   colour of, 147–148
   connectors for, 133, 135, 136
   description of, 126
   flexible, 128, 136
   induction heating and, 160
   installation of, 127–129
   for service entrance, 126
   in signal systems, 374
   for suspended ceiling lighting fixtures, 189, 190, 320–322
   types of, 126
   to wire grounding-type receptacles, 155
ASU. *See* American standard unit (ASU)
Attics
   branch circuits in, 101, 281–287
   conductors in, 116, 117
   exhaust fans in, 279–280
   installing cable in, 131–132, 134
   temperature and, 116, 117
Authorization for connection, 5
Automatic heat sensor/detector, 287, 356
Autotransformer dimmer control, 225, 227, 228
AWG. *See* American wire gauge (AWG)

B

Ballasts
   dimmer, 228–229
   noise from fluorescent, 179
   thermal protection for, 191, 192
Baseboard heaters, 291, 292
   circuit requirements for, 294–295, 296, 297
   colour of conductors for, 298
   electric furnace and, 297
   kitchen fans and, 252
   outlets near, 91, 204
   relay for, 294
   typical, 294, 295
Basement
   cable installation in, 333
   calculating load for, 25–26, 29–30
   conduit installation in, 46, 418

non-metallic sheathed cable in unfinished, 127
   receptacles in, 108, 110, 167
Bathroom/washroom
   bonding for, 241–243
   branch circuits for, 237–246
   ceiling exhaust fans in, 279–280
   ceiling heater circuits in, 243–245
   defined, 240, 243
   GFCI circuit breakers for, 167
   GFCI receptacle outlets for, 166, 172, 237, 241
   hanging fixtures in, 241
   humidity control in, 280–281
   lighting branch circuit for, 211, 212, 237–246
Bathtubs. *See* Hot tubs; Hydromassage tub
Bedroom, lighting branch circuit for, 201–202, 203, 204, 210
Bell wiring, 113, 374
Blueprints, 12. *See also* Plans
Boathouses, GFCI receptacle outlets and, 166
Boiler, 306
Bonding, 27, 28, 47–49, 50
   bathroom circuit and, 241–243
   cable assemblies and, 124
   cable television, 393
   cables and, 42, 43, 155
   combination heater and, 243–244
   connections of, 43
   cooktop units and, 264
   defined, 48
   dryers and, 264, 279
   EMT for equipment, 134
   flexible metal conduit and, 138.140
   food waste disposals and, 270
   GFCIs and, 169, 171
   hydromassage tub and, 145–146
   jumpers and, 53
   kitchen and, 257
   methods of, 51–53
   multiple meters and, 54
   non-metallic sheathed cable and, 123, 124, 126, 257
   non-metallic water pipes and, 49
   outlets and, 94
   portable dishwasher and, 271
   ranges and, 264, 279
   receptacle replacement and, 173–176
   at receptacles, 155–158
   recessed lay-in fixtures and, 322
   to reduce fire hazard, 53
   satellite television and, 397
   spas/hot tubs and, 385–386
   submersible pumps and, 338, 339
   swimming pools and, 381–384
   underground wiring and, 417–418
   wall boxes and, 203
   washer head screws and, 158
   water heater and, 345
Bonding jumpers, 52, 53
Bonding lug, 53

Bonding screw, 42, 43, 155, 203
  multiple meters and, 54
  outlets and, 94
Box fill, 91
Boxes, 79. *See also* Outlets; Receptacles;
      Switch/device boxes; Wall boxes
  in bathroom, 243
  bonding and, 155
  cable clamps for, 133, 136
  ceiling light, 116
  chime transformer and, 374
  common, 84
  for conduit wiring, 83, 87
  covers on, 77, 84, 90, 392
  for dimmer controls, 225
  extension ring/covers on, 77, 90
  for extra-low-voltage relay, 365–366
  flush-mounted, 81
  grounded, 155
  grounding wires attached to, 158
  for incandescent lighting, 191
  junction (*See* Junction boxes)
  in kitchen, 257
  knockout/pryouts on, 133, 136
  in master bedroom, 210–211
  metallic, 73, 158–159
  multioutlet assembly mounted on, 335
  non-metallic, 73, 78, 82, 391, 392
  number of conductors in, 86–91, 203
  octagonal, 73, 77, 78, 84, 158
  for outdoor lighting, 413–414
  paddle fans and, 211, 213, 214
  plaster rings and, 83, 90
  position of, 85
  raised covers on, 84, 90, 392
  rodents and, 88
  square, 83, 84
  for suspended ceiling lighting fixtures, 189–191
  for television, 391, 392
  for track lighting, 224
  trade sizes of, 6
  types of, 79, 84
  utility, 84
  with vapour boot, 75, 77
  various wall/ceiling materials and, 85
Branch circuits, 195–196. *See also* Lighting
      branch circuits
  ampacity rating of, 100, 113, 114
  for attics, 281–287
  for bathroom heater, 243–245
  cable runs and, 196–197
  for central vacuum system, 420
  code requirements for (table), 107
  for convenience receptacles in kitchen, 253–256
  for counter-top cooking units, 266
  dedicated, 167
  defined, 100
  electric furnace and, 298
  GFCIs and, 167, 168, 169
  heating by separate, 292
  for hydromassage tub, 245–246
  for laundry, 276–280
  layout for, 107–110
  load estimation for, 198–199, 223
  neutral conductor and multiwire, 147
  number/location of, 99–110
  for oil/gas heating, 313
  for ovens, 266
  overcurrent devices for, 42, 43, 55–58, 100, 345
  for ranges, 266
  receptacle, 106–110
  separate, for heating, 292
  for separate built-in cooking units, 267
  short-circuit protection and, 287
  for smoke alarms, 357
  swimming pools and, 385
  track lighting, 222–223, 224, 225
  TVSS on, 179
  for utility room, 335–346

voltage and, 103–104
  voltage drop and, 119–120
  for water heaters, 345
  wiring diagram of, 108, 199–202
  for workshop, 331–335
Breaker main/breaker branches, 61–63
British thermal unit (Btu), 265
Building codes, 1
  attics and, 283
  branch circuits and, 196
  cable wiring and, 129, 132
  lighting fixtures and, 185, 320
  non-metallic piping and, 49
  remote-control switches and, 364
Building views, 11–12
Building wire
  cross-sectional area for, 119
  temperature and, 117
Bushings, 51–52, 53, 134
BX. *See* Armoured cable (BX)

C

Cable clamps, 133, 136
Cable layout, 199–202
  for bathroom, 238
  for entry-porch circuit, 230
  for front bedroom, 202
  for garage, 412
  for kitchen, 250
  for laundry/washroom/attic, 277
  for living room, 221
  for master bedroom, 210
  for recreation room, 321
  for remote-control extra-low-voltage system, 368
  for study/bedroom, 211, 212
  for telephone, 398
  for television, 396
  typical, 201
  for workshop, 332
Cable runs, 196–197
  estimating length of, 201
  longer than 20 m, 323
  maximum length of, 121–124
  short, 320
  in signal system, 374
Cable television, 392, 393
Cables. *See also* Armoured cable (BX); Non-metallic
      sheathed cable (Romex)
  in attics, 131–132, 134, 282–283, 284, 285
  for baseboard heaters, 298
  in basement, 333
  bonding and, 155
  bundled, 126
  chimney and, 333
  coaxial, 392, 393
  colour coding of, 147–148, 149, 152–155, 367
  conductor count with, 92
  connectors for, 133, 135, 136
  cross-sectional area for, 119
  data, 401–402
  diagrams of, 199–202
  electrical schedules for, 20
  EMT *vs.*, 91
  extra-low-voltage, 367, 368
  "fished," 129
  heating, 292
  heating ducts and, 125, 132–133, 135, 333
  methods of protecting, 129–131, 134
  multiconductor, 374
  outlet boxes and, 78, 82
  pryouts for, 133, 136
  for remote-control systems, 367–368
  for service entrance, 126
  shielded coaxial, 392
  for signal systems, 374
  "station wire," 397
  for submersible pumps, 338
  for telephone, 397–401
  for television, 391–392, 395–397

temperature and, 116, 117
  through concrete walls, 418
  through wood/metal framing, 129–131
  in underground service, 37, 40
  vs. EMT for recreation room, 327
Calculations. *See also* Loads
  cable run, 121–123
  electric furnace, 298
  fault current, 53, 60
  floor area, 24–30
  load (*See* Loads)
  service entrance (*See* Service entrance calculations)
  voltage drop, 118–119
Canadian Building Code, 99, 353
  Parts I and II, 186
Canadian Electrical Code (CEC), 1–3
  appendices in, 2–3
  code terminology in, 3, 35
  diagrams in, 2
  local standards *vs.,* 5, 23
  measurements in, 6
  sections of, 2
  standards steering committee for, 2
  tables in (*See* Tables, *CEC*)
Canadian Fire Alarm Association (CFAA), 354
Canadian Standards Association (CSA), 2, 4
Carbon monoxide detectors, 358
Cartridge fuses, 56, 57, 59
CATV. *See* Community antenna television (CATV)
CEC. See Canadian Electrical Code (CEC)
Ceiling fixtures
  in bathroom, 241
  exhaust fan, 286 (*See also* Exhaust/ventilation fan)
  in front entry, 230, 232
  in garage, 411–413
  in hallways, 241
  for kitchen, 251
  paddle fans with, 211, 213 (*See also* Paddle fans)
  suspended (*See* Suspended ceiling lighting fixtures)
  in workshop, 331, 332
Ceiling heater circuits, 243–245
Ceiling outlet, 85, 116
  for fan, 73
  insulation and, 190, 191
  non-metallic sheathed cable and, 128
  single-pole switch and, 152
  symbols for, 74
Central vacuum system, 332, 420
CFAA. *See* Canadian Fire Alarm Association (CFAA)
Chime transformer, 331, 370–373
Chimes, 370–374
  additional to door (extension), 373, 374
  bell wire and cable for, 373–374
  clock-, 374
Chimneys, 126, 333
Circuit breakers, 57–63
  AFCI, 177, 178, 202
  for air conditioners, 300
  branch circuits and, 100, 102
  colour coding and, 149
  GFCI (*See* GFCI circuit breakers)
  GFCI combined with AFCI in, 177
  for hydromassage bathtubs, 386–387
  interrupting ratings for, 58–63, 60, 168
  jet pump and, 337
  in kitchen, 256–257
  for main disconnect, 42
  service entrance and, 27, 42, 44
  for spas/hot tubs, 386
  swimming pools and, 381
  thermal-magnetic, 287
Circuit directory, 37, 40
Circuit loading guidelines, 197–198
Circuit schedule, 37, 40, 41, 44
Circuits
  branch (*See* Branch circuits)
  Class 1 or 2, 299, 313–316
  closed or closed loop, 360
  four-wire, 148

Circuits, *continued*
  "hot," 165
  minimum number of, 28, 99–110, 198–199
  series, 156
  three-wire, 148, 253–256, 256–257
  two-wire, 147–148
Circular mil area (CMA), 7, 113, 121
Clamps, 133, 136
Class 1 or 2 circuits
  electric heating/cooling and, 299
  gas/oil heating and, 313–316
Class 2 transformer, 366, 370, 371
Class A GFCI, 166, 167, 172, 242
  for hydromassage tub, 245
  receptacle replacement and, 176
Class J fuses, 57, 59
Class P ballasts, 191, 192
Clips, 189
Clocks
  chimes and, 374
  receptacles for, 167, 252, 253
  refrigerator circuit and, 253
  symbol for outlet for, 76
Closed circuits, 360
Closed loop circuits, 360
Closets, 202, 204, 205–208, 233
Clothes dryers. *See* Dryer load; Dryers
CMA. *See* Circular mil area (CMA)
Codes, 1–3. *See also* Canadian Electrical Code
  (CEC); Standards
Cold resistance, 149–150
Colour coding, 147–149, 152–155
  baseboard heaters and, 298
  of conductors, 146–149, 152–155, 298, 367, 374
  of extra-low-voltage cables, 367
  of telephone cables, 398, 399
  of termination jacks, 403
  on wiring diagram, 199, 201
Combination bushings, 53
Combination lighting and receptacle circuit, 106
Combination panel, 34
Community antenna television (CATV), 393
Compression connectors, 115
Computer, 401–405. *See also* Data systems
  home automation and, 405–406
  lighting branch circuit for, 211
Condominium
  interrupting ratings and, 168
  receptacles in, 167, 168
Conductor insulation, 116, 147, 160, 177
  arcing due to failure of, 177
  induction heating and, 160
  shock due to breakdown of, 165
  temperature and, 116, 187
  test for, 178
  thermoplastic, 116
Conductors
  aluminum *vs.* copper, 27, 114, 116, 121
  ampacity and, 27, 101, 120–121
  apartment dwelling and, 30
  applications of different sizes of, 113
  approved combinations of equipment
    and (table), 116
  arcing and, 177
  for baseboard heaters, 295, 298
  bonding (*See* Bonding)
  branch circuits and, 100–103, 187, 190
  for cable television, 393
  colours of, 146–149, 152–155, 298, 367
  in conduits, 116
  connecting of, 114–116
  consumer service, 37
  in data systems, 403, 404
  dryers and, 278–279
  ELC, 316
  for electric furnaces, 296, 297
  EMT/rigid metal conduit and, 134
  for gas/oil heating systems, 315–316
  grounded circuit, 146, 147

grounding electrode, 29, 34
"hot" (*See* "Hot" conductor)
identifying, 146–149, 298
induction heating and, 159–161
mast service and, 35–36
neutral (*See* Neutral conductors)
number of, in box, 86–91, 203
overcurrent protection for, 37
for pump motor, 336
rating/setting of overcurrent protection of, 104
for remote-control extra-low-voltage systems, 367
resistance of, 121
sizing relationship and, 121
swimming pools and, 381–384, 385
for telephone, 398–401
temperature and, 116–117, 118, 120–121, 263
toggle switches and, 150
for underground wiring, 416, 417–418
for water heater, 345
wire size for, 112–113
on wiring diagrams, 199, 201
Conduit, 28, 36–37, 38
  bonding and, 155, 156
  boxes for, 83, 86–91, 87, 92
  colour coding of, 148, 154
  conductors for, 116
  connectors for, 136
  empty, 334
  EMT/rigid metal, 134
  flexible (*See* Flexible conduit)
  junction boxes with, 320 (*See also* Junction boxes)
  knockouts for, 133, 136
  Liquidtight flexible, 140–142
  for outdoor lighting, 414–419
  rigid PVC (*See* Rigid PVC conduit)
  in service entrance, 134
  smaller size, 120–121
  "stub-up," 414
  supporting threaded, 415
  for telephone, 401
  temperature and, 138
  through concrete walls, 418
  trade sizes of rigid steel, 6
  in workshop, 334
Conduit fill, 44–45
Connection permit, 5
Connectors, 114–116
  for cable, 133, 135, 136
  for track lighting, 224
Construction, shock due to defective, 165
Construction drawings, 18–20. *See also* Drawings
  measurements on, 5–6, 18
  orthographic projection for, 11
  specifications and, 20
  symbols on, 12, 13–17
  as two-dimensional, 11, 19
  typical set of, 19–20
Construction Health & Safety Manual, 3
Construction Safety Association of Ontario, 3
Construction sites, GFCIs for, 176, 177
Consumer Product Safety Commission (CPSC), 342
Consumer service
  ampacity rating for, 27, 30, 42, 54
  CEC rules for, 37
  defined, 33
  more than one service box for, 40
  PVC for, 134
Consumer service conductors, 37, 46
Continuous load, 101–102
Control circuits, for oil/gas heating, 313–316
Convenience receptacles, 253–256
  multi-outlet assemblies and, 334
  for television, 392
Cooking units
  counter-mounted, 263–264
  separate built-in, 262, 267
  surface-heating elements for, 264
Cool white (CW) lamps, 240
Cool white deluxe (CWX) lamps, 240, 250, 320

Cooling system, heat pump for, 298
Copper conductors, 27, 112
  allowable ampacities for, 117
  aluminum *vs.*, 27, 114
  connectors with, 114–116
  grounding and, 48, 49, 50
  in non-metallic sheathed cable, 123, 124
  resistance of, 121
Cord connections (of fixed appliances), 271
Correction factors, 101, 104, 126, 282
Cost estimates. *See also* Economics
  lighting fixtures included in, 72–73
  outlets in, 71
  specifications and, 20
Counter-mounted cooking unit circuit, 263–264
Countertop receptacles, 253–256
Coxial cable, 392, 393
CPSC. *See* Consumer Product Safety Commission
  (CPSC)
Crimp-type connectors, 115
CSA. *See* Canadian Standards Association (CSA)
Current permit, 5
CW. *See* Cool white (CW) lamps
CWX. *See* Cool white deluxe (CWX) lamps

D

Data systems, 401–405
  cable for, 401–402
  lighting branch circuit and, 211
  termination jacks in, 403, 404
  wireless data networks and, 403, 405
DC. *See* Direct-current (DC) circuits
"Decora" switches, 154
Decorative appliances, 312
Decorative lighting, 230, 231, 414
Decorative pools, 379, 387
Dedicated branch circuits, 167
Demand watts
  in electric furnace calculations, 298
  in service entrance calculations, 23, 24,
    26, 27
Demarcation point, 397
Derating, ampacity, 101
"Designer" switches, 154
Details (in drawings), 19
Device boxes. *See* Switch/device boxes
Diagrams
  CEC, 2
  one-line, 20
  riser, 20
Dimensions (on drawings), 12, 18
Dimmer switches/controls, 86, 90, 223, 225–229
  autotransformer, 225, 227
  home automation and, 405
  single-pole, 223, 225
  solid-state, 224–225, 226
  three-way, 225, 249
Dining room, lighting fixtures for, 251
Direct-current (DC) circuits
  rectifier to change AC to, 367
  switches for, 149
"Dirty" power, 178
Discharge lighting, 191
Disconnect switches, 154, 155
  for electric air conditioner, 301
  for oil/gas heating systems, 309
  for water heater, 345
Disconnecting means
  for air conditioner, 301
  for food waste disposal, 268–269
  main, 29, 42
  service-entrance (*See* Service-entrance
    disconnecting means)
  for spas/hot tubs, 386
  for water heater, 345
Dishwasher, 243, 270–271, 342
Distribution panel, chime transformer and, 374
Door chimes, 370–374
Doorjamb switch, 233

Double-insulated tools/appliances, 158, 242–243, 271
Double-pole GFCI circuit breaker, 167
Double-pole pressure switch, 337
Double-pole switches, 154–155, 199
   symbol for, 80
Drawings, 9–20. *See also* Construction drawings
   architectural, 18
   dimensions on, 12, 18
   lines on, 9–10
   notations on, 12, 13–17, 73, 83
   pictorial, 10–11
   shop, 20
   specifications and, 20
   symbols, 12, 13–17 (*See also* Symbols)
   technical, 9–10
   working, 18–20
Dry-niche fixtures, 381, 384
Dryer load, 26, 29, 30, 118. *See also* Small
     appliance circuits
Dryers, 243, 264, 276–279
   symbol for outlet for, 75
Dual-Element Fuse, 55–57, 59
Dual receiver satellite system, 396
Duplex receptacle outlet, 79, 103
   for central vacuum system, 332
   GFCI for, 170
   in hallways, 241
   for kitchen countertop, 254
   multi-outlet assemblies and, 334
   single *vs.*, 253
   split-wired, 256–257
   symbol for, 75, 79

E

Economics
   CEC and, 2
   GFCIs and, 171, 172
   lighting branch circuits and, 204
   and mounting of chime transformer, 374
ELC conductors, 316
Electric furnace, 292. *See also* Furnace
   air conditioner/heat pump and, 299
   baseboard heaters and, 297–298
   circuit requirements for, 296–298
   code requirements for, 293
   "separate circuit" rule for, 292
Electric heating, 291–298, 299. *See also* Heaters
   baseboard heaters for (*See* Baseboard heaters)
   control of, 292–294
   furnace for (*See* Electric furnace; Furnace)
   "separate circuit" rule for, 292, 298
   of swimming pool decks, 384–385
   types of, 292
Electric service, defined, 33
Electric utility. *See* Utility companies
Electrical contractors
   codes/standards and, 1
   lighting fixtures and, 72–73
   specifications and, 20
   underground service and, 37
Electrical drawings, 19–20. *See also* Construction
     drawings
Electrical energy, cost of, 63–65
Electrical equipment. *See also* Equipment
   approved, 3–5
   in bathroom, 237
   CSA label on, 3–4
   in kitchen, 257
   for spas/hot tubs, 385
   swimming pools and, 381
   symbols for, 12
   ULC label on, 3–5
Electrical hazards, 47, 165–166. *See also* Arcing;
     Fire; Shock
   CEC and, 2
   Class 1 or 2 circuit wiring and, 313–315
   codes/standards and, 1
   fault current calculations to prevent, 60

from lightning, 398
   overcurrent devices and, 59
   from swimming pools, 379–381
Electrical inspection, 5
   interrupting ratings and, 168
   lighting fixtures and, 320
   neutral conductor size and, 117
   outlets and, 94
Electrical layout, 107–110
Electrical metallic tubing (EMT), 129, 130,
     134, 136
   box fill and, 91
   empty raceways of, 334
   grounding of, 47
   insulation and, 53
   outlets and, 78
   support of, 136
   trade sizes of, 6
Electrical schedules, 20, 41, 44
Electrical specifications. *See* Specifications
Electrical symbols. *See* Symbols
Electrical Wiring: Commercial, 53, 59, 60, 62, 118
Electricians
   branch circuits planned by, 99
   codes/standards and, 1
   colour coding and, 148–149
   GFCIs and, 171
   interrupting ratings and, 168
   lighting fixtures and, 185, 197
   underground service and, 37
   wiring diagrams by, 199
   working drawings and, 19
Electrocution, 380
Electromagnetic interference (EMI), 179
Electronic air cleaning, 292
Electronic appliances/equipment
   noise from, 179
   TVSS for, 178–179
Electronic solid-state dimmer controls, 223, 225
Electrostatic air cleaners, 292
Elevations (in drawings), 11, 13, 14, 19
EMI. *See* Electromagnetic interference (EMI)
Empty conduits, 334
EMT. *See* Electrical metallic tubing (EMT)
Energy conservation, 292
Engineering scale, 18
Entry, front, branch circuits for, 229–233
Equipment. *See also* Electrical equipment
   arcing due to failure of, 177
   electronic (*See* Electronic appliances/equipment)
   mechanical, 20
   service entrance (*See* Service entrance equipment)
   shock due to misuse of, 165
   symbols for, 12
Equipment-grounding conductor, 156, 381
Estimating loads. *See* Loads
Ethernet, 401, 402
Exhaust/ventilation fan, 243, 244, 252
   in attic, 279–280, 281–287
   ceiling, 286, 331, 332
   control of, 284–287
   in gable of house, 283, 285
   in laundry/washroom, 279–280
   motor of, 287
   operation of, 284
   overload protection for, 287
   range hood, 252
   in recreation room, 327
Exit light, symbol for, 76
Expansion joints, 138, 140
Extension chimes, 373, 374
Extension cords, 197
Extra-low-voltage multiconductor cables,
     358, 360
Extra-low-voltage power circuits
   for chimes, 370–374
   electric heating/cooling and, 299
   "hot," 374
   oil/gas heating and, 313–316

remote-control (*See* Remote-control low-energy
     circuits)
   for swimming pools, 384
Extra-low-voltage switches, 364–365, 368

F

Faceplates, 93–94, 392
Facial saunas, 243
Fans
   exhaust (*See* Exhaust/ventilation fan)
   for kitchen, 252
   paddle (*See* Paddle fans)
   symbol for, 76
Fault currents, 52–53, 60
Feeders, 100
   for submersible pumps, 338
   track lighting and, 223
   voltage drop and, 119–120
Field-assembled grounding electrode, 52
Field-installed series-rated system, 62–63
Field wiring, for oil/gas heating systems, 306, 315
Fire. *See also* Arcing
   baseboard heaters and, 296
   bonding and, 53
   caused by extension cords, 197
   CEC and, 2
   Class 1 or 2 circuit wiring and, 313, 314
   closet lighting and, 204
   codes/standards and, 1
   fault current calculations to prevent, 60
   heating ducts and, 132
   lighting fixtures and, 186
   service entrance conductors and, 37
   smoke detectors, 353
Fire detectors, 360
First floor living area, 24–25, 105
Fixture drops, 320, 322
Fixture studs, 86, 89
Fixture tails, 190
Fixture whips, 190
Fixtures. *See* Lighting fixtures
FLA. *See* Full load amperes (FLA)
Flexible conduit, 128, 136, 138, 140–142. *See also*
     Non-metallic flexible conduit
   metallic, 138, 140–142
   non-metallic, 140, 142
   support of, 141
   for suspended ceiling lighting fixtures, 189–190,
     320–322, 322
Flexible connectors, 138
Floor area, calculating
   for apartment dwelling, 29–30
   for single-family dwelling, 24–26
Floor-mounted electric heating units, 292
Floor plans, 11, 19
   electrical outlets on, 71, 83
   GFCIs on, 168
   symbols on, 12, 17
Fluorescent lighting, 185, 187, 191
   circuit loading and, 198
   closets and, 204
   colour and, 240, 250
   dimmers for, 227, 228–229
   in kitchen, 250
   noise from, 179
   plug-in relays and, 366
   in recreation room, 320–322
   valance, 211, 212
   for workbench, 331
Flush-mounted chime, 372
Flush-mounted outlet box, 81
Food waste disposal, 267–270
   overcurrent protection and, 268
   receptacles for, 167, 256
   turning on and off, 269–270
Fountains, 387
Four-way switch, 152, 154, 159, 161
   for living room, 220, 221
   symbol for, 80

Four-wire circuit, 148
Fourth Canadian Edition, 62
Freestanding range, 263–264, 266–267
Freezer. *See* Refrigerator/freezers
Front entry/porch, branch circuits for, 229–233
Full load amperes (FLA), 281
Furnace. *See also* Electric furnace
    gas, 296, 298
    oil/gas heating systems and, 306–312
Furnace load, 26, 27, 30, 118, 296–298
Furring strips, 327
Fuse rejecters, 55
Fused main/breaker branches, 60–61
Fused main/fused branches, 58, 60
Fuseholders, 55
Fuses
    for air conditioners, 300
    branch circuits and, 100
    cartridge, 56, 57, 59
    characteristics of low-voltage, 56
    Class J, 57, 59
    Dual-Element, 55–57, 59
    interrupting ratings for, 58–63, 60, 168
    jet pump and, 336, 337
    overload element of, 56
    plug, 55, 56, 59
    service entrance and, 27, 44, 55–57
    short-circuit element of, 56–57
    swimming pools and, 381
    time-delay (*See* Time-delay fuses)
    types of, 55

G

Gang-type switch (device) box, 78, 82–83, 90, 126, 128
Garage
    attached *vs.* detached, 413
    central vacuum system in, 420
    lighting branch circuit for, 411–413
    receptacles in, 109, 167, 413
    underground wiring for, 414–419
Garage door opener, 419–420
    home automation and, 405
    receptacles for, 413, 419, 420
Gas fireplaces, 312
Gas furnace, 296, 298
Gas heating system, 292, 306–312
Gas pipes, grounding and, 47–49
GCFIs. *See* Ground-fault circuit interrupters (GFCIs)
General lighting, 196, 198
General-use switch, 149–150, 281
GFCI circuit breakers, 167, 169, 172
    AFCI combined in, 177
    for bathroom, 167
    for hydromassage tubs, 386
    for spas/hots, 386
GFCI receptacle outlets, 90, 158, 159
    protecting downstream receptacles, 173
    testing/recording of, 172, 173
    warning signs by, 172, 173
Glassbreak detectors, 359
Ground bushing, 51–52, 53
Ground clamps, 53, 54
Ground-fault circuit interrupters (GFCIs), 166–177.
    See also GFCI receptacle outlets
    for bathroom, 237, 240
    for branch circuit outside receptacles, 202–203
    code requirements for, 166–168
    for construction sites, 176, 177
    current-withstand rating for, 168
    feedthrough, 168, 171–172, 176
    for hydromassage bathtubs, 386–387
    indentifying, 172, 173
    internal components and connections of, 170
    for kitchen, 256
    nusance tripping of, 168
    operation of, 169
    for outdoor wiring, 413
    as part of receptacle outlet, 169
    precautions for, 168

reset ("R") button for, 172, 174
    in residence circuits, 168–171
    for spas/hot tubs, 385, 386
    swimming pools and, 381, 384, 385
    test ("T") button for, 172, 174
    time/current curve and, 166
    warning signs for, 172, 173
    for water pumps, 387
    in wet bar, 327
Ground plate, 49, 52
Ground rod, 49, 52
Grounded circuit conductor, 146, 147, 257
Grounded neutral conductor, 47–49, 51
Grounded outlet boxes, 155
Grounding
    of appliances, 242–243
    and bonding at receptacles, 155–158
    of cable television, 393
    defined, 48
    of dishwasher, 271
    of food waste disposal, 270
    low-impedance path for, 47, 48
    methods of, 49–51
    of multiple services, 54
    noise from improper, 179
    purpose and requirements of, 47
    of satellite television, 395, 397
    of spas/hot tubs, 385–386
    of swimming pools, 381–384
    of telephone, 401
    of water pump, 338
Grounding conductor
    connecting method for, 157
    defined, 48
Grounding electrode conductors, 29, 34, 47–49,
        50, 51
    common types of, 52
    "in-situ," 49, 52
    multiple meters and, 54
Grounding lugs, 53
Grounding terminals, 155, 157
Grounding-type receptacles, 157
    and bonding wall boxes, 203
    GFCIs and, 168, 169, 171
    parts of, 158
    receptacle replacement and, 176
Guard strips, 131, 132, 282, 283
Guy wire, 37

H

Hallways
    lighting fixtures in, 232, 241
    receptacles in, 241
Hand-held tools/appliances, 243
    double-insulated, 271
    nongrounding-type receptacles and, 158–159
"Hard-wired" appliances, 271, 385
Hazards. *See* Electrical hazards
Heat detectors/sensors, 287, 356
Heat pumps, 292, 298, 300
    "separate circuit" rule for, 292, 298
    terminology for, 299–301
Heat sensor, 287
Heaters, 118. *See also* Heating systems
    baseboard (*See* Baseboard heaters)
    bathroom ceiling, 243–244
    combination, 243, 244
Heatex, 298
Heating, induction, 159–161
Heating cable, 292, 385
Heating ducts, cable and, 125, 132–133, 135, 333
Heating load, 26, 29, 30, 291
    air conditioning and, 301
Heating systems. *See also* Electric heating
    control of, 292–294, 313–316
    gas, 292, 306–312
    home automation and, 405
    with hot water, 306
    millivolt, 312–313

oil, 292, 306–312
    self-generating, 312–313
    "separate circuit" rule for, 292, 298
    supply circuit wiring for, 313
    with warm air, 306
Heating, ventilating, and air conditioning (HVAC), on
        floor plans, 19
Home automation, 405–406
Home theatre room, 406
"Hot" circuit, 165
"Hot" conductor, 117, 118, 147
    appliances and, 243
    changing colours and, 148, 149
    GFCIs and, 168, 170
    in three-wire cables, 327
"Hot" extra-low-voltage circuit, 374
Hot resistance, 150
"Hot spot," 117
Hot tubs, 245–246, 379, 385–386
    hazard associated with, 381
    leakage current collectors for, 387
    spas *vs.,* 385
Hot water boilers, 306–312
Hub, 45
Human needs, 4
Humidifiers, 292
Humidistat, 280–281
Humidity control, 280–281, 285, 292
HVAC. *See* Heating, ventilating, and air conditioning
        (HVAC)
Hydromassage tub, 237, 245–246, 379, 386–387

I

IBEW, 354
IEEE. *See* Institute of Electrical and Electronics
        Engineers (IEEE)
Impulse transients, 178
"Inboard/outboard" connection, 229
Incandescent lighting fixtures, 185, 187
    in closet, 204, 208
    colour and, 240, 250
    dimmer controls for, 227–228
    inrush currents and, 226–228
    in kitchen, 250
    outlet boxes for, 191
    symbol for, 74
Induction heating, 159–161
Inductive loads, 149, 150, 197
Infinite-position temperature controls, 264, 265
Infinite-speed switch, 287
Input percentage, 264
Inrush current, 149–150, 226–228
Institute of Electrical and Electronics Engineers
        (IEEE), symbols and, 12
Insulation
    on ceiling, 190, 191
    for chimes, 374
    on conductors (*See* Conductor insulation)
    electric heating and, 292
    on extra-low-voltage system, 370, 374
    kitchen fans and, 252
    on remote-control extra-low-voltage
        system, 370
Insulation test, 178
International standards, 4
International System of Units (SI), 5–7
Interrupting ratings, 168
Inverse time characteristic, 47
Ionization type smoke detector, 76, 354
Isolated ground receptacle, 179, 180
Isolation transformers, GFCIs and, 168, 169

J

Jet pump, 335–337
Joint compound, 27
Joules, 178
Jumpers
    bonding and, 52, 53, 155–157, 158
    grounding and, 47–48

Junction boxes, 75, 77–78, 81
  for appliances, 262–263
  fixture drops and, 320, 322
  for hydromassage tub, 145–146
  paddle fans and, 213
  for suspended ceiling lighting fixtures, 188, 190, 191, 320–322, 322
  swimming pools and, 381

K

Kilovolt-ampere (kVA) value, 59–60
Kilowatt-hour (kWh) measurement, 63–65
Kilowatt (kW) measurement, 63–64
Kilowatt (kW) rating, 24, 197
Kitchen
  bonding for, 257
  convenience receptacles in, 253–256
  fan outlets in, 252
  layout for receptacles in, 109
  lighting branch circuits in, 249–253
  lighting fixtures for, 251
  multiwire circuits in, 256–257
  small appliance circuits in, 253–257
  special-purpose outlets in, 255
  split-wired receptacles in, 105, 256–257
  track lighting in, 249, 251
  undercabinet lighting in, 249–250, 252
Knockouts, 136
kVA. See Kilovolt-ampere (kVA) value
kW. See Kilowatt (kW) rating
kWh. See Kilowatt-hour (kWh) measurement

L

Lamps. See also Lighting fixtures
  colour and, 240, 250
  types of kitchen, 250
  Ultralume 3000, 250
Laundry, 276–280
  ceiling exhaust fans in, 279–280
  GFCIs and, 172
  humidity control in, 280–281
  receptacles in, 109, 166, 279
  special-purpose outlets in, 255
Lawn mower/trimmer, 166, 243
Lawn sprinklers, 405
Lay-in fluorescent fixtures, 320–322
Lead-in wire, 392
Leakage current collectors, 387
LED. See Light-emitting diode (LED)
Leviton Decora Home Control, 405–406
Light-emitting diode (LED), 179
Lighting branch circuits, 104–106, 107–110. See also Lighting fixtures
  for attics, 281–287
  for bathroom, 237–246
  dimmer controls for, 223, 225–229
  fans and, 252
  for front bedroom/study, 201–202, 203
  for front entry/porch, 229–233
  for garage, 411–413
  for hall, 232
  for kitchen, 249–253, 256–257
  for laundry, 279–280
  for living room, 202, 220–229
  for master bedroom, 204, 210
  number of, 198–199
  outdoor, 202–203
  overview of, 220–221
  for paddle fans, 211, 213–214
  for recessed fixtures, 187, 190, 191
  for recreation room, 320–327
  for study/main bathroom, 211, 212
  track, 222–223
  for valances, 211, 212
  for washroom, 279–280
  for workshop, 331–332
Lighting fixtures, 185–193. See also Lighting branch circuits
  for bedroom, typical ones, 209

ceiling (See Ceiling fixtures)
  for closets, 204, 205–208
  conductor wire size for, 113
  for dining room, 251
  discharge, 191
  electrical schedules for, 20
  fluorescent vs. incandescent, 204 (See also Fluorescent lighting; Incandescent lighting)
  for garage, 411–413
  for halls, 232, 241
  home automation and, 405
  for kitchen, 251, 257
  labels on, 186, 191, 193
  load for single-family dwelling, 25
  location of, 99
  magnetic low-voltage, 227
  neon, 191, 192
  outdoor, 229–233, 418–419
  outlets and, 72–75, 79
  prewired, 191
  receptacle layout and, 197
  recessed (See Recessed lighting fixtures)
  for recreation room, 320–327
  for spas/hot tubs, 385
  surface-mounted, 73, 79, 185
  suspended ceiling, 185, 320–322
  for swimming pools, 380, 381, 384
  symbols for, 12, 79
  track, 106, 222–223, 224, 225
  type IC, 190
  types of, 73, 79, 196
  undercabinet, 249–250, 252
  voltage limitations on, 191, 193
  wall brackets for, 232
  in wet locations, 230, 233
  for workshop, 331–332
Lighting outlets, 71
  layout for, 107–110
  location of, 94
  number/location of, 99–110
  receptacle combined with, 106
  symbols for, 71–76, 79
  in typical dwelling, 101
Lightning surge, 398
Line conductors, meters and, 46
Line-to-ground fault, 168
Line-to-ground surge, 178
Line-to-line load, 118
Line-to-line surge, 178
Line-to-neutral load, 118, 168
Line-to-neutral surge, 178
Lines, in drawings, 9–10
Liquidtight flexible conduit, 140–142
Living room, 24–25
  branch circuits for, 220–229
  receptacles in, 202, 222
Load centres
  chime transformer and, 374
  service entrance equipment and, 42–44, 60–63
  typical, 43
Loads
  actual, 105
  for apartment dwelling, 29–30
  balance of, 199
  for bathroom, 243
  for bedroom/study, 203, 211
  branch circuits and, 100–103, 197–199, 204, 223
  continuous, 101–102
  dividing, 199
  for front entry/porch, 231
  for garage, 413
  guidelines for, 197–198
  heating/air conditioning, 301 (See also Air conditioning load; Heating load)
  inductive, 149, 150, 197
  for kitchen, 250
  for laundry, 277
  line-to-line, 118
  line-to-neutral, 118

for living room, 222
  for master bedroom, 204
  method for estimating, 197–199
  motor, 150
  for multioutlet assemblies, 334
  paddle fans and, 214
  for recreation room, 322, 325–327
  resistive, 149, 150, 197
  for service entrance (See Service entrance calculations)
  short-circuit current and, 59–60
  for single-family dwelling, 24–26
  for track lighting, 223
  tungsten filament lamp, 149, 150
  for water heater, 345
  for workshop, 331, 332
Locator switch, 156
Locknuts, 53
Loomex, 123
Low-energy power circuits, gas/oil heating and, 313–316
Low-voltage fuses, 56
Low-voltage lighting fixtures, magnetic, 227
Low-voltage wiring. See also Extra-low-voltage power circuits
  for central vacuum system, 420
  for garage door opener, 420
Luminaires. See Lighting fixtures
Lutron dimmers, 227–228

M

Maestro dimmers, 228
Magnetic contacts, 359
Magnetic low-voltage lighting fixtures, 227
Main panel, 42
Main service bonding conductor, 28
Main service conduit size, 28
Main service disconnect, 29
  circuit breaker for, 42
  location of, 37, 40, 42
  meters and, 46
Main service fuse/circuit breaker, 27
Main service ground, 27–28
Main service panel ampacity, 28–29
Main service switch, 27, 28–29
Maslow, Abraham, 3
Mast-type service entrance, 35–37
  clearance for, 35–36
  service mast as support in, 36–37
Master amplifier distribution system, 391–392
Master selector switch, 364–365, 368
MCM, 113
Measurements, 5–7
  conversion table of, 6
  as dimensions on drawings, 12, 18
Mechanical equipment, electrical schedules for, 20
Mechanical floor plan, 19
Megger, 178
Metal conduit
  bonding and, 155
  swimming pools and, 381
  underground, 417
Metal oxide varistors (MOVs), 179
Metal wall plates, 91–92
Metal water pipe, grounding and, 34, 49, 50–51
Metallic outlet boxes, 73, 159–161
Metallic raceways
  induction heating and, 159–161
  multiple meters and, 54
  recessed lighting fixtures and, 187
  underground wiring and, 417
Meter base, rigid PVC conduit and, 134
Meters
  ground clamp for, 54
  jumpers around, 47
  location of watt-hour, 33–34
  multiple, 53–54
  overhead service and, 38, 45–47
  reading of, 63, 65
  single-phase watt-hour, 63, 64

Meters, *continued*
   typical sockets for, 45
   underground service and, 37
Metric measurements, 5–7, 18
Microwave oven, 255, 266
Mil, defined, 113
Millivolt system, 312–313
Minimum cover, 415
Modular plug, for telephone, 398, 400
Motion detectors, 359, 405
Motor branch circuits, 100
Motor circuit switch, 280–281
Motor loads, switches and, 150
Motorized master control, 365
Motors, 243
   for central vacuum systems, 420
   electrical schedules for, 20
   of exhaust fans, 287
   for garage door openers, 419
   for jet pumps, 336–337
   noise from, 179
   of paddle fans, 214
   for spas/hot tubs, 386
MOVs. *See* Metal oxide varistors (MOVs)
Multiconductor cables, for signal systems, 374
Multifamily dwelling, meters for, 53–54
Multigang meter socket, 30, 54
Multioutlet assembly, 75, 331, 332, 333–334, 335
Multiset coupler, 391–392
Multiswitch control, 363
Multiwire circuits. *See also* Three-wire circuit
   in kitchen, 256–257
   neutral conductor and, 147
   in recreation room, 322–327
   split-wired receptacles and, 256–257

N

National Building Code of Canada, 99, 358
National Standards of Canada (NSC), 4
Neon lights, 191, 192, 282, 283
Network interface card (NIC), 402
Neutral conductors
   connections for, 43
   defined, 147
   GFCIs and, 168, 169
   grounding and, 47–49, 51
   identifying, 146, 147
   multiple meters and, 54
   open, in three-wire circuit, 326, 327
   service entrance calculations and, 24, 27, 29
   service entrance equipment and, 43, 47–49, 54
   size of, 117–118
   split-wired receptacles and, 256
   three-wire circuits and, 322, 324
   vs. "hot" phase conductors, 117, 118
NIC. *See* Network interface card (NIC)
Night-light, 243, 244
NMD90 cable, 123, 124, 125, 277, 280–281
   for attic exhaust fan circuit, 283
   in recreation room, 320, 322, 323
   for water heater, 344, 345
   in workbench lighting, 331
NMSC cable, 126, 128, 298, 320, 2103
NMW cable, 123, 124, 125
NMWU cable, 123, 124, 125, 414, 415
   for submersible pump, 338
No-niche fixtures, 384
Noise, 179, 252
Non-metallic flexible conduit, 140, 142. *See also*
      Flexible conduit
Non-metallic outlet boxes, 73, 78, 82, 391, 392
Non-metallic raceways, 45
   multiple meters and, 54
   underground wiring and, 417
Non-metallic sheathed cable (Romex), 116
   in attics, 131–132
   bonding and, 155–156, 158, 257
   "bundled" together, 125, 126
   clamps for, 133, 136

colour of, 147–148, 149
connectors for, 133, 135, 136
description of, 123–124
furring strips and, 327
GFCIs and, 168, 169
installation of, 124–125, 127, 128, 135
metal boxes and, 126
for outdoor lighting, 414–419
outlet boxes and, 78, 82
in signal systems, 374
temperature and, 125, 126
to wire grounding-type receptacles, 155, 157
Non-metallic water pipe, 49, 52. *See also* Plastic
      (PVC) piping
Nongrounding-type receptacles, 158–159
   GFCIs and, 168, 169, 171
Notations, 12, 13–17, 73, 83
NSC. *See* National Standards of Canada (NSC)
Nuisance tripping, 168, 169, 172

O

"O" ring expansion joint, 138, 140
Occupational Health and Safety Act (OSHA), 176
OCD. *See* Overcurrent device (OCD)
Ohm's law, 325–327, 346, 380
Oil heating system, 292, 306–312
   "separate circuit" rule for, 292, 298
Orthographic projection, 11
OSHA. *See* Occupational Health and Safety Act (OSHA)
Outdoor installations, 413–419
   fixtures marked for, 418–419
   master selector switch for, 368
   minimum cover in, 415
   remote-control extra-low-voltage systems and, 367
   of spas/hot tubs, 385
   in vehicular *vs.* non-vehicular areas, 415
Outdoor receptacles
   GFCIs for, 168, 170
   kitchen and, 255
   lighting branch circuit and, 202–203, 229–333
   for spas/hot tubs, 385
Outlets, 71–94. *See also* Boxes; Receptacles
   bonding and, 94
   branch circuits and (*See* Branch circuits)
   defined, 71
   grouping, 107, 195–196, 197
   junction boxes and, 75, 77–78, 81
   for kitchen fan, 252
   lighting fixtures and, 72–75 (*See also* Lighting
      outlets)
   lighting *vs.* receptacle, 71
   location of, 91, 93, 204
   for outdoor lighting, 413–414
   special-purpose (*See* Special-purpose outlets)
   symbols for, 71–76, 79, 83, 199, 200
   for television, 391, 392
   toggle switches *vs.*, 71
   TVSS for, 179, 180
   on wiring diagrams, 199–202
Oven. *See* Range/oven
Overcurrent device (OCD). *See also* Overcurrent
      protection
   attics and, 287
   branch circuit ampacity rating and, 113, 114
   for chime transformer, 371, 373
   electric furnace and, 298
   GFCI as part of branch-circuit, 167
   interrupting rating for, 168
   for jet pump motor, 336–337
   for spas/hot tubs, 386
   swimming pools and, 381, 385
   switches with, 287
Overcurrent protection, 44. *See also* Ground-fault
      circuit interrupters (GFCIs)
   for bathroom, 237
   branch-circuit devices for, 42, 43, 100–103,
      336–337
   for dimmer controls, 225
   for dryers, 277, 278, 279

for food waste disposal, 268
fuses for, 55–57
for gas/oil heating systems, 316
grounding conductor and, 48
interrupting ratings for, 58–63
inverse time characteristic in, 47
for jet pump, 336–337
panelboards and, 42
rating/setting of, 104
service entrance conductors and, 37, 47–51,
   55–57
time/current curve and, 166
for water heaters, 345
Overhead service, 33, 34–35, 38, 39
Overload protection
   for electric heating, 292
   of fan motor, 287
   for jet pump motor, 337

P

Pad-mounted transformer, 34, 35, 37
Paddle fans, 73, 211, 213–214
   dimmers with, 227
   symbol for, 76
Panel schedule, GFCIs on, 168
Panelboards
   branch circuits and, 101
   in combination panel, 34
   electrical schedules for, 20
   overcurrent protection and, 42
   series-rated, 62–63
   swimming pools and, 381
Panels
   chime transformer and, 374
   in recreation room, 324–327
   service entrance (*See* Service entrance panels)
   temporary power, 176
   three-wire circuit and, 324–327
PAR lighting fixtures, 414
Parallel arc, 177
Pendant kit assembly, for track lighting, 223, 224
Pentrox, 27
Personnel safety, 3
Photocontrols, home automation and, 405
Photoelectronic type smoke detector, 76, 354
Pictorial drawings, 10–11, 199, 200
Pilot lights, 155, 156, 157, 281–287
   for extra-low-voltage switches, 364
   symbol for, 80
Piping symbols, 15
Plans (in drawings), 11, 13
Plastic bushing, 53
Plastic-case breakers, 61–63
Plastic (PVC) conduit, 134, 137, 138
   bending, 137, 138
   fitting for, 139
   support of, 136, 139
   underground, 418
Plastic (PVC) piping, 49. *See also* Non-metallic
      water pipe
Plate electrode, 49, 52
Plot/site plans, 9, 11
Plug fuses, 55, 56
Plug-in relays, 366
Plumbing code, water heaters and, 341–342
Plumbing floor plan, 19
Plumbing symbols, 15
Porch. *See* Front entry/porch
Portable dishwasher, 271
Portable GFCI devices, 176, 177
Portable tools, multioutlet assembly for, 334
Post lamps, 414, 418, 419
   master selector switch for, 368
   underground wiring for, 414, 418, 419
Pressure cable connector, 114
Pressure switch, 337
Pressure terminal connector, 114
Pryouts, 136
Pulsating DC supply, 367

Pump load, 26, 27, 30. *See also* Submersible pump; Sump pump; Water pump
Pump motors, swimming pools and, 381
Push buttons
  for garage door opener, 420
  symbol for, 371
PVC. *See entries beginning* Plastic (PVC)

R

Raceways. *See also* Metallic raceways; Non-metallic raceways
  bushings and, 53
  Class 1 circuits in, 313, 315, 316
  electrical schedules for, 20
  empty EMT, 334
  neutral conductor and, 27
  non-metallic, 45
  in signal systems, 374
  temperature and, 117
  through wood/metal framing, 129–131
  trade sizes of, 6
  underground wiring and, 417
Radio, lighting branch circuit and, 202
Radio frequency interference (RFI), 179
Range/oven. *See also* Cooking units
  branch circuits for, 266
  freestanding, 263–264, 266–267
  load for, 26, 29, 118 (*See also* Small appliance circuits)
  microwave, 255, 266
  self-cleaning, 265
  symbol for outlet for, 75
  typical surface unit of, 264
  wall-mounted, 26, 262–266
Rated-load current, 299–300
Receptacles, 71, 72. *See also* Outlets
  for air conditioners, 299
  baseboard heaters and, 295–296, 297
  in basement, 108, 110
  in bedroom/study, 201–202
  bonding at, 155–158
  code requirements for (table), 102, 107
  convenience, 253–256
  countertop, 253–256
  duplex, 79, 103
  faceplates for, 93–94
  in garage, 413
  GFCI (*See* GFCI receptacle outlets)
  grounding-type, 157
  in hallways, 241
  home automation and, 405
  in laundry, 279
  layout for, 107–110, 197
  lighting combined with, 106
  in living room, 202
  location of, 91, 93, 204, 254–255
  multioutlet assemblies and, 334
  non-metallic boxes for, 78
  nongrounding-type, 158–159
  number/location of branch, 99–110
  position of, 91–93
  removing tab on "hot" side of, 100, 103
  replacing, 173–176
  for small appliances, 99–100
  for spas/hot tubs, 385
  special-purpose, 83, 90
  split-switched (*See* Split-switched receptacle wiring)
  split-wired, 105
  symbols for, 73, 75, 79
  toggle switches and, 149
  TVSS for, 179, 180
  wet bar and, 322–327
  on wiring diagrams, 199–202
  in workshop, 331–332
Recessed clock outlet, 252, 253
Recessed lighting fixtures, 73, 79, 105, 185
  for accent lighting, 220
  code requirements for, 187–191, 193
  fluorescent, 320–322

  for front entry closet, 233
  lay-in, 320–322
  for recreation room, 320–327
  for spas/hot tubs, 385
  symbol for, 74, 79
  temperature and, 116, 186, 191
  thermal cutout for, 186, 187
  "through" wiring for, 196
Recreation room, 320–327
  lay-in recessed fixtures for, 320–322
  wet bar in, 322–327
Rectifier, 367
Refrigerator/freezers, 243, 253
  GFCIs and, 172
  receptacles for, 167
  separate circuit for, 331
Relay, extra-low-voltage, 363–368
Remote-control low-energy circuits, 317, 363–370
  code requirements for, 368, 370
  conductor wire size for, 113, 367
  home automation and, 405
  oil/gas heating and, 313–316
  relay used in, 363–368
  switches for, 364–365
  transformers for, 366–367, 370
  typical, 368
REN. *See* Ringer equivalence number (REN)
Residential electrical wiring
  codes/standards for, 1–3
  lighting in, 191, 193, 196
Resistance-type heating cables, 292
Resistive loads, 149, 150, 197
RFI. *See* Radio frequency interference (RFI)
Rigid metal conduit, 134, 417
Rigid PVC conduit, 134, 137, 138
  bending, 137, 138
  fitting for, 139
  support of, 136, 138, 139
  in utility room, 336
Rigid steel conduit, 6, 134, 136
Ring wave transients, 178
Ringer equivalence number (REN), 400
Rod electrodes, 49, 52
Romex. *See* Non-metallic sheathed cable (Romex)
Room air conditioners, 298–299, 300
Routers, 403, 405
Runs. *See* Cable runs

S

Safety, 1. *See also* Electrical hazards
  CEC and, 2
  of electrical equipment, 3
  of gas/oil heating systems, 306
  GFCIs and, 171
  of lighting fixtures, 185
  personnel, 3
  telephone and, 401
Satellite antennas, 393–394
Satellite dish, 394–395
Scales
  on construction drawings, 18
  as measuring instruments, 6, 18
SCC. *See* Standards Council of Canada (SCC)
Scene controllers, 406
Schedules, 19–20, 41, 44. *See also* Panel schedule
Sconce, 79
Scotchlok, 114, 115
Screw terminal, 114
Second-floor balcony, receptacles on, 167
Sectional views (in drawings), 11–12, 13, 19
Security lighting, 196, 358–360
Security systems, 358–360
  home automation and, 405
Self-generating system, 312–313
Separate circuits
  for built-in cooking units, 267
  for freezers, 331
  for heating, 292
  for sump pump, 332

Series arc, 177
Series circuit, 156
Series-connected system, 62–63
Series-rated system, 62–63
Service box, 27, 40, 42
  for apartment dwelling, 30
  in combination panel, 34
  defined, 33
  multiple meters and, 54
Service conduit, 28
Service entrance
  armoured cable for, 126
  conduit wiring for, 134
  interrupting ratings for, 168
  multioutlet assemblies and, 334
  rigid PVC for, 134
  track lighting and, 223
Service entrance calculations, 23–30
  for apartment, 29–30
  for conductors, 23–24, 27, 28
  conduit and, 28, 36–37, 38
  for disconnecting means, 23–24, 27, 29
  grounding conductor in, 28
  minimum code requirements in, 24
  for neutral conductors, 24, 27, 29
  panels and, 28–29
  for single-family dwelling, 24–29
  subpanel and, 29
Service-entrance disconnecting means, 23–24, 27
  location of, 34, 37, 40, 42
  main, 29
  service boxes for, 40, 42
Service entrance equipment, 33–65
  bonding of, 51–53
  branch circuit overcurrent devices and, 55–58
  circuit breakers and, 57–63
  conductors and, 35–36, 37, 40
  conduit sizing and, 44–45
  and cost of electrical energy, 63–65
  fuses and, 55–57
  grounding of, 47–51
  load centre and, 42–44, 60–63
  mast-type, 35–37
  meters and, 45–47, 63–65
  multiple meters and, 53–54
  neutral conductors and, 43, 47–49, 54
  overhead service, 33, 34–35
  panels and, 60–63
  short-circuit currents and, 59–60
  for single-family dwelling, 51
Service entrance panels, 28–29, 34, 37
  circuit directory on, 40, 44
  fuse, 55
  typical main panel, 42
Service panel ampacity, 28–29
Sheet metal duct work symbols, 16
Shielded coaxial cable, 392
Shock, 47, 48, 165–166
  baseboard heaters and, 296, 298
  bonding wall boxes and, 203
  CEC and, 2
  codes/standards and, 1
  construction sites and, 176
  GFCIs and, 166, 168
  hydromassage tub and, 245
  portable dishwashers and, 271
  receptacle replacement and, 176
  from swimming pools, 379–381
Shop drawings, 20
Short-circuit currents, 59–60
  boxes and, 88
  fan motors and, 287
SI. *See* International System of Units (SI)
Side views (in drawings), 11
Signal circuits, 317
  chimes and, 370–374
  oil/gas heating and, 313–316
Signal system (chimes), 370–374

Single-circuit grounding, multi-outlet assemblies and, 334
Single-family dwelling
  minimum ampacity rating for, 42
  service entrance calculations for, 24–26
  service entrance equipment for, 51
Single-phase watt-hour meter, 63, 64
Single-pole, double-throw momentary contact switch, 364, 365
Single-pole AFCI circuit breaker, 177, 178
Single-pole dimmer switch, 223, 225, 227, 228
Single-pole GFCI circuit breaker, 167
Single-pole switch, 150–151, 152
  in bathroom, 237
  in front entry/porch, 230, 231, 233
  in recreation room, 320, 327
  symbol for, 80
  in workshop, 331
Single receptacle outlet
  duplex *vs.*, 75, 253
  symbol for, 75
Sinks, receptacle outlets near, 166, 167
Sirens, 359
Site/plot plans, 11, 19
Sliding glass doors, 210
Small appliance circuits, 99–110
  bonding for, 257
  for kitchen, 253–257
  receptacles and, 99–100
Smoke detectors/alarms, 353–357, 359
  interconnected, 357
  receptacles for, 167
  security system and, 360
  symbols for, 76
  types of, 354
SNI. *See* Standard network interface (SNI)
Snow blowers, 243
Soffit lighting, 240, 241, 242
Solder-type lugs, 114
Solderless connectors, 114, 115, 155
Solid-state dimmer switches/controls, 224–225, 226
Sone (of fan noise), 252
Spas, 379, 385–386
  hazard associated with, 381
  hot tubs *vs.*, 385
  leakage current collectors for, 387
Special-purpose outlets, 262–271
  for hydromassage bathtub, 83, 245
  in kitchen/laundry, 255
  symbol for, 75, 76
  in utility room, 335, 336
Specifications, 20
  lighting fixtures and, 72–73
  outlets and, 72–73, 83
Speed-control switch, 285, 286
Speed of recovery, 341
Spikes, 178
Splices, 155–156, 199
Split-bolt connectors, 114, 115
Split-circuit receptacles, 255
  multi-outlet assemblies and, 334
Split-phase motors, 419
Split-switched receptacle, symbol for, 75
Split-switched receptacle wiring, 104, 202, 204, 210, 211
  diagram of, 221
Split-wired receptacle, 105
  for kitchen, 254, 256–257
  multiwire circuits and, 256–257
Standard network interface (SNI), 398
Standards. *See also* Codes
  CEC, 3
  CSA and, 4
  international, 4
  local, 5
  SCC and, 4
  for symbols, 12
  TIA-570A, 3
  ULC and, 4–5

Standards Council of Canada (SCC), 4
  symbols and, 12, 72
Standards steering committee *(CEC)*, 2
"Station wire," 397
Structural flow plan, 19
Study, lighting branch circuit for, 201–202, 203, 211, 212
Submersible pump, 337–338
Subpanel, in service entrance, 29, 34
Sump pump, 243
  receptacles in, 166
  separate circuit for, 332
Supply service
  CEC rules for, 37
  defined, 33
  overhead, 33, 34–35, 38, 39
  underground, 33, 34, 37, 40
Surface-mounted chime, 372
Surface-mounted lighting fixtures, 73, 79, 185, 385
Surges
  lightning, 398
  TVSS for, 178–180, 211
Suspended ceiling lighting fixtures, 185, 186, 187–191
  junction boxes for, 188, 190, 191, 320, 322
  for recreation room, 320–327
  support of, 188–189, 320–322
Swimming pool, 24, 36, 245, 379–385
  code-defined, 381
  electric heating of deck area of, 384–385
  GFCIs for, 168, 169
  permanently installed *vs.* storable, 381
  receptacles for, 167
Swimming pools, lighting fixtures in, 380, 381, 384
Switch/device boxes, 75, 77–78, 84.
  See also Boxes
  in bathroom, 244
  cable clamps for, 133, 136
  extra support for, 83, 89
  ganged, 82–83, 90
  grounding wires attached to, 158
  induction heating and, 161
  knockout/pryouts on, 133, 136
  location of, 91, 93
  with metal mounting strips, 83, 86
  number of conductors in, 86–91
  rodents and, 88
  for television, 391
  for toggle switches, 150
Switches, 73, 75, 77–78, 146
  AC/DC general-use, 150
  for attic lighting, 281–287
  for bathroom, 243
  in bathroom, 237
  colour coding and, 147–149, 152–155
  dimmer (*See* Dimmer switches/controls)
  disconnect, 154, 155
  doorjamb, 233
  for exhaust fans, 284–287
  extra-low-voltage, 364–365, 368
  for garage door openers, 420
  general-use, 149–150, 281
  GFCIs and, 172
  for humidity control, 280–281, 285
  infinite heat control, 265
  infinite-speed, 287
  junction boxes and, 78
  for kitchen fans, 252
  location of, 93
  locator, 156
  master selector (*See* Master selector switch)
  motor circuit, 280–281
  with overcurrent devices, 287
  paddle fans and, 214
  with pilot lights (*See* Pilot lights)
  remote-control, 364–365
  series circuits and, 156
  speed-control, 285, 286

  standard types of, 80
  study/bedroom, 212
  symbols for, 80
  T-rated, 150, 227
  timer, 90, 245, 285, 286
  toggle (*See* Toggle switches)
  on wiring diagrams, 199–202
Symbols, 12, 13–17
  ANSI, 12
  architectural drafting, 13–14
  for chimes, 371
  for electrical equipment, 12
  legend of, 19, 72
  miscellaneous, 76
  non-standard, 72
  notations and, 12, 73
  outlet, 71–74, 79, 83, 199, 200
  plumbing/piping/valve, 15
  for push buttons, 371
  SCC and, 12, 72
  sheet metal duct work, 16
  for switches, 80
  for telephone, 397
  for television, 391
System grounding, 47–51
System International (SI), 5–7

T

T-rated switches, 150, 227
T-slot receptacles, 256–257
Tables, *CEC*, 2, 3
  Table 1, 113
  Table 2, 24, 27, 29, 101, 104, 114, 116, 117, 120, 187, 263, 265, 345, 414
  Table 3, 113
  Table 4, 27, 114, 116
  Table 5A, 101, 113, 282, 283, 315
  Table 5C, 101, 104, 113, 125, 315
  Table 6, 44, 134
  Table 8, 24, 28, 44, 45
  Table 9, 24, 28, 44, 45
  Table 10, 24, 28, 44, 45, 176
  Table 11, 116
  Table 12, 113
  Table 13, 24, 27, 58, 101, 104, 279
  Table 16, 24, 54, 124, 156, 381, 382, 386, 387, 417
  Table 17, 24, 27, 29, 49, 50, 51, 54
  Table 19, 116, 123, 124, 191, 263, 315, 401, 414
  Table 22, 86
  Table 23, 6, 86, 88, 90, 203
  Table 29, 336–337
  Table 41, 24, 28, 53
  Table 43, 49, 50
  Table 45, 287, 336
  Table 53, 415, 416
  Table 57, 315
  Table D1, 191, 414
  Table D3, 121, 122, 123, 323
  Table D16, 287, 336
Tables (referred to in construction drawings), 12
Tap-off units, 392
Tape measure, 5–6
Task lighting, 196, 222
Technical drawings, 9–10
TECK90 cable, 126
Telephone, 397–401
  colour coding of cables for, 398, 399
  conductors for, 398–401
  lighting branch circuit and, 202, 211
  modular plug for, 398, 400
  security systems and, 358
Television, 391–397
  CATV, 393
  code rules for cable, 393
  lighting branch circuit and, 202
  satellite, 393–395
  toaster and, 326–327

Temperature
  conductor ratings for, 116, 263
  conductors and, 116–117, 118, 120–121
  recessed lighting fixtures and, 187
Temperature control
  exhaust fans and, 287
  infinite-position, 264, 265
  by surface heating elements, 264, 265
Temporary lighting, 176
Temporary power panels, 176
Termination jacks, in data systems, 403, 404
Terminations, arcing due to poor, 176–177
Terminology, 3, 35
Therapeutic pools, 379
Thermal insulation, lighting fixtures and, 190
Thermal-magnetic circuit breaker, 287
Thermostat (for electric heating), 292–294
  for baseboards, 295, 297–298
  for wall heating units, 295
  for water pump, 340, 341, 342
Thermostat (for exhaust fan), 285
Thermostat (for oil/gas systems), 306, 309, 311–312
  Class 2 circuit and, 315
  mercury, 312
  self-generating systems and, 313
Thermostat (for oven), 265–266
Thermostat symbol, 76
Thermostat wire, 374
Threaded boss, 45
Three-way switch, 151, 152, 153, 154
  dimmer type, 225, 227, 249
  for front entry/porch, 231, 233
  symbol for, 80
  track lighting and, 249
  on wiring diagram, 199–202
Three-wire cable, 159, 220–221
  "hot" conductor in, 327
  induction heating and, 159–161
  in recreation room, 324–327
  remote-control, 364
  in signal system, 374
Three-wire circuit, 148. *See also* Multiwire circuits
  correct *vs.* incorrect connection of, 325
  for countertop receptacles, 253–256
  problems with, 326
  in recreation room, 322–327
  split-wired receptacles and, 256–257
  vs. two two-wire circuits, 322, 324
Three-wire grounding-type attachment plug, 158
TIA-570A, 3
Time/current curve, 166
Time-delay fuses, 55–57, 287
  jet pump and, 336, 337
Timer switches, 90
  for exhaust fans, 285, 286
  home automation and, 405–406
  for hydromassage tub, 245
Toaster, television and, 326–327
Toggle switches, 71, 149–155
  in attic, 282
  flush, 149
  for front entry, 231
  ratings of, 149–150
  symbols for, 80
  types of, 150–152
Track lighting, 106, 222–223, 224, 225
  for kitchen, 249, 251
  symbol for, 74
Trade sizes, conversions for, 6–7
Transformers
  chime, 331, 370–373
  device boxes and, 89–90
  GFCIs and isolation, 169
  low-impedance, 59–60
  for oil/gas heating systems, 312, 316
  for remote-control extra-low-voltage systems, 366–367, 370
  short-circuit currents and, 59–60

swimming pools and, 381, 384
  symbol for, 76
Transient voltage surge suppression (TVSS), 178–180, 211
Travellers, colour coding for, 152–154
Tubular support, in mast-type service entrance, 37
Tungsten filament lamps, 149, 150
  inrush currents and, 226–228
  plug-in relays and, 366
TVSS. *See* Transient voltage surge suppression (TVSS)
Two-wire cable, 159, 161
  for baseboard heaters, 298
  for remote-control circuits, 364
  in signal system, 374
Two-wire circuits, 147–148
  maximum cable run for, 121–123
  one three-wire circuit *vs.* two, 322, 324
  for smoke alarms, 357
"Typical," 12

U

ULC. *See* Underwriters' Laboratories of Canada (ULC)
Undercabinet lighting, 249–250, 252
Underground meter base, 46
Underground service, 33, 34, 37, 40
Underground wiring, 414–419
Underwater lighting fixtures, 380, 384
Underwriters' Laboratories of Canada (ULC), 3, 4–5, 353–354
Underwriters' Laboratories (UL), 316
Unshielded twisted pair (UTP) cable, 401–402
Utility companies, 5, 27, 29
  charges by, 63–65
  meter reading and, 63
  neutral conductor size and, 117
  open neutral conductor and, 327
  residential subdivisions and, 37
  service entrance equipment and, 33
  underground service and, 37
  water heaters and, 342, 344
Utility room, 335–346, 374
UTP. *See* Unshielded twisted pair (UTP) cable

V

Vacuum system, central, 332, 420
Valance lighting, 211
Values, 4
Valve symbols, 15
Vanity lighting, 237, 239, 240, 243
Vapour barriers
  electric heating and, 292
  for kitchen outlets, 252
Vapour boot, 75, 77
Volt-amperes, 197
Voltage, 103–104
  of chime transformers, 371
  lighting fixture limitations of, 191, 193
  surges in, 178–180, 211, 398
  swimming pools and, 380
  telephone and hazardous, 398
  for television, 392
  water heaters and, 346
Voltage drop, 118–120, 121
  causes of, 118
  code references to, 119–120
  conductor temperature and, 118, 120
  electric furnace and, 296
  formula for, 118–119
  Table D3 and, 121–123
  three-wire circuits and, 322, 323, 324
Voltage transients, 178–179

W

W. *See* White (W) lamps
Wading pools, 379
Wall boxes
  in bathroom, 244
  bonding, 203

dimmer units for, 225
  for doorjamb switch, 233
  lighting branch circuits and, 204
  for porch/entry lighting, 231
  in recreation room, 327
  sizing, 203
Wall heating units, thermostat for, 295
Wall lighting outlet, symbol for, 79
Wall-mounted electric heating units, 292
Wall-mounted lighting fixture, 79
  for porch/entry, 230, 232
  symbol for, 74
Wall-mounted oven, 26, 262, 263–266
Wall plates, 400
Warm-air systems, 306–312
Warm white deluxe (WWX) lamps, 240, 250, 320
Warm white (WW) lamps, 240, 250
Warning signs, GFCI, 172, 173
Washing machines, 172, 243, 279
Washroom. *See* Bathroom/washroom
Water heater load, 26, 27, 30, 118
Water heater meters, 53
Water heaters, 338–346
  corrosion of, 340
  dishwaters and, 342
  heating elements of, 340–341
  scalding from, 341–342
  for spas/hot tubs, 386
  speed of recovery for, 341
  symbol for, 76
Water pipes
  ground clamp for, 54
  grounding and, 47–49, 50–51, 52
  metal, 34, 49
  non-metallic, 49, 52
Water pump, 335–338, 387
Watt-hours, 33–34, 63, 64, 178
Wattage rating, 197, 340, 346
Watts loss, three-wire circuits and, 322, 324
Weatherproof receptacles, 413, 414
Well casings, ground clamp for, 54
Wet bar, 322–327
Wet locations, 417, 418
Wet-niche fixtures, 381, 384
Whirlpool bath, 245–246, 386–387
White (W) lamps, 240
Wing nut, 114, 115
Wire connectors, 114–116
Wire gauge, 6, 7, 112
Wire nut, 114, 115
Wire sizing, 100–103, 112–113, 121
Wireless data networks, 403, 405
Wireless key, 359
Wireless keypad, 359
Wireless panic button, 359
Wiring
  cable (*See* Cable layout; Cable runs; Cables)
  conduit (*See* Conduit)
  least expensive, 124
  load centre and, 42–44
  temporary, 176
  "through," 191, 196
  underground, 414–419
Wiring diagrams, 108, 199–202. *See also* Cable layout
Workbench lighting, 331
Working drawings, 18–20
Workshop, 331–335
WW. *See* Warm white (WW) lamps
WWX. *See* Warm white deluxe (WWX) lamps

X

X-10 system technology, 405–406
X-ray equipment, noise from, 179

Z

Zone control, 291, 292